Lecture Notes in Computer Science

Lecture Notes in Artificial Intelligence   14127

Founding Editor

Jörg Siekmann

Series Editors

Randy Goebel, *University of Alberta, Edmonton, Canada*
Wolfgang Wahlster, *DFKI, Berlin, Germany*
Zhi-Hua Zhou, *Nanjing University, Nanjing, China*

The series Lecture Notes in Artificial Intelligence (LNAI) was established in 1988 as a topical subseries of LNCS devoted to artificial intelligence.

The series publishes state-of-the-art research results at a high level. As with the LNCS mother series, the mission of the series is to serve the international R & D community by providing an invaluable service, mainly focused on the publication of conference and workshop proceedings and postproceedings.

Davide Calvaresi · Amro Najjar ·
Andrea Omicini · Reyhan Aydogan ·
Rachele Carli · Giovanni Ciatto · Yazan Mualla ·
Kary Främling
Editors

Explainable and Transparent AI and Multi-Agent Systems

5th International Workshop, EXTRAAMAS 2023
London, UK, May 29, 2023
Revised Selected Papers

 Springer

Editors
Davide Calvaresi ⓘD
University of Applied Sciences and Arts
Western Switzerland
Sierre, Switzerland

Andrea Omicini ⓘD
Alma Mater Studiorum, Università di
Bologna
Bologna, Italy

Rachele Carli ⓘD
Alma Mater Studiorum, Università di
Bologna
Bologna, Italy

Yazan Mualla ⓘD
Université de Technologie de
Belfort-Montbéliard
Belfort Cedex, France

Amro Najjar ⓘD
Luxembourg Institute of Science
and Technology
Esch-sur-Alzette, Luxembourg

Reyhan Aydogan ⓘD
Ozyegin University
Istanbul, Türkiye

Giovanni Ciatto ⓘD
Alma Mater Studiorum, Università di
Bologna
Bologna, Italy

Kary Främling ⓘD
Umeå University
Umeå, Sweden

ISSN 0302-9743 ISSN 1611-3349 (electronic)
Lecture Notes in Artificial Intelligence
ISBN 978-3-031-40877-9 ISBN 978-3-031-40878-6 (eBook)
https://doi.org/10.1007/978-3-031-40878-6

LNCS Sublibrary: SL7 – Artificial Intelligence

This Springer imprint is published by the registered company Springer Nature Switzerland AG
The registered company address is: Gewerbestrasse 11, 6330 Cham, Switzerland

Preface

AI research has made several significant breakthroughs that has boosted its adoption in several domains, impacting our lives on a daily basis. Nevertheless, such widespread adoption of AI-based systems has raised concerns about their foreseeability and controllability and led to initiatives to "slow down" AI research. While such debates have mainly taken place in the media, several other research works have emphasized that achieving trustworthy and responsible AI would necessitate making AI more transparent and explainable.

Not only would eXplainable AI (XAI) increase acceptability, avoid failures, and foster trust, but it would also comply with relevant (inter)national regulations and highlight these new technologies' limitations and potential.

In 2023, the fifth edition of the EXplainable and TRAnsparent AI and Multi-Agent Systems (EXTRAAMAS) continued the successful track initiated in 2019 in Montreal and followed by the 2020 to 2022 editions (virtual due to the COVID-19 pandemic circumstances). Finally, EXTRAAMAS 2023 was held in person and proposed bright presentations, a stimulating keynote (titled "Untrustworthy AI" given by Jeremy Pitt, Imperial College London), and engaging discussions and Q&A sessions.

Overall, EXTRAAMAS 2023 welcomed contributions covering areas including (i) XAI in symbolic and subsymbolic AI, (ii) XAI in negotiation and conflict resolution, (iii) Explainable Robots and Practical Applications, and (iv) (X)AI in Law and Ethics.

EXTRAAMAS 2023 received 26 submissions. Each submission underwent a rigorous single-blind peer-review process (three to five reviews per paper). Eventually, 16 papers were accepted and collected in this volume.

Each paper was presented in person (with the authors' consent, they are available on the EXTRAAMAS website[1]). Finally, The Main Chairs would like to thank the special track chairs, publicity chairs, and Program Committee for their valuable work, as well as the authors, presenters, and participants for their engagement.

June 2023

Davide Calvaresi
Amro Najjar
Andrea Omicini
Kary Främling

---

[1] https://extraamas.ehealth.hevs.ch/.

Organization

General Chairs

Davide Calvaresi   University of Applied Sciences and Arts Western Switzerland, Switzerland

Amro Najjar   Luxembourg Institute of Science and Technology, Luxembourg

Andrea Omicini   Alma Mater Studiorum, Università di Bologna, Italy

Kary Främling   Umeå University, Sweden

Special Track Chairs

Reyhan Aydogan   Ozyegin University, Turkiye

Giovanni Ciatto   Alma Mater Studiorum, Università di Bologna, Italy

Yazan Mualla   UTBM, France

Rachele Carli   Alma Mater Studiorum, Università di Bologna, Italy

Publicity Chairs

Yazan Mualla   UTBM, France

Benoit Alcaraz   University of Luxembourg, Luxembourg

Rachele Carli   Alma Mater Studiorum, Università di Bologna, Italy

Advisory Board

Tim Miller   University of Melbourne, Australia

Michael Schumacher   University of Applied Sciences and Arts Western Switzerland, Switzerland

Virginia Dignum   Umeå University, Sweden

Leon van der Torre   University of Luxembourg, Luxembourg

Program Committee

Andrea Agiollo	University of Bologna, Italy
Remy Chaput	University of Lyon 1, France
Paolo Serrani	Universià Politecnica delle Marche, Italy
Federico Sabbatini	University of Bologna, Italy
Kary Främling	Umeå University, Sweden
Davide Calvaresi	University of Applied Sciences and Arts Western Switzerland, Switzerland
Rachele Carli	University of Bologna, Italy
Victor Contreras	University of Applied Sciences and Arts Western Switzerland, Switzerland
Francisco Rodríguez Lera	University of León, Spain
Bartłomiej Kucharzyk	Jagiellonian University, Poland
Timotheus Kampik	University of Umeå, Sweden, Signavio GmbH, Germany
Igor Tchappi	University of Luxembourg, Luxembourg
Eskandar Kouicem	Université Grenoble Alpes, France
Mickaël Bettinelli	Université Savoie Mont Blanc, France
Ryuta Arisaka	Kyoto University, Japan
Alaa Daoud	INSA Rouen-Normandie, France
Avleen Malhi	Bournemouth University, UK
Minal Patil	Umeå University, Sweden
Yazan Mualla	UTBM, France
Takayuki Ito	Kyoto University, Japan
Lora Fanda	University of Geneva, Switzerland
Marina Paolanti	Università Politecnica delle Marche, Italy
Arianna Rossi	University of Luxembourg, Luxembourg
Joris Hulstijn	University of Luxembourg, Luxembourg
Mahjoub Dridi	UTBM, France
Thiago Raulino	Federal University of Santa Catarina, Brasil
Giuseppe Pisano	Univeristy of Bologna, Italy
Katsuhide Fujita	Tokyo University of Agriculture and Technology, Japan
Jomi Fred Hubner	Federal University of Santa Catarina, Brazil
Roberta Calegari	University of Bologna, Italy
Hui Zhao	Tongji University, China
Niccolo Marini	University of Applied Sciences Western Switzerland, Switzerland
Salima Lamsiyah	University of Luxembourg, Luxembourg
Stephane Galland	UTBM, France
Giovanni Ciatto	University of Bologna, Italy

Matteo Magnini	University of Bologna, Italy
Viviana Mascardi	University of Genoa, Italy
Giovanni Sileno	University of Amsterdam, The Netherlands
Sarath Sreedharan	Arizona State University, USA

Contents

Cross-Domain Applied XAI

Explainable Agents and Multi-Agent Systems

Mining and Validating Belief-Based Agent Explanations

Ahmad Alelaimat[✉], Aditya Ghose, and Hoa Khanh Dam

Decision Systems Lab, School of Computing and Information
Technology University of Wollongong, Wollongong 2522, Australia
{aama963,aditya,hoa}@uow.edu.au

Abstract. Agent explanation generation is the task of justifying the decisions of an agent after observing its behaviour. Much of the previous explanation generation approaches can theoretically do so, but assuming the availability of explanation generation modules, reliable observations, and deterministic execution of plans. However, in real-life settings, explanation generation modules are not readily available, unreliable observations are frequently encountered, and plans are non-deterministic. We seek in this work to address these challenges. This work presents a data-driven approach to mining and validating explanations (and specifically belief-based explanations) of agent actions. Our approach leverages the historical data associated with agent system execution, which describes action execution events and external events (represented as beliefs). We present an empirical evaluation, which suggests that our approach to mining and validating belief-based explanations can be practical.

Keywords: Explainable agents · Mining explanations · BDI agents

1 Introduction

Explainable agents have been the subject of considerable attention in recent literature. Much of this attention involves folk psychology [1], which seeks to explain the actions of an agent by citing its mental state (e.g., the beliefs of the agent, and its goals and intentions). Roughly speaking, when an explainee requests explanations about a particular action, two common explanation styles might be adopted: (1) a goal-based explanation and (2) a belief-based explanation [2–4]. This paper focuses on the latter style[1] in the context of the well-known Belief-Desire-Intention (BDI) paradigm [5]. Fundamentally, belief-based explanations help answer the following question: *What must have been known for the agent to perform a particular action over another?* Arguably, a sufficiently

---

A. Ghose-Passed away prior to the submission of the manuscript. This is one of the last contributions by Aditya Ghose.

[1] We submit that goal-based explanations are also of great value to develop explainable agents, and we believe that an extension of the techniques presented in this work can address these but are outside the scope of the present work.

© The Author(s), under exclusive license to Springer Nature Switzerland AG 2023
D. Calvaresi et al. (Eds.): EXTRAAMAS 2023, LNAI 14127, pp. 3–17, 2023.
https://doi.org/10.1007/978-3-031-40878-6_1

detailed explanation (e.g., one that justifies an action with extensive information about the agent reasoning) will require no additional information to answer this question. Nevertheless, in certain settings (e.g., time-constrained environments), explanations are more useful when they are relatively unfaithful [6,7]. Explaining by beliefs can also help solve a range of problems, such as encouraging behaviour change [4], enhancing human-agent teaming [8], and transparency [9].

Although there is a large and growing body of work on explanation generation in the field of autonomous agents [1], much of this work has traditionally assumed the availability of explanation generation modules, reliable observations, and deterministic execution of plans. However, in real-life settings, autonomous agents are not explainable by design, unreliable observations are frequently encountered, and plans are non-deterministic. We seek in this work to address these challenges by proposing techniques that mine the historical data associated with agent system execution to generate belief-based explanations of agents' past actions. We shall refer to this problem as *mining belief-based explanations*. Our proposal relies on the notion of audit logging. Many of the existing MAS frameworks (e.g., JACK framework [10]) support logging different aspects of agent behaviour. Of these, we are interested in two particular aspects: (1) a behaviour log that records the creation and completion of the past executed actions, and (2) a belief log that describes the belief set activity of the agent during the recording of (1). One such implementation of audit logging is the tracing and logging tools for JACK Intelligent Agents [10]. Our approach to mining belief-based explanation involves two steps:

1. We leverage these two audit logs in chronological order to generate one sequence database taken as input by a sequential pattern miner. The intuition behind this is to identify commonly occurring patterns of action execution events preceded by sequences of beliefs. Here, we intend to mine the enabling beliefs of each action referred to in the behaviour log.
2. We define a validation technique that leverages a state update operator (i.e., an operator that defines how the specification of a belief state is updated as a consequence of the agent's perception of the environment), agent's past experiences provided by the above-cited sequence database to compute the soundness and completeness of the mined belief-based explanations.

Mining and validating belief-based explanations can be outlined as follows: Given as inputs (1) a behaviour log of past executed actions, (2) a belief log, (3) a plan or plans that execution generated these logs, (4) a state update operator, compute: the belief-based explanations of every action referred to in the behaviour log. While inputs (1) and (2) are used for mining belief-based explanations, inputs (3) and (4) are used for validating the mined explanations. As we show later in this work, inputs (3) and (4) can also be used to generate detailed belief-based explanations.

The remainder of this work is organized as follows. Section 2 introduces our running example and some required preliminaries. Section 3 describes our approach to updating belief-based explanations, which sits at the core of this work.

Section 4 describes our approach to mining belief-based explanations. In Sect. 5, we describe how the mined explanations can be validated. Section 6 reports an empirical evaluation of this work. Related work is discussed in Sect. 7 before we conclude and outline future work in Sect. 8.

2 Preliminaries and Running Example

Agents with BDI architecture are designed to imitate human practical reasoning using beliefs, desires and intentions. The beliefs represent what the agent knows about the environment, the desires are goals (i.e., objectives) that the agent would like to bring about, and the intentions are plans that the agent is committed to executing. With these anthropomorphic handles, the BDI agent derives its actions and, consequently, can explain them. As the literature (e.g., [11]) suggests, this is an elegant means to explain agent systems with considerable underlying complexity. Although an explainable BDI agent must faithfully reflect its reasoning cycle (i.e., including its beliefs, goals, plans, and intentions), explaining by beliefs can provide value in ways other handles cannot. To illustrate this, we study two scenarios for Engine failure on take-off (EFTO): (1) accelerate-stop and (2) accelerate-go, as described in [12].

Example 1. As shown in Fig. 1, the pilot agent has two plans written on the basis of Jason agent programming language [13] to handle EFTO in large twin-engine jet aircraft: (p1) accelerate-stop and (p2) accelerate-go. p1 involves reducing thrust, applying speed breaks and notifying the tower of the emergency. p2 begins with ensuring full power is applied (including mixture, throttle, landing gear and flaps), followed by liftoff, and then notifying the tower of the emergency and the intended landing. To handle an EFTO successfully, two critical speeds must be calculated before each take-off, namely V1 (the speed at which the pilot can abort the take-off safely) and V2 (the speed at which the pilot can take-off safely).

```
@p1 +efto(aircraft):  V1 < Speed < V2
   <- idle(throttle);
      deploy(brakes);
      .send(tower,tell,stop(accelerate)).
@p2 +efto(aircraft):  V1 < Speed < V2
   <- increase(mixture);
      increase(throttle);
      take_up(flap);
      pull(yoke);
      take_up(gear);
      .send(tower,tell,go(accelerate));
      .send(tower,tell,return(landing)).
```

Fig. 1. Jason plans for handling EFTO

A particularly stressful situation for the pilot is when EFTO occurs between V1 and V2 (i.e., the aircraft is going too fast to accelerate-stop but too low to accelerate-go). For this particular situation, the two plans (p1 and p2) are applicable (i.e., the plan selection is based on some conditions not represented in the agent code).

Existing explanation generation techniques can be summarized as follows: A BDI agent triggers an action with respect to its goals and beliefs, which can be represented in terms of a Goal Hierarchy Tree (GHT) [2–4]. A GHT is a tree structure representing a high-level abstraction of an agent's reasoning. At the root of the tree, the agent's main goal is placed. A link from the top-level goal to one or more sub-goals means that these sub-goals must be achieved as part of the top-level goal. Tree leaves represent actions that the agent can execute. For the agent to execute an action, certain beliefs placed directly above the action must be true. What existing explanation algorithms do is select the beliefs and goals that are directly above the selected action node in the GHT to design explanation patterns. We argue that such explanation generation techniques can provide irrational explanations in many settings. To illustrate this consider the following scenarios.

Example 2. Assume that the pilot decided to accelerate-stop at one instance and accelerate-go at another instance in the past. Now, consider the following queries that may arise by an aviation candidate:

– *Why did the pilot agent pull the throttle lever to idle immediately after the EFTO?*
– *Why did the pilot agent move the mixture knob to rich after the EFTO?*

Following the current norms of explanation generation, the two queries in Example 2 can be readily answered using the goal efto(aircraft) and the belief that V1 < Speed < V2. Another way to say that both queries are explained by the same goal and same belief. It is clear that these explanations are inaccurate - they do not accurately describe how the pilot agent came to its decision to accelerate-stop at the first instance nor to accelerate-go at the later instance.

2.1   Audit Logs

During agent systems execution, a wide variety of data on changes in the agent's mental attitude and environment can be represented in the form of audit logs. Collecting such data can be implemented using audit logging tools such as Mind Inspector in Jason platform [13] and Design Tracing Tool (DTT) in JACK platform [10]. We are interested in two modes of audit logging: (1) behaviour logs and (2) belief logs. A behaviour log describes the historical execution of plans as sequences of events where each event refers to some action. A behaviour log can be represented as a set of triples $\langle$p_label, t_i, $a_i\rangle$, where the value of t_i refers to the starting time of the action a_i, which has been executed as part of a plan labelled p_label. An excerpt of the behaviour log associated with the plans in our running example during two different flights is recorded in Table 1.

Table 1. A behaviour log of the pilot agent.

Flight No.	Plan	Timestamp	Action
2065	p1	t75	idle(throttle)
2065	p1	t77	deploy(brakes)
2065	p1	t80	send(tower, msg)
2072	p2	t1027	increase(mixture)
2072	p2	t1029	increase(throttle)
2072	p2	t1031	take up(flap)
2072	p2	t1033	pull(yoke)
2072	p2	t1035	take up(gear)
2072	p2	t1037	send(tower, msg)
2072	p2	t1038	send(tower, msg)

A belief log records the history of the external events perceived by the target agent. It consists of a set of couples $\langle t_i, q_i \rangle$, where t_i value indicates the time when the agent added the belief q_i to its belief base. Note that belief logs[2] record new beliefs as they are added to the belief base but do not record persistent beliefs. Determining which beliefs hold at a certain point of system execution, therefore, requires updating machinery (e.g., the state update operator described in Subsect. 3). Table 2 illustrates an excerpt of a belief log describing external events perceived by the pilot agent during two different flights.

Table 2. A belief log of the pilot agent

Flight No.	Timestamp	Beliefs	Flight No.	Timestamp	Beliefs
2065	t70	runway(dry)	2072	t1024	efto
2065	t71	wind(cross)	2072	t1025	V1 = 156
2065	t72	efto	2072	t1025	V2 = 166
2065	t73	V1 = 129	2072	t1025	Flaps = 15
2065	t73	V2 = 145	2072	t1026	Speed = 161
2065	t73	Flaps = 15	2072	t1028	escalating(fuel flow)
2065	t74	Speed = 135	2072	t1030	accelerating(thrust)
2065	t76	decelerating(thrust)	2072	t1032	Flaps = 0
2065	t78	steady(aircraft)	2072	t1034	liftoff(aircraft)
2072	t1022	runway(wet)	2072	t1036	up(gear)
2072	t1023	wind(head)	2072	t1040	liftoff(aircraft)

---

[2] One can leverage JACK capability methods to make belief set activities available at agent level [14]. This manipulation allows, in turn, to store of enabling beliefs based on the user-defined data structure.

Normally, it is more convenient to represent beliefs in first-order sentences to maintain consistency and constraints. To avoid handling different groundings of the variables as distinct beliefs, we need to neglect the precise grounding of valuables. Nevertheless, there are settings where we need precise instantiations of the variables.

3   Updating Belief-Based Explanations

Updating belief-based explanations is useful for at least two reasons. First, it could be used to contextualise explanations (i.e., providing users with detailed explanations). Another way to say that updated belief-based explanations can help answer the following question for any step of plan execution: *What must have been known in detail for the agent to perform a particular action over another?* As a second reason, it could also be used to validate the mined explanations, which we describe in detail in later sections.

At each action step in a plan execution, we derive the updated belief-based explanation of an action by combining the enabling beliefs of the preceding actions with the enabling beliefs of the action we are at. For the purpose of this work, we ignore other constructs appearing in a plan body (e.g., achievement and test goals). We assume that each action in a plan is associated with enabling beliefs (i.e., no provisional execution of actions) written as conjunctive normal-form sentences using a state description language that might involve propositional variables (i.e., variables that can be true or false) and non-Boolean variables (i.e., new value assignments). We allow the updated belief-based explanations to be non-deterministic for two reasons (1) in any plan with OR branching, one might arrive at an action through multiple trajectories, and (2) much of the existing state update operators resolve inconsistencies in multiple different ways. Among the two well-known state update operators in the literature - the Possible Worlds Approach (PWA) [15], and the Possible Models Approach (PMA) [16] - our work relies on the PWA. More precisely, we use the state update operator $\oplus$ as defined below, assuming the presence of a background knowledge base KB.

Definition 1 ($\oplus$ Operator). For the two belief states s_i, s_j, and the knowledge base KB, the state update operator $\oplus$ can be defined as follows:

$$s_i \oplus s_j = \{s_j \cup s_i' \mid (s_i \wedge s_i' \cup s_j \cup KB \not\models \perp) \wedge (\nexists\ s_i''\ \text{such that}$$
$$s_i' \subset s_i'' \subseteq s_i \wedge s_i'' \cup s_j \cup KB \not\models \perp)\},$$

in which if $s_j \cup s_i$ is consistent, then the resulting updated explanation is $s_j \cup s_i$. Otherwise, we need to define $s_i' \subseteq s_i$ such that $s_j \cup s_i'$ is consistent and there is no exists s_i'' such that $s_i' \subset s_i'' \subseteq s_i$ and $s_j \cup s_i''$ is consistent. Note that we might need to refer to a general version of the state update operator, i.e., if $S = \{s_1, \ldots, s_n\}$ is a finite set of belief states, then $S \oplus s = \{s_i \oplus s \mid s_i \in S\}$. Note also that the output of the state update operator is not always unique state specifications. Actually, the output might be a set of non-deterministic possible belief states. For the purpose of illustrating why this might be the case, we consider the following example.

Example 3. Consider the following knowledge base

$$KB = r \rightarrow \neg (d \wedge q)$$

representing a rule for the pilot agent, where the propositional letter r can be read as there is an EFTO, the letter d as the thrust is accelerating, and the letter q as the aircraft is ascending. Now, let $(d \wedge q)$ hold in some previous belief state, and r came to be held in the belief state where we are at. Applying $\oplus$, the generated two alternative scenarios describing the updated belief states are

1. $\{d \wedge r\}$ and
2. $\{q \wedge r\}$

which is to say, the rule in KB expresses that whenever the pilot agent believes that there is an EFTO, then it is believed that either the thrust is accelerating or the aircraft is ascending (i.e., the thrust cannot accelerate unless descending after engine failure).

With the intention of obtaining complete detailed belief-based explanations of the agent behaviour, we need to apply the state update operator over each pair of actions in the behaviour log repeatedly, with the previous updated belief-based explanations associated with the former action as the first argument and the current enabling beliefs associated with the later action as the second argument.

4 Mining Belief-Based Explanations

Mining belief-based explanations starts with transforming the observations in the audit logs into observation sequences, each of which involves the execution of an action and the manifestations of its enabling beliefs. Note that logging tools are usually designed to record one mode of observation per log. That is to say, action execution and external events manifestation are recorded in separate logs. To that end, we also need to start a correlation between the two logs to obtain an observation log that serves as a sequential database, which will be mined to extract belief-based explanations using a sequential rule miner. We define this correlation as follows.

Definition 2 (Observation sequence, and log). Let A be an actions space, B be a belief states space, $a_0, \ldots, a_n \in A$, and $b_0, \ldots, b_n \in B$. An observation instance (t) is an alternating sequence of the form $b_0, a_1, \ldots, b_n, a_n$. D_{All} is an observation log, such that $D_{All} \in 2^T$, where T is the set of all observation instances.

Mining belief-based explanations relies on the two following premises: (1) that the beliefs observed in the belief log immediately before executing an action can be the enabling beliefs of that action, and (2) that the persistent beliefs observed a long time before the execution of an action are typically not the enabling beliefs of that action, but may be of that action plus some others. Hence, we use the basic relation "direct successor" [17] over the actions and beliefs in the D_{All} as follows:

Definition 3 (Direct successor). Let D_{All} be an observation sequence over T. Let $b, a, b' \in T$.

1. Direct predecessor state: $b >_D a$ *iff* $\langle b, a \rangle$ is a subsequence of T, and
2. Direct successor state: $a >_D b'$ *iff* $\langle a, b' \rangle$ is a subsequence of T.

Relation $>_D$ describes which external events directly follow/precede a given action. Direct predecessor relation over D_{All} would offer learning entries of the form $\langle b, a \rangle$, where a is an action applicable at the belief state b. For our purposes, we do not distinguish between $\langle q, p, a \rangle$ and $\langle p, q, a \rangle$, because we are only interested in relating actions to their enabling beliefs but not in the relation amongst enabling beliefs. Against this background, we create O_{Direct}, which is a sequence of the following form

$$\langle \langle \langle b_{11}, .., b_{1n} \rangle, a_1 \rangle \rangle, .., \langle \langle b_{i1}, .., b_{im} \rangle, a_i \rangle \rangle, .., \langle \langle b_{p1}, .., b_{pk} \rangle, a_p \rangle \rangle \rangle$$

where each $\langle a_{i-1}, a_i \rangle$ represents an ordered pair of actions, and each $\langle b_{i1}, .., b_{im} \rangle$ represents the observed beliefs before the execution of action a_i and after action a_{i-1} execution. An exception is required for the first recorded action in D_{All}, as there is no preceding action. In this case, we use the timestamp of the initial high-level event as the start time of the system execution. Given D_{All}, we view the problem of mining belief-based explanations as finding all the sequences $\langle b, a \rangle$ that satisfy some predefined measures of interestingness, assuming unique activity execution (i.e., there is no concurrent execution of actions).

Again, we are interested in discovering all the beliefs that are observed always, or most of the time, directly before the execution of each action referred to in the behaviour log. Association rule learning can be an effective means for discovering regularities between beliefs and actions. Fundamentally, given two itemsets X and Y, the sequential rule $X \rightarrow Y$, states that if the elements in X occur, then it will be followed by the elements in Y. For a sequential rule to be an interesting one, it must satisfy minimum support (how frequently the rule appears in D_{All}) and minimum confidence (the accuracy of the rule). Such rules are considered in our context as belief-based explanations. We use the well-known CMRules algorithm [18] to discover this form of rules.

Example 4. Continuing with our running example, Table 3 shows the results of applying the CMRules algorithm to D_{All} obtained from Table 1 and Table 2.

CMRules algorithm can discover all the interesting association rules from D_{All}. Nevertheless, additional post-processing is required, where we rule out any association rule that its consequent label is not a single action name or its antecedent label is not beliefs.

Table 3. Mined belief-based explanations

plan	action	belief-based explanations
p1	idle(throttle)	runway(dry) ∧ wind(cross) ∧ efto ∧ Flaps = 15 ∧ V1 > Speed > V2
p1	deploy(brakes)	decelerating(thrust)
p1	send(tower, msg)	steady(aircraft)
p2	increase(mixture)	runway(wet) ∧ wind(head) ∧ efto ∧ Flaps = 15 ∧ V1 > Speed > V2
p2	increase(throttle)	escalating(fuel_flow)
p2	take_up(flap)	accelerating(thrust)
p2	pull(yoke)	liftoff(aircraft)

5 Validating the Explanation Process

Our guiding intuition here is that the state update operator and the available data used to update belief-based explanations can also be leveraged to validate the mined explanations. Our validation approach involves some mechanisms that take as inputs (1) A D_{All}, (2) A state update operator, and compute the soundness and completeness of the mind explanations. We keep assuming unique action execution through this section.

First, we need inputs (1) and (2) to generate a sequence (denoted hereafter as $D_{Updated}$) that associates each action in D_{All} to the set of updated beliefs of all actions executed up to that point. Each object in $D_{Updated}$ is a pair of the form $\langle \beta_i, a_i \rangle$, where β the set of updated beliefs of all actions executed up to a_i. $D_{Updated}$ can be simply obtained using $\oplus$ over D_{All} assuming the presence of a KB defined in the same language as that in which the beliefs are described. Table 4 illustrates the results of applying the operator $\oplus$ to input D_{All}.

Table 4. Updated belief-based explanations of plan p1

plan	action	belief-based explanations
p1	idle(throttle)	runway(dry) ∧ wind(cross) ∧ efto ∧ Flaps = 15 ∧ V1 > Speed > V2
p1	deploy(brakes)	decelerating(thrust) ∧ runway(dry) ∧ wind(cross) ∧ efto ∧ Flaps = 15 ∧ V1 > Speed > V2
p1	send(tower, msg)	decelerating(thrust) ∧ steady(aircraft) ∧ runway(dry) ∧ wind(cross) ∧ efto Flaps = 15 ∧ V1 > Speed > V2

It should be noted that a single action in $D_{Updated}$ can be associated with a set of sets of beliefs due to the non-determinism nature of the $\oplus$ operator. Now, to validate the mined explanations, it is useful to establish the following:

– **Soundness.** A sound belief-based explanation is one that is mined correctly (i.e., observed in D_{All}). Another way to say that a detailed belief-based explanation to a given point in the plan execution should contain the mined belief-based explanation updated using $\oplus$ at that point in the plan execution. Formally, for each plan execution sequence in $D_{Updated}$ and each action a_i the following condition must hold: $\beta_i \cup KB \models b$ for some $b \in b_{a_1}, \ldots, b_{a_i}$, where b_{a_i} is the mined belief-based explanation of action a_i .
– **Completeness.** A complete belief-based explanation requires that all the enabling beliefs of a given action are mined. Actually, this can be viewed as a reversal of the above-cited entailment relation (i.e., $b \cup KB \models \beta_i$)

Where it is possible for the mined explanations to be unsound or incomplete, further post-processing may be required for more reliable results. We overtake this problem by seeking more observations and/or re-mining with lower support and confidence thresholds.

6   Evaluation

Agent explanation generation can be evaluated on several grounds. Commonly, the first is a human-rated evaluation of the explainability of the generated explanations. However, we consider this more as a part of explainable agent development, so our focus in this section is on evaluating our mining technique, which includes the performance measures of the mined explanations (i.e., precision and recall).

This section presents the evaluation of our approach to mining belief-based explanations. First, we present the setup and implementation for our experiments. Next, our evaluation is detailed.

6.1   Data and Implementation

We implemented our approach as a plugin for our Toolkit XPlaM[3] [19] and evaluated it using a synthetic log of 1000 execution instances, representing firefighter agent past behaviour as described in [3]. We also used a plan library consisting of 15 plans[4].

---

[3] The source code for XPlaM Toolkit (including the code for the approach presented here) has been published online at https://github.com/dsl-uow/xplam.
[4] We published the datasets supporting the conclusions of this work online at https://www.kaggle.com/datasets/alelaimat/explainable-bdi-agents.

6.2   Performance Results

Our goal of the evaluation is to establish that the proposed approach can generate generally reliable explanations. To that end, we recorded the precision (i.e., number of correctly mined explanations over the total number of mined explanations) and recall (i.e., number of correctly mined explanations over the total number of actual explanations) obtained from applying the CMRules algorithm and the above-cited validation technique. We consider minimum confidence (*min conf*) and minimum support (*min supp*) of the mined rules as the most important factors for mining belief-based explanations. The results are depicted in Table 5 and are summarized below.

Table 5. Performance for different *min conf* and *min supp*

min conf	1.00	0.95	0.9	0.85	0.8
Precision	0.8948	0.8910	0.8731	0.8487	0.8235
Recall	0.5678	0.6387	0.7265	0.7854	0.8547
min supp	0.5	0.4	0.3	0.2	0.1
Precision	0.9216	0.8741	0.8254	0.7703	0.6987
Recall	0.2257	0.3458	0.5361	0.6966	0.8815

min conf:

1. Confidence threshold considerably impacts performance results, i.e., higher *min conf* leads to higher precision but lower recall. For example, our approach achieved the highest precision of 0.89 with the lower recall of 0.56 for *min conf* of 1.00.
2. It is necessary, thus, to find a trade-off between precision and recall when varying *min conf* threshold. Hence, we use *min conf* of 0.9 for testing the impact of varying *min supp* threshold.

min supp

1. Similar to the *min conf*, varying *min supp* has a significant impact on the performance results. For example, we achieved the highest precision of 0.92 with an insignificant recall of 0.22 for *min supp* of 0.5.
2. Note that extracting association rules related to infrequent events needs *min supp* to be low. However, this, in turn, sacrifices the precision of the mined explanations.

We noticed some insightful observations through our experiments. First, the size of D_{All} does not impact the performance results of the mined explanations *iff* D_{All} includes all possible behaviours of the observed agent. For example, we achieved very close precision values for D_{All} of 200 and 800 execution instances, which were 0.836 and 0.842, respectively. Second, varying *min conf* and *min supp* have a significant impact on the performance of the mined explanations and, consequently, come with several limitations (e.g., time and effort). Finally, the premise that the beliefs observed in the belief log immediately before executing an action can be the enabling beliefs of that action can be deceptive sometimes (i.e., the enabling beliefs of a late action can hold before plan execution). It is necessary, therefore, for the user to determine precise *min conf* and *min supp* thresholds.

7   Related Work

Although a large and growing body of work exists on developing explainable agents and robots [1], few works focus on generating explanations for intelligent agents.

Harbers et al. [3] described four algorithms to design explainable BDI agents: one using parent goals, one using top-level goals, one using enabling beliefs, and one using the next action or goal in the execution sequence. They found that goal-based explanations were slightly preferable to belief-based expansions to explain procedural actions (i.e., a sequence of actions and sub-goals) based on users' evaluation. Nevertheless, belief-based explanations were preferable in explaining conditional and single actions. Similar explanation algorithms were proposed by Kaptein et al. [2], but to investigate the difference in preference of adults and children for goal-based and belief-based explanations. They found that both adults and children preferred goal-based explanations. Related, but in a different ontology, is the work presented in Kaptein et al. [20], in that the agent explains its action in terms of its beliefs, goals and emotions.

Sindlar et al. [21] proposed an abductive approach to infer the mental states of BDI agents in terms of beliefs and goals. To that end, they described three explanatory strategies under three perceptory presumptions: complete, late, and partial observations. Sindlar et al. extend their work to an explanation approach that takes into account three organizational principles: roles, norms, and scenes in [22]. The extended work proposes an approach to how the observed behaviour of game players can be explained and predicted in terms of the mental state of virtual characters. Related, but in a different domain, is the work presented in [23]. Sequeira and Gervasio propose a framework for explainable reinforcement learning that extracts relevant aspects of the RL agent interaction with its environment (i.e., interestingness elements) in [23]. They suggested four dimensions of analysis: frequency, execution certainty, transition-value, and sequence analysis to extract the interestingness elements, which are used to highlight the behaviour of the agent in terms of short video clips.

All the previous approaches presented in this paper can theoretically generate explanations of agents' actions, but assuming reliable observations, availability of an explanation generation module and deterministic execution of plans. On the other hand, we leverage the past execution experiences of the target agent, which allows performing various techniques to handle unreliable observations (e.g., measures of interestingness). Although historical data might be hard to be obtained, our approach still can be effective in settings where no more than common patterns (i.e., plans), information that other players would observe or expect the target agents to be aware of (i.e., external events) and the external behaviour of the target agent are available. Much of the work done on agent explanation generation shares the common judgment that relatively short explanations are more useful to explainees. Nevertheless, as shown in our running example, detailed explanations are critical in some cases (e.g., explaining BDI plan selection). Finally, and in contrast with the literature, our explanation generation mechanism allows explanations to be non-deterministic through the updating of belief-based explanations, as described in Sect. 3.

8   Conclusion

In this paper, we addressed the problem of agent explanation mining (and specifically belief-based explanations) in the context of the well-known BDI paradigm. This problem was formulated as follows: "Given the past execution experiences of an agent and an update operator, generate the belief-based explanation of each action referred to in the agent's past execution experiences". We presented an update operator that is able to generate detailed explanations. Through examples, we also showed that detailed explanations could be useful in explaining BDI plan selection. We have tackled the problem of explanation generation in non-deterministic settings. At this point in time, we are trying to extend the application of the proposed approach to other BDI handles.

References

1. Anjomshoae, S., Najjar, A., Calvaresi, D., Främling, K.: Explainable agents and robots: results from a systematic literature review. In: 18th International Conference on Autonomous Agents and Multiagent Systems (AAMAS 2019), Montreal, Canada, 13–17 May 2019, pp. 1078–1088. International Foundation for Autonomous Agents and Multiagent Systems (2019)
2. Kaptein, F., Broekens, J., Hindriks, K., Neerincx, M.: Personalised self-explanation by robots: the role of goals versus beliefs in robot-action explanation for children and adults. In: 2017 26th IEEE International Symposium on Robot and Human Interactive Communication (RO-MAN), pp. 676–682. IEEE (2017)
3. Harbers, M., van den Bosch, K., Meyer, J.-J.: Design and evaluation of explainable bdi agents. In: 2010 IEEE/WIC/ACM International Conference on Web Intelligence and Intelligent Agent Technology, vol. 2, pp. 125–132. IEEE (2010)
4. Abdulrahman, A., Richards, D., Bilgin, A.A.: Reason explanation for encouraging behaviour change intention. In: Proceedings of the 20th International Conference on Autonomous Agents and MultiAgent Systems, pp. 68–77 (2021)
5. Georgeff, M., Rao, A.: Modeling rational agents within a bdi-architecture. In: Proceedings of 2nd International Conference on Knowledge Representation and Reasoning (KR 1991), pp. 473–484. Morgan Kaufmann (1991)
6. Harbers, M., van den Bosch, K., Meyer, J.-J.C.: A study into preferred explanations of virtual agent behavior. In: Ruttkay, Z., Kipp, M., Nijholt, A., Vilhjálmsson, H.H. (eds.) IVA 2009. LNCS (LNAI), vol. 5773, pp. 132–145. Springer, Heidelberg (2009). https://doi.org/10.1007/978-3-642-04380-2_17

7. Calvaresi, D., Mualla, Y., Najjar, A., Galland, S., Schumacher, M.: Explainable multi-agent systems through blockchain technology. In: Calvaresi, D., Najjar, A., Schumacher, M., Främling, K. (eds.) EXTRAAMAS 2019. LNCS (LNAI), vol. 11763, pp. 41–58. Springer, Cham (2019). https://doi.org/10.1007/978-3-030-30391-4_3

8. Verhagen, R.S., Neerincx, M.A., Tielman, M.L.: A two-dimensional explanation framework to classify AI as incomprehensible, interpretable, or understandable. In: Calvaresi, D., Najjar, A., Winikoff, M., Främling, K. (eds.) EXTRAAMAS 2021. LNCS (LNAI), vol. 12688, pp. 119–138. Springer, Cham (2021). https://doi.org/10.1007/978-3-030-82017-6_8

9. Mualla, Y.: Explaining the behavior of remote robots to humans: an agent-based approach. PhD thesis, Université Bourgogne Franche-Comté (2020)

10. Winikoff, M.: Jack^{TM} intelligent agents: an industrial strength platform. In: Bordini, R.H., Dastani, M., Dix, J., El Fallah Seghrouchni, A. (eds.) Multi-Agent Programming. MSASSO, vol. 15, pp. 175–193. Springer, Boston, MA (2005). https://doi.org/10.1007/0-387-26350-0_7

11. Abdulrahman, A., Richards, D., Ranjbartabar, H., Mascarenhas, S.: Belief-based agent explanations to encourage behaviour change. In: Proceedings of the 19th ACM International Conference on Intelligent Virtual Agents, pp. 176–178 (2019)

12. Multi-engine aeroplane operations and training. Technical report, Civil Aviation Safety Authority (July 2007)

13. Bordini, R,H., Hübner, J.M., Wooldridge, M.: Programming multi-agent systems in AgentSpeak using Jason. John Wiley & Sons (2007)

14. Howden, N., Rönnquist, R., Hodgson, A., Lucas, A.: Jack intelligent agents-summary of an agent infrastructure. In: 5th International Conference on Autonomous Agents, vol. 6 (2001)

15. Ginsberg, M.L., Smith, D.E.: Reasoning about action i: a possible worlds approach. Artifi. intell. **35**(2), 165–195 (1988)

16. Winslett, M.S.: Reasoning about action using a possible models approach, pp. 1425–1429. Department of Computer Science, University of Illinois at Urbana-Champaign (1988)

17. Maruster, L., Weijters, A.J.M.M.T., van der Aalst, W.M.P.W., van den Bosch, A.: Process mining: discovering direct successors in process logs. In: Lange, S., Satoh, K., Smith, C.H. (eds.) DS 2002. LNCS, vol. 2534, pp. 364–373. Springer, Heidelberg (2002). https://doi.org/10.1007/3-540-36182-0_37

18. Fournier-Viger, P., Faghihi, U., Nkambou, R., Nguifo, E.M.: Cmrules: mining sequential rules common to several sequences. Knowl.-Based Syst. **25**(1), 63–76 (2012)

19. Alelaimat, A., Ghose, A., Dam, K.H.: Xplam: a toolkit for automating the acquisition of BDI agent-based digital twins of organizations. Comput. Industry **145**, 103805 (2023)

20. Kaptein, F., Broekens, J., Hindriks, K., Neerincx, M.: The role of emotion in self-explanations by cognitive agents. In: 2017 Seventh International Conference on Affective Computing and Intelligent Interaction Workshops and Demos (ACIIW), pp. 88–93. IEEE (2017)

21. Sindlar, M.P., Dastani, M.M., Dignum, F., Meyer, J.-J.C.: Mental state abduction of BDI-based agents. In: Baldoni, M., Son, T.C., van Riemsdijk, M.B., Winikoff, M. (eds.) DALT 2008. LNCS (LNAI), vol. 5397, pp. 161–178. Springer, Heidelberg (2009). https://doi.org/10.1007/978-3-540-93920-7_11

22. Sindlar, M.P., Dastani, M.M., Dignum, F., Meyer, J.-J.C.: Explaining and predicting the behavior of BDI-based agents in role-playing games. In: Baldoni, M., Bentahar, J., van Riemsdijk, M.B., Lloyd, J. (eds.) DALT 2009. LNCS (LNAI), vol. 5948, pp. 174–191. Springer, Heidelberg (2010). https://doi.org/10.1007/978-3-642-11355-0_11
23. Sequeira, P., Gervasio, M.: Interestingness elements for explainable reinforcement learning: understanding agents' capabilities and limitations. Artif. Intell. **288**, 103367 (2020)

Evaluating a Mechanism for Explaining BDI Agent Behaviour

Michael Winikoff[1]([✉])(iD) and Galina Sidorenko[2](iD)

[1] Victoria University of Wellington, Wellington, New Zealand
michael.winikoff@vuw.ac.nz
[2] Halmstad University, Halmstad, Sweden
galina.sidorenko@hh.se

Abstract. Explainability of autonomous systems is important to supporting the development of appropriate levels of trust in the system, as well as supporting system predictability. Previous work has proposed an explanation mechanism for Belief-Desire-Intention (BDI) agents that uses folk psychological concepts, specifically beliefs, desires, and valuings. In this paper we evaluate this mechanism by conducting a survey. We consider a number of explanations, and assess to what extent they are considered believable, acceptable, and comprehensible, and which explanations are preferred. We also consider the relationship between trust in the specific autonomous system, and general trust in technology. We find that explanations that include valuings are particularly likely to be preferred by the study participants, whereas those explanations that include links are least likely to be preferred. We also found evidence that single-factor explanations, as used in some previous work, are too short.

Keywords: Explainable Agency · Belief-Desire-Intention (BDI) · Evaluation

1 Introduction

"Explainability is crucial for building and maintaining users' trust in AI systems." [16]
"Automated systems should provide explanations that are technically valid, meaningful and useful to you and to any operators or others who need to understand the system" https://www.whitehouse.gov/ostp/ai-bill-of-rights/, published 4 October 2022.

It is now widely accepted that explainability is crucial for supporting an appropriate level of trust in autonomous and intelligent systems (e.g. [12,16,29]). However, explainability is not just important to support (appropriate) trust. It also makes a system understandable [34], which in turn allows systems to be challenged, to be predictable, to be verified, and to be traceable [34].

In this paper we focus on *autonomous agents*: software systems that are able to act autonomously. This includes a wide range of physically embodied systems

---

The authors were at the University of Otago, New Zealand, when most of the work was done.

D. Calvaresi et al. (Eds.): EXTRAAMAS 2023, LNAI 14127, pp. 18–37, 2023.
https://doi.org/10.1007/978-3-031-40878-6_2

(e.g. robots) and systems that do not have physical embodiment (e.g. smart personal assistants) [25, 26, 28]. Although autonomous systems use AI techniques, not all AI systems are autonomous, e.g. a system may be simply making recommendations to a human, rather than taking action itself.

Explainability is particularly important for autonomous systems [20, 36], since, by definition, they take action, so, depending on the possible consequences of their actions, there is a need to be able to trust these systems appropriately, and to understand how they operate. One report proposes to include "...*for users of care or domestic robots a why-did-you-do-that button which, when pressed, causes the robot to explain the action it just took*" [32, Page 20]. It has also been argued that explainability plays an important role in making autonomous agents accountable [8].

However, despite the importance of explainability of autonomous systems, most of the work on explainable AI (XAI) has focused on explaining machine learning (termed "data-driven XAI" by Anjomshoae *et al.* [4]), with only a much smaller body of work focusing on explaining autonomous agents (termed "goal-driven XAI" by Anjomshoae *et al.* [4], and "explainable agency" by Langley *et al.* [20]). Specifically, a 2019 survey [4] found only 62 distinct published papers on goal-driven XAI published in the period 2008–2018.

In order to develop a mechanism for an autonomous agent to be able to answer in a useful and comprehensible way questions such as "*why did you do X?*", it is useful to consider the social sciences [23]. In particular, we draw on the extensive (and empirically-grounded) work of Malle [21]. Malle argues that humans use folk psychological constructs in explaining their behaviour[1]. Specifically, in explaining their behaviour, humans use the concepts of beliefs, desires[2], and valuings[3].

Prior work [38] has used these ideas to develop a mechanism that allows Belief-Desire-Intention (BDI) agents [5, 6, 27] (augmented with a representation for valuings, following [9]) to provide explanations of their actions in terms of these concepts.

In this paper we conduct an empirical human subject evaluation of this mechanism, including an evaluation of the different component types of explanations (e.g. beliefs, desires, valuings). Such evaluations are important in assessing the effectiveness of explanatory mechanisms. For example, are explanations using beliefs seen as less or more preferred than explanations that use desires, or that use valuings? Empirical evaluation can answer these questions, and by answering them, guide the development and deployment of explanation mechanisms for autonomous agents. Specifically, the key research question we address[4] is: **What forms of explanation of autonomous agents are preferred?**.

---

[1] There is also empirical evidence that humans use these constructs to explain the behaviour of robots [13, 33].

[2] Terminology: we use "goal" and "desire" interchangeably.

[3] Defined by Malle as things that "*directly indicate the positive or negative affect toward the action or its outcome*". Whereas values are generic (e.g. benevolence, security [30]), valuings are about a given action or outcome. Valuings can arise from values, but can also be directly specified without needing to be linked to higher-level values. In our work we represent valuings as preferences over options [38, Sect. 2].

[4] We also consider (Sect. 4.4) the question: "to what extent is trust in a given system determined by a person's more general attitudes towards technology, and towards Artificial Intelligence?".

An earlier evaluation of this explanation mechanism has been conducted [37] (the results of which are also briefly summarised in [38]). However, this paper differs from the earlier evaluation in that: (i) we use a different scenario, (ii) we use different patterns of explanations, including links (which were not included in the earlier evaluation), (iii) we also include questions on trust in technology, and (iv) we conduct a deeper and more sophisticated analysis, including an assessment of the effects of the different explanatory component types, and of the correlation between trust in the autonomous system and more general trust in technology.

We propose a number of hypotheses, motivated by existing literature (briefly indicated below, and discussed in greater length in Sect. 5). Our hypotheses all relate to the *form* of the explanation. Since the explanation we generate has four types of explanatory factors, we consider for each of these types how they are viewed by the user (H1–H3). Furthermore, since including more types of explanatory factors results in longer explanations, we also consider the overall effect of explanation length (H4).

H1: Explanations that include valuings are more likely to be preferred by users over other forms of explanations (that do not include valuings). This hypothesis is based on the finding of [37].

H2: Explanations that include desires are more likely to be preferred by users over explanations that include beliefs. This hypothesis is based on the findings of [7, 15, 17] (discussed in detail in Sect. 5).

H3: Explanations that include links are *less* likely to be preferred by users over other forms of explanations (that do not include links). This hypothesis is based on the findings of [15].

H4: Shorter explanations are more likely to be preferred by users. This hypothesis is based on the arguments of (e.g.) [17]. Note that they argued that explanations ought to be short, and therefore only evaluated short explanations. In other words, their evaluation did not provide empirical evidence for this claim.

The remainder of this paper is organised as follows. We begin by briefly reviewing the explanation mechanism that we evaluate (Sect. 2). Next, Sect. 3 presents our methodology, and then Sect. 4 presents our results. We finish with a review of related work (Sect. 5), followed by a brief discussion (Sect. 6) summarising our findings, noting some limitations, and indicating directions for future work.

2   Explanation Mechanism

We now briefly review the explanation mechanism. For full details, we refer the reader to [38]. In particular, here we focus on the *form* of the explanations, omitting discussion of *how* the explanations are generated.

We use the following scenario: *Imagine that you have a smart phone with a new smart software assistant, SAM. Unlike current generations of assistants, this one is able to act proactively and autonomously to support you. SAM knows that usually you use one of the following three options to get home: (i) Walking, (ii) Cycling, if a bicycle is available, and (iii) Catching a bus, if money is available (i.e. there is enough credit on your card). One particular afternoon, you are about to leave to go home, when the*

E1: A bicycle was not available, money was available, the made choice (catch bus) has the shortest duration to get home (in comparison with walking) and I believe that is the most important factor for you, I needed to buy a bus ticket in order to allow you to go by bus, and I have the goal to allow you to catch the bus.

E2: A bicycle was not available, money was available, and the made choice (catch bus) has the shortest duration to get home (in comparison with walking) and I believe that is the most important factor for you.

E3: The made choice (catch bus) has the shortest duration to get home (in comparison with walking) and I believe that is the most important factor for you.

E4: A bicycle was not available, and money was available.

E5: A bicycle was not available, money was available, and I have the goal to allow you to catch the bus.

Fig. 1. Explanations E1–E5

phone alerts you that SAM has just bought you a ticket to catch the bus home. This surprises you, since you typically walk or cycle home. You therefore push the "please explain" button.

An explanation is built out of four types of building blocks: desires, beliefs, valuings, and links.

– A **desire (D)** explanation states that the agent having a certain desire was part of the reason for taking a certain action. For example, that the system chose to buy a bus ticket because it desired to allow you to catch the bus.

– A **belief (B)** explanation states that the agent having a certain belief was part of the reason for taking a certain action. For example, that the system chose to buy a ticket because it believed that a bicycle was not available.

– A **valuing (V)** explanation states that the agent chose a certain option (over other options) because it was *valued*. For example, that the system chose to select catching a bus because it was the fastest of the available options, and that getting home more quickly is valued.

– Finally, a **link (L)** explanation states that a particular action was performed in order to allow a subsequent action to be done. For example, that the agent bought the ticket in order to allow the user to then catch the bus (which requires having a ticket).

A full explanation may use a number of each of these elements, for example: *A bicycle was not available (B), money was available (B), the made choice (catch bus) has the shortest duration to get home (in comparison with walking) and I believe that is the most important factor for you (V), I needed to buy a bus ticket in order to allow you to go by bus (L), and I have the goal to allow you to catch the bus (D).*

3   Methodology

We surveyed[5] participants[6], who were recruited using advertisements in a range of undergraduate lectures within the Otago Business school, by email to students at institutions of two colleagues, Frank and Virginia Dignum, with whom we were collaborating on related work, and by posting on social media. New Zealand based participants were given the incentive of being entered into a draw for a NZ$100 supermarket voucher.

The scenario used the software personal assistant ("SAM") explained in Sect. 2.

Each participant is presented with five possible explanations (see Fig. 1) which are given in a random order, i.e. each participant sees a different ordering. The explanations combine different elements of the explanation mechanism described earlier in this paper. Specifically, there are four types of elements that can be included in an explanation: beliefs, valuings, desires, and links. Explanation E1 includes all four elements, explanation E2 filters out the desires and links, E3 includes only valuings, E4 includes only beliefs, and E5 includes only beliefs and desires.

For each of the five explanations E1–E5 participants were asked to indicate on a Likert scale of 1–7[7] how much they agree or disagree with the following statements: "This explanation is Believable (i.e. I can imagine a human giving this answer)", "This explanation is Acceptable (i.e. this is a valid explanation of the software's behaviour)", and "This explanation is Comprehensible (i.e. I understand this explanation)". Participants were also asked to indicate whether they would like further clarification of the explanation given, for instance, by entering into a dialog with the system, or providing source code.

Once all five explanations were considered, participants were asked to rank the explanations from 1 (most preferred) to 5 (least preferred). They were also asked to indicate the extent to which they agreed with the statement "I trust SAM because it can provide me a relevant explanation for its actions" (7 point Likert scale).

Next, the survey asked a number of questions to assess and obtain information about general trust in technology, including attitude to Artificial Intelligence. The 11 questions consisted of 7 questions that were adopted from McKnight *et al.* [22, Appendix B]. Specifically, we used the four questions that McKnight *et al.* used to assess faith in general technology (item 6 in their appendix), and the three questions that they used to assess trusting stance (general technology, item 7). We also had four questions that assessed attitudes towards Artificial Intelligence. Finally, the respondents were asked to provide demographic information.

4   Results

We received 74 completed responses to the online survey. The demographic features of the respondents are shown in Table 1.

---

[5] The survey can be found at: https://www.dropbox.com/s/ec6fg3u1rqhytcb/Trust-Autonomous-Survey.pdf.

[6] Ethics approval was given by University of Otago (Category B, D18/231).

[7] Where 1 was labelled "Strongly Disagree", 7 was labelled "Strongly Agree", and 2–6 were not labelled.

Table 1. Selected demographic characteristics of respondents (percentage distributions; percentages may not sum to 100% due to rounding)

Characteristic		Percentage
Gender	male	55.4
	female	41.9
	not answered	2.7
Age	18–24	39.2
	25–34	27.0
	35–44	14.9
	45–54	14.9
	55–64	4.0
Education	High school graduate	17.6
	Bachelor/undergraduate degree	44.6
	PhD degree/Doctorate	36.5
	not answered	1.4
Ethnicity	New Zealander (non Māori)	31.1
	Māori	2.7
	European	46.0
	Other	20.3

4.1 Analysis of Believability, Acceptability and Comprehensibility of Explanations

We begin by analysing how participants assessed each of the explanations E1-E5 on three characteristics: Believability, Acceptability and Comprehensibility. Each explanation was assessed on its own (in random order), i.e. the participants in this part of the survey were not asked to compare explanations, but to assess each explanation in turn.

The descriptive statistics regarding the Believability, Acceptability and Comprehensibility of the five Explanations are shown below (recall that 1 is "strongly disagree" and 7 is "strongly agree", so a higher score is better).

We used paired Wilcoxon-signed rank tests to estimate differences in means. The results are given in Table 2. These results show that most of the differences between pairs of explanations in terms of their Believability, Acceptability, and Comprehensibility are statistically significant[8] with $p < 0.005$.

Characteristic	Explanation	Mean	Std. Dev.	Median
Believability	E1	3.90	1.78	4
	E2	4.80	1.50	5
	E3	5.08	1.34	5
	E4	3.73	1.87	4
	E5	3.76	1.72	4
Acceptability	E1	5.12	1.70	5
	E2	5.14	1.52	5
	E3	4.57	1.74	5
	E4	3.76	1.95	4
	E5	4.45	1.81	5
Comprehensibility	E1	5.55	1.38	6
	E2	5.77	1.03	6
	E3	5.62	1.04	6
	E4	4.99	1.63	5
	E5	4.85	1.64	5

Figure 2 depicts the relationships in Table 2. For believability (top left of Fig. 2) explanations E3 and E2 are statistically significantly different to explanations E1, E4 and E5 (in fact E3 and E2 are better than E1, E4 and E5 since they have a higher median). However, E3 and E2 are not statistically significantly different to each other, nor are there statistically significant differences amongst E1, E4 or E5. For acceptability (bottom of Fig. 2) the situation is a little more complex: explanations E1 and E2 are statistically significantly different to the other three explanations[9] (but not to each other), and E3 and E5 are both statistically significantly better than E4 (but E3 and E5 are not statistically significantly different). Finally, for comprehensibility (top right of Fig. 2), explanations E2, E3 and E1 are statistically significantly different to explanations E4 and E5, but for each of the two groups of explanations there are not statistically significant differences within the group.

Overall, considering the three criteria of believability, comprehensibility, and acceptability, these results indicate that E2 is statistically significantly better than E4 and E5 according to all criteria, and is statistically significantly better than E1 (Believability only), and E3 (Acceptability only). Explanation E3 was statistically significantly better than E4 (all criteria), E5 (Believability and Comprehensibility), and

---

[8] We use a significance level of 0.005 rather than 0.05 to avoid type II errors, given the number of tests performed. The significance level is calculated as $\sqrt[10]{0.95} = 0.9948838$, giving a threshold for significance of around 0.005.

[9] Although for E1-E3 it is only at $p = 0.0273$.

Table 2. Statistical Significance of Differences in means for Believability, Acceptability and Comprehensibility. Bold text indicates statistical significance with $p < 0.005$ and "***" indicates $p < 0.0001$.

Characteristic	Explanation	E1	E2	E3	E4	E5
Believability	E1	–	***	***	0.6006	0.6833
	E2	***	–	0.2015	***	***
	E3	***	0.2015	–	***	***
	E4	0.6006	***	***	–	0.9808
	E5	0.6833	***	***	0.9808	–
Acceptability	E1	–	0.7357	0.0273	***	***
	E2	0.7357	–	**0.0041**	***	***
	E3	0.0273	**0.0041**	–	**0.0003**	0.6481
	E4	***	***	**0.0003**	–	**0.0002**
	E5	***	***	0.6481	**0.0002**	–
Comprehensibility	E1	–	0.1275	0.7370	**0.0040**	**0.0022**
	E2	0.1275	–	0.1510	***	***
	E3	0.7370	0.1510	–	**0.0005**	**0.0005**
	E4	**0.0040**	***	**0.0005**	–	0.6060
	E5	**0.0022**	***	**0.0005**	0.6060	–

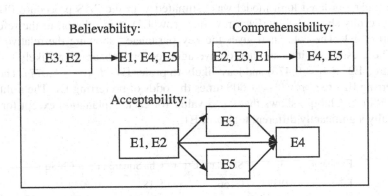

Fig. 2. Visual representation of the significance results in Table 2 where an arrow indicates a statistically significant difference (arrow is directional from better to worse)

E1 (Believability). Explanation E1 was statistically significantly better than E4 and E5 (Comprehensibility and Acceptability), and E3 (Acceptability). Finally, E5 is better than E4 (Acceptability only).

So, overall E2 can be seen as the best explanation since it is ranked statistically significantly differently to all other explanations (with a higher median) on at least one of the three characteristics (Believability, Acceptability, and Comprehensibility), but no other explanation is better than it on any characteristic. Next are E1 and E3 which are statistically different (specifically better) than E4 and E5 on some characteristics (for E1 Comprehensibility and Acceptability but not Believability, and for E2 Believability and Acceptability, but not Comprehensibility).

4.2   Analysis of Rankings of Explanations

The analysis below relates to the part of the survey where respondents were asked to rank a set of five explanations from 1 (most preferred) to 5 (least preferred).

To analyse the ranked data we employed a general discrete choice model (linear mixed model), using a ranked-ordered logit model which is also known as an exploded logit [3].

A discrete choice model is a general and powerful technique for analysing which factors contributed to the outcome of a made choice. It is required in this case because each of the five explanations being ranked represented a combination of explanatory factor types. The ranked-ordered logit is used to deal with the fact that the data represents a ranking: after selecting the most preferred explanation, the next selection is made out of the remaining four explanations. This means that the selections are not independent.

The ranked-ordered logit is based on a multistage approach where the standard logit [3] is applied to the most preferred choice J_1 in the set of all alternatives $(J_1, \ldots, J_K)$, then to the second-ranked choice J_2 in the set $(J_2, \ldots, J_K)$ after the first-ranked item was removed from the initial choice set and so on.

The ranked-ordered logit model was estimated with the SAS procedure PHREG, yielding results shown below. Each row (e.g. row E2) is in relation to the reference explanation, E1. The column β gives the key parameter, showing the relative likelihood. These estimates indicate that, on average, respondents are most likely to prefer explanation E2 ($\beta_{E2} = 0.475$) and least likely to prefer E4 ($\beta_{E4} = -1.077$). The odds of preferring E2 are $exp^{0.475} = 1.608$ times the odds of preferring E1. The right-most column ("Pr > ChiSq") shows that the β value for each explanation except for E3 is statistically significantly different to that of E1.

Explanation	β	Standard Error	Chi-Square	Pr > ChiSq
E2	0.475	0.166	8.18	0.0042
E3	−0.154	0.165	0.878	0.3488
E4	−1.077	0.17	40.016	<.001
E5	−0.887	0.168	28.034	<.0001

We also calculated the Wald chi-square for all the possible pairs or coefficients (see below). All but two of the tests were statistically significant[10], with p-values less than 0.005 (actually less than 0.001). The two non-significant pairs were E1–E3 and E4–E5, which were not significant at the 0.005 level.

---

[10] As before, we use a significance level of 0.005 rather than 0.05 to avoid type II errors, given the number of tests performed.

Label	Wald Chi-Square	Pr > ChiSq
$\beta_{E2} - \beta_{E3}$	14.1768	0.0002
$\beta_{E2} - \beta_{E4}$	77.3522	<.0001
$\beta_{E2} - \beta_{E5}$	61.7307	<.0001
$\beta_{E3} - \beta_{E4}$	28.8808	<.0001
$\beta_{E3} - \beta_{E5}$	18.9785	<.0001
$\beta_{E4} - \beta_{E5}$	1.3091	0.2526
$\beta_{E2} - \beta_{E1}$	8.1801	0.0042
$\beta_{E3} - \beta_{E1}$	0.8780	0.3488
$\beta_{E4} - \beta_{E1}$	40.0157	<.0001
$\beta_{E5} - \beta_{E1}$	28.0341	<.0001

This analysis therefore allows us to conclude that, based on participants ranking of the explanations, E2 is most preferred, followed by E1 and E3, which are not significantly differently ranked, and then E4 and E5 (also not statistically significantly different in ranking). In other words, we have three tiers: E2 (most preferred), E1 and E3 (less preferred than E2), and E4 and E5 (least preferred). This is consistent with the results of the previous section.

In order to provide additional confidence in the logit analysis, we also performed a series of comparisons between pairs of items using a Wilcoxon signed rank test. This also found that all differences were significant at the 0.005 level, except for the two pairs that were not significantly different at this level according to the regression analysis. Thus, the exploded logit model gives results that are qualitatively the same as those obtained by a standard nonparametric method.

We also investigated whether there are differences between males and females in their ranking of explanations. Using the same exploded logit model and new dummy variable for gender, we computed the Wald chi-square statistic for the null hypothesis that differences between gender-dependent coefficients are zero, which had p-value 0.95. Thus, there is no evidence for a difference between men and women in ranking explanations. A similar analysis was made for age-dependent groups of respondents

Table 3. The construction of the explanations.

Component	E1	E2	E3	E4	E5
B(eliefs)	1	1	0	1	1
V(aluings)	1	1	1	0	0
D(esires)	1	0	0	0	1
L(inks)	1	0	0	0	0
Length in words:	63	36	27	9	20
Length in characters:	318	206	152	54	101

and found no significant difference in ranking of explanations in relation to age (p-value 0.158).

4.3 Effects of Explanation Components

Next, we investigated the effects of explanation components (e.g. beliefs, desires, valuings) and how they affect ranking. There were four possible components: beliefs, valuings, desires and links. The constructed explanations are shown in Table 3 where ones indicate the presence of respective components and zeros indicate their absence. For example, the first column indicates that explanation E1 has all four components, whereas the second column shows that E2 has only the beliefs and valuings components.

As shown in Table 4, all except one of the coefficients of the exploded logit model are significantly different from zero at level $p = 0.005$ and the only exception β_D, corresponding to desires, is significant at the 0.05 level. A positive coefficient indicates that this component is more preferred, whereas a negative coefficient indicates that the component is less preferred. Thus, respondents prefer explanations that have V, B, and D components. They are reluctant to prefer explanations that have links. The magnitudes of coefficients in Table 4 can be interpreted as follows. The presence of V components in the explanation has produced $100 \times (\exp^{\beta} -1) = 100 \times (\exp^{2.4} -1) = 1002.3$ percent increase in the odds of preferring this explanation to the one where V is absent, controlling for other components. The presence of beliefs in the explanation has produced $100 \times (\exp^{0.82} -1) = 127$ percent increase in the odds of preferring this explanation to the one where B is absent, controlling for other components. The presence of desires in the explanation has produced $100 \times (\exp^{0.54} -1) = 71.6$ percent increase in the odds of preferring this explanation to the one where D is absent, controlling for other components. For links we have $100 \times (\exp^{-1.16} -1) = -68.65\%$, which implies that the odds of preferring explanation with links over the one where L is absent goes down by 68.65%.

Table 4. Respondents' Preferences in Ranking Components V,B,D,L: Analysis of Maximum Likelihood Estimates

Parameter	Parameter Estimate (β)	Standard Error	Chi-Square	Pr > ChiSq
V	2.402	0.224	115.28	<.0001
B	0.821	0.176	21.661	0.0001
D	0.543	0.224	5.88	0.0153
L	−1.164	0.285	16.6224	0.0001

As before, we also calculated the Wald chi-square for all the possible pairs or coefficients. We found that the difference between preferring B and D is not statistically significant ($p = 0.33$), whereas the difference among all others components is significant (see Table 5).

Table 5. Statistically Significant Differences in regression coefficients

Label	Wald Chi-Square	Pr > ChiSq
$\beta_V - \beta_B$	52.2652	<.0001
$\beta_V - \beta_D$	90.3121	<.0001
$\beta_V - \beta_L$	56.0711	<.0001
$\beta_B - \beta_D$	0.9473	0.3304
$\beta_B - \beta_L$	27.2910	<.0001
$\beta_D - \beta_L$	12.5446	0.0004

This analysis shows that of the four factors that are included in the explanations, the presence of V components most strongly (and significantly) correlates with higher preference for the explanation. In other words, explanations including valuings are more likely to be preferred.

4.4 Analysis of Overall Trust in SAM

Our final analysis considered the relationship between overall trust in a specific autonomous system (SAM), and broader trust in technology in general, and AI specifically. The question being addressed here is: to what extent is trust in a given system, such as SAM, determined by a person's more general attitudes towards technology, and towards Artificial Intelligence?

As noted earlier, the survey included 11 questions that assessed three dimensions of attitudes [22]: faith in technology (4 questions), general attitude to technology (3 questions), and attitude to Artificial Intelligence (4 questions).

We conducted a reliability analysis to assess the internal consistency of these blocks of questions. The results (see Table 6) show that the Cronbach's alpha coefficients ranged from[11] 0.73 to 0.85. We also considered all of the questions taken together ("Merged" in Table 6), which yielded a higher alpha. This meant that the questions forming the components of the scale were sufficiently intercorrelated to allow the dimensions to be merged. We therefore merged the three dimensions into a single item that measured each participant's attitude to technology in general (including AI).

In order to assess the extent to which broader background attitudes to technology influenced trust in SAM we compared the calculated background trust measure (average of the ten questions) against each participant's response to the question "I trust SAM because it can provide me a relevant explanation for its actions" (Likert response on a 1–7 scale).

To estimate the correlation between background trust in technology and trust in SAM, we calculated Spearman's coefficient. The coefficient value of 0.46 confirms that

---

[11] For the AI group of questions, the analysis indicated that dropping the third question would improve the alpha from 0.69 to 0.79, which was done, meaning that we used a total of 10 questions. The dropped question was: "I think that current problems with use of AI (bias, breach of privacy, etc.) will be solved in the short term".

Table 6. Analysis of dimensions of background trust to technology

Characteristic	Cronbach's alpha
Faith	0.73
General attitude to technology	0.85
Attitude to AI	0.79
Merged	0.91

there appears to be a positive correlation between the two variables ($\rho_S = 0.46$, $n = 74$, $p = 3.8 \times 10^{-5}$). Thus, high values of background trust in technology are associated with high "trust in SAM" scores.

Interestingly, although the correlation is clearly significant ($p = 3.85 \times 10^{-5}$), it is not that strong ($\rho_S = 0.46$, which is considered a moderate strength correlation). In other words, knowing that a person has, say, a high level of trust in technology in general, does not allow one to confidently predict that they will therefore have a high level of trust in an autonomous system (see Fig. 3). In other words, trust in autonomous systems is not purely determined by background trust in technology more broadly.

We also assessed the effects of gender. A Wilcoxon test performed for two independent groups (men and women) showed no evidence for a difference in means for SAM score (W = 551.5, p-value = 0.33). So, we can conclude that there is no evidence that men and women give different scores to SAM.

5   Related Work

As noted in the introduction, there is comparatively little work on goal-driven XAI. Focusing specifically on approaches that use beliefs and desires, and that conduct an evaluation, there are a number of papers.

Harbers *et al.* [7, 14, 15] consider an explanation mechanism that is similar to the one we evaluate in that it uses explanation templates that correspond to our explanatory components of beliefs, desires, and links. However, they do not have a corresponding template for valuings. Furthermore, their explanations do not take into account possible alternatives, i.e. they explain why X was done solely in terms of what *enabled* X to be done, rather than considering why X was *selected* from amongst the available options. In general, X may be enabled, but whether it is selected can depend also on the availability of other options. For example, choosing to catch a bus because a bicycle is not available, so cycling (which otherwise would be preferred) is not an option. An explanation in terms of what enabled us to catch a bus (having money), is not useful. A useful explanation in this scenario is that the preferred option (cycling) was not available due to the lack of a bicycle being available.

Turning to the evaluations, Broekens *et al.* [7] report on an evaluation using a cooking domain. They had 30 participants who were randomly allocated to one of the three explanation types. Participants were asked to score an explanation for each action in

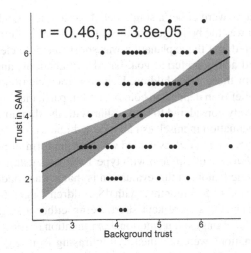

Fig. 3. Correlation between "trust in SAM" score and background trust.

terms of naturalness[12] and usefulness[13]. They found that, in general, goal-based explanations were preferred. However, the specific preferred explanation depended on the action and its context. For example, where an action is an "or" (i.e. its parent goal requires a single child to be selected), then a belief-based explanation is more helpful.

Harbers *et al.* [15] report on an evalation using a fire-fighting domain, with 20 participants who were not experts in the domain. For each action, they asked participants which of four explanations was preferred: the parent goal (in the goal hierarchy tree[14]), the parent's parent goal, the beliefs, and a link explanation. Similarly to Broekens *et al.* they found that the choice depended on the action and its context. However, in general, links were barely selected as preferred, and while goals were well-received, for "or" actions beliefs were preferred.

These results are consistent with ours in that we also found that links were not preferred. One difference is that while their explanations consisted of a single type (e.g. belief or goal or link), we considered more complex explanations that mixed elements. And, of course, they did not consider valuings, so our key finding, that valuings are more preferred than either belief-based or goal-based explanations, was not able to be identified by their work.

Kaptein *et al.* [17,19] considered explanations in the context of an e-health application. In earlier work [17] they evaluated user preferences for explanations in the context of a personal assistant that worked with a fictitious child ("jimmy") who has type 1 diabetes mellitus. Participants (19 adults and 19 children) were provided with a number of scenarios, and asked to select their preferred explanation for each one. The expla-

---

[12] Explained as: "With a natural explanation we mean an explanation that sounds normal and is understandable, an explanation that you or other people could give.".

[13] Explained as: "Indicate how useful the explanations would be for you in learning how to make pancakes.".

[14] The tree of goals, beliefs, and actions.

nations given as options were either a single belief, or a single goal. In both cases the explanation provided was the belief/goal immediately above the action in the goal hierarchy tree. This ensured that the explanation was short (a single element). They found that both children and adults preferred goal-based explanations, and that adults had a stronger preference for these than children. However, they caution that the preference between goals and beliefs can depend on context, and in particular, that in their work the participants were already considered to be familiar with the domain, since the children participating in the evaluation themselves had type 1 diabetes.

In later work [19] Kaptein et al. evaluated whether the form of the explanation provided affected the *behaviour* of children with type 1 diabetes using an e-health support system. A distinguishing feature of this evaluation is that it was conducted "in the wild" over a longer time period (2.5–3 months), with 48 children[15] aged 6–14. As in the previous evaluation, explanations were kept short, being either a single belief or single goal ("cognitive" explanations), or an emotional explanation ("affective" explanations). The emotional explanations were obtained by rephrasing from e.g. "I want to …" to "It would make me happy if you …". They found only a single statistically significant result, which was counter-intuitive: providing explanations (either cognitive or affective) correlated with children following the tasks *less* often. The authors hypothesised a number of possible explanations for this behaviour, for example, that children read the explanation, and if the aim of the task is to teach them something that they already believe they know, then they are therefore less likely to select that task.

Again, these results are consistent with ours, in that we found varying preference between beliefs and desires. However, as noted for Harbers et al., their explanations did not mix explanation types, and they did not consider valuings. On the other hand, they included affective explanations, which were not part of our evaluation.

More recently, Abdulrahman et al. [1,2] conducted an empirical human subject study to assess explanations provided by an intelligent virtual advisor. Their study was limited to university students (mostly under 20 years old), with 91 participants. It concerned a virtual assistant ("Sarah") that was designed to give advice to help students manage stress. Like us, they drew inspiration from Malle, but they did not include valuings in their explanations. They considered explanations that contained beliefs only, desires only, and both beliefs and desires[16]. The key question they consider is to what extent "… *do explanations that refer to the user's beliefs or goals influence the user's intention to change the behaviours recommended by the agent?*". They did not find a difference between belief-only and goal-only explanations, but found that belief-and-goal explanations did not lead to a significant change in intentions to join a study group (the recommendation from the agent), which they ascribe to the explanation being longer.

Mualla et al. [24] propose an explanation mechanism focussed on parsimony, which requires balancing brevity and adequacy of the explanation. They use contrastive explanations and different forms of filtering to attempt to provide parsimonious explanations. Their evaluation, which is done using a scenario involving understanding UAV operations, hypothesises that using contrastive rather than only normal explanations, and

---

[15] One child was excluded from the data analysis due to a data glitch.

[16] Since their virtual assistant was only providing advice, rather than performing a sequence of actions, it did not make sense to have link explanations.

adaptive rather than static filtering, both improve understandability of explanations. They divided participants into three groups: normal explanations and static filtering (SF), normal explanations and adaptive filtering (AF), and adaptive filtering with both normal and contrastive explanations (AC). Comparing survey results for these groups they found that while adaptive filtering on its own was not necessarily better (AF vs. SF), the combination of adaptive filtering and contrastive explanation did make a significant difference (SF vs. AC). They also evaluated trust, but did not find any statistically significant relationship regarding the effect of explanation type on trust. This last point can perhaps be explained by our finding that trust is to some extent influenced by background trust in technology: if the effect of explanations on trust is only partial (since trust is also influenced by other factors, such as trust in technology), then we might expect to see that the effect on trust of changing the *form* of the explanation would not be statistically significant. Our findings regarding the length of explanations support their argument for parsimony: our most preferred explanation was neither the longest nor the shortest. Finally, we note that their explanation mechanism does not include valuings, and that our results suggest that it should.

6 Discussion

We have conducted a human participant empirical evaluation of explanations of BDI agents, where the explanations consist of different types of explanatory components: beliefs, desires, valuings, and links.

We found that participants assess the different explanations somewhat differently for Believability, Acceptability, and Comprehensibility, and that most of the differences between the assessment of different explanations were statistically significant (Sect. 4.1). Overall, considering both assessing each explanation on its own (Sect. 4.1) and explicitly ranking the explanations (Sect. 4.2), we have a consistent preference for E2 (which has belief and valuing explanatory components), followed by E1 (all component types) and E3 (valuing only), which are not distinguishable from each other. The least preferred explanations were E4 (belief only) and E5 (belief and desire), which are also not distinguishable from each other in terms of preferences.

Analysing the data to assess preferences for the different types of explanatory components (beliefs, desires, valuings, links; see Sect. 4.3), we found that the presence of valuing components make an explanation significantly more likely to be preferred, and that the presence of belief and/or desire components also makes an explanation more likely to be preferred, but less so than valuings. On the other hand, the presence of a link component makes an explanation less likely to be preferred.

Finally (Sect. 4.4), there is statistically significant correlation between trust in SAM and trust in technology in general ($p = 3.85 \times 10^{-5}$), but the correlation has moderate strength ($\rho_S = 0.46$). Since our survey assessed trust in technology before participants were introduced to SAM, we have that trust in technology cannot be influenced by anything related to SAM. Therefore, the correlation can be interpreted as indicating that while trust in technology in general (including AI) influences trust in SAM (as might be expected), it does not *determine* it. This is an encouraging finding: if we had found that preexisting trust in technology and AI in general strongly affected (or even determined) trust in a given autonomous system, then there would be a limited (or no) role for explanations to affect the level of trust.

Returning to our hypotheses, we have that:

H1: Explanations that include valuings are more likely to be preferred by users over other forms of explanations (that do not include valuings). This hypothesis is confirmed by our findings (Sect. 4.1, 4.2 & 4.3).

H2: Explanations that include desires are more likely to be preferred by users over explanations that include beliefs. This hypothesis is **not** confirmed: we did not find a statistically significant difference between preferences for beliefs and desires (Sect. 4.3).

H3: Explanations that include links are less likely to be preferred by users over other forms of explanations (that do not include links). This hypothesis is confirmed by our findings (Sect. 4.3).

H4: Shorter explanations are more likely to be preferred by users. Interestingly, this hypothesis is **not** confirmed: explanations E1 (the longest, with all four types of explanatory factors) and E3 (with only a single factor) did not have a statistically significant difference in preference (Sect. 4.2). Indeed, E1 was considered more acceptable than E3, whereas E3 was considered more believable than E1 (Sect. 4.1). Furthermore, there was not a significant difference in their comprehensibility (Sect. 4.1). Indeed, the two least-preferred explanations (E4 and E5) were the shortest!

Based on these findings, we provide the following advice to guide the development of explanations.

Firstly, it is clear that valuings are valued. Explanations that included a valuing component (E1, E2 and E3) were significantly more likely to be preferred. This is consistent with the findings of the previous evaluation [37], which also found that valuings were valued[17]. We therefore recommend that when developing explanation mechanisms based on this framework, that valuing explanatory factors are included in explanations.

Secondly, we found that explanations including link components were less likely to be preferred. The evaluation by Harbers *et al.* [15] also found that link explanations were barely selected as preferred. However, we exercise a note of caution: we only had one explanation that included links (E1), and it may also be that the lower preference for this explanation reflects its length. We therefore do not recommend excluding link explanatory components at this point, but rather suggest that further evaluation would help to clarify whether they are indeed seen as less preferred.

Thirdly, we did not find that users prefer short explanations. The most preferred explanation (Sect. 4.2) was E2, which is longer than E3 and E4. On the other hand, the longest explanation (E1) was not the least preferred. Although the length of an explanation clearly can play a role, with too-long explanations being less useful, our findings do not support the approach taken by previous work to limit explanations to a single belief or a single goal. We therefore recommend that when providing explanations, the explanations are not limited to only single factors. Furthermore, when evaluating forms

---

[17] Specifically, their explanations corresponding in structure with our E2 (valuing and belief) and E3 (valuing) were most preferred.

of explanation, longer explanations should also be considered and included in the evaluation.

There is scope for further evaluation, with different scenarios, and with different forms of explanations. Two specific forms of explanation that would be good to consider are emotions, and interactive explanations. Keptein *et al.* [18] argue that explanations should include emotions. This is an interesting idea, and one that would be good to investigate further. It would also be good to consider other evaluation metrics such as relevance and the extent to which explanations relate to what the user already knows. Finally, our evaluation only considered explanations that were presented to the user all at once. It would also be good to consider explanations that are presented in the form of a dialogue, with an initial reason being given, and then additional information being provided as the user interacts with the system (See e.g. [10, 11, 31, 35]).

Acknowledgements. We would like to thank Dr Damien Mather, at the University of Otago, for statistical advice. This work was supported by a University of Otago Research Grant (UORG).

References

1. Abdulrahman, A., Richards, D., Bilgin, A.A.: Reason explanation for encouraging behaviour change intention. In: Dignum, F., Lomuscio, A., Endriss, U., Nowé, A. (eds.) AAMAS 2021: 20th International Conference on Autonomous Agents and Multiagent Systems, Virtual Event, United Kingdom, 3–7 May 2021, pp. 68–77. ACM (2021). https://www.ifaamas.org/Proceedings/aamas2021/pdfs/p68.pdf
2. Abdulrahman, A., Richards, D., Bilgin, A.A.: Exploring the influence of a user-specific explainable virtual advisor on health behaviour change intentions. Auton. Agents Multi Agent Syst. **36**(1), 25 (2022). https://doi.org/10.1007/s10458-022-09553-x
3. Allison, P.D., Christakis, N.A.: Logit models for sets of ranked items. Sociol. Methodol. **24**, 199–228 (1994). https://www.jstor.org/stable/270983
4. Anjomshoae, S., Najjar, A., Calvaresi, D., Främling, K.: Explainable agents and robots: results from a systematic literature review. In: Elkind, E., Veloso, M., Agmon, N., Taylor, M.E. (eds.) Proceedings of the 18th International Conference on Autonomous Agents and MultiAgent Systems, AAMAS 2019, Montreal, QC, Canada, 13–17 May 2019, pp. 1078–1088. International Foundation for Autonomous Agents and Multiagent Systems (2019). https://dl.acm.org/citation.cfm?id=3331806
5. Bratman, M.E., Israel, D.J., Pollack, M.E.: Plans and resource-bounded practical reasoning. Comput. Intell. **4**, 349–355 (1988)
6. Bratman, M.E.: Intentions, Plans, and Practical Reason. Harvard University Press, Cambridge (1987)
7. Broekens, J., Harbers, M., Hindriks, K., van den Bosch, K., Jonker, C., Meyer, J.-J.: Do you get it? user-evaluated explainable BDI agents. In: Dix, J., Witteveen, C. (eds.) MATES 2010. LNCS (LNAI), vol. 6251, pp. 28–39. Springer, Heidelberg (2010). https://doi.org/10.1007/978-3-642-16178-0_5
8. Cranefield, S., Oren, N., Vasconcelos, W.W.: Accountability for practical reasoning agents. In: Lujak, M. (ed.) AT 2018. LNCS (LNAI), vol. 11327, pp. 33–48. Springer, Cham (2019). https://doi.org/10.1007/978-3-030-17294-7_3
9. Cranefield, S., Winikoff, M., Dignum, V., Dignum, F.: No pizza for you: value-based plan selection in BDI agents. In: Proceedings of the Twenty-Sixth International Joint Conference on Artificial Intelligence, IJCAI-17, pp. 178–184 (2017). DOI: https://doi.org/10.24963/ijcai.2017/26

10. Dennis, L.A., Oren, N.: Explaining BDI agent behaviour through dialogue. In: Dignum, F., Lomuscio, A., Endriss, U., Nowé, A. (eds.) AAMAS 2021: 20th International Conference on Autonomous Agents and Multiagent Systems, Virtual Event, United Kingdom, 3–7 May 2021, pp. 429–437. ACM (2021), https://www.ifaamas.org/Proceedings/aamas2021/pdfs/p429.pdf
11. Dennis, L.A., Oren, N.: Explaining BDI agent behaviour through dialogue. Auton. Agents Multi Agent Syst. **36**(1), 29 (2022). https://doi.org/10.1007/s10458-022-09556-8
12. Floridi, L., et al.: Ai4people–an ethical framework for a good AI society: opportunities, risks, principles, and recommendations. Minds Mach. **28**(4), 689–707 (2018). https://doi.org/10.1007/s11023-018-9482-5
13. de Graaf, M.M.A., Malle, B.F.: People's explanations of robot behavior subtly reveal mental state inferences. In: 14th ACM/IEEE International Conference on Human-Robot Interaction, HRI 2019, Daegu, South Korea, 11–14 March 2019, pp. 239–248. IEEE (2019). https://doi.org/10.1109/HRI.2019.8673308
14. Harbers, M.: Explaining agent behavior in virtual training. SIKS dissertation series no. 2011–35, SIKS (Dutch Research School for Information and Knowledge Systems) (2011)
15. Harbers, M., van den Bosch, K., Meyer, J.C.: Design and evaluation of explainable BDI agents. In: Huang, J.X., Ghorbani, A.A., Hacid, M., Yamaguchi, T. (eds.) Proceedings of the 2010 IEEE/WIC/ACM International Conference on Intelligent Agent Technology, IAT 2010, Toronto, Canada, 31 August–3 September 2010, pp. 125–132. IEEE Computer Society Press (2010). https://doi.org/10.1109/WI-IAT.2010.115
16. High-Level Expert Group on Artificial Intelligence: The assessment list for trustworthy artificial intelligence (2020). https://digital-strategy.ec.europa.eu/en/library/assessment-list-trustworthy-artificial-intelligence-altai-self-assessment
17. Kaptein, F., Broekens, J., Hindriks, K.V., Neerincx, M.A.: Personalised self-explanation by robots: the role of goals versus beliefs in robot-action explanation for children and adults. In: 26th IEEE International Symposium on Robot and Human Interactive Communication, RO-MAN 2017, Lisbon, Portugal, 28 August–1 September 2017, pp. 676–682. IEEE (2017). https://doi.org/10.1109/ROMAN.2017.8172376
18. Kaptein, F., Broekens, J., Hindriks, K.V., Neerincx, M.A.: The role of emotion in self-explanations by cognitive agents. In: Seventh International Conference on Affective Computing and Intelligent Interaction Workshops and Demos, ACII Workshops 2017, San Antonio, TX, USA, 23–26 October 2017, pp. 88–93. IEEE Computer Society (2017). https://doi.org/10.1109/ACIIW.2017.8272595
19. Kaptein, F., Broekens, J., Hindriks, K.V., Neerincx, M.A.: Evaluating cognitive and affective intelligent agent explanations in a long-term health-support application for children with type 1 diabetes. In: 8th International Conference on Affective Computing and Intelligent Interaction, ACII 2019, Cambridge, United Kingdom, 3–6 September 2019, pp. 1–7. IEEE (2019). https://doi.org/10.1109/ACII.2019.8925526
20. Langley, P., Meadows, B., Sridharan, M., Choi, D.: Explainable agency for intelligent autonomous systems. In: Singh, S., Markovitch, S. (eds.) Proceedings of the Thirty-First AAAI Conference on Artificial Intelligence, San Francisco, California, USA, 4–9 February 2017, pp. 4762–4764. AAAI Press (2017). https://aaai.org/ocs/index.php/IAAI/IAAI17/paper/view/15046
21. Malle, B.F.: How the Mind Explains Behavior: Folk Explanations, Meaning, and Social Interaction. The MIT Press, Cambridge (2004). ISBN 0-262-13445-4
22. Mcknight, D.H., Carter, M., Thatcher, J.B., Clay, P.F.: Trust in a specific technology: an investigation of its components and measures. ACM Trans. Manag. Inf. Syst. **2**(2), 12:1–12:25 (2011). https://doi.org/10.1145/1985347.1985353
23. Miller, T.: Explanation in artificial intelligence: insights from the social sciences. Artif. Intell. **267**, 1–38 (2019). https://doi.org/10.1145/1824760.1824761

24. Mualla, Y., et al.: The quest of parsimonious XAI: a human-agent architecture for explanation formulation. Artif. Intell. **302**, 103573 (2022). https://doi.org/10.1016/j.artint.2021.103573
25. Müller, J.P., Fischer, K.: Application impact of multi-agent systems and technologies: a survey. In: Shehory, O., Sturm, A. (eds.) Agent-Oriented Software Engineering, pp. 27–53. Springer, Heidelberg (2014). https://doi.org/10.1007/978-3-642-54432-3_3
26. Munroe, S., Miller, T., Belecheanu, R., Pechoucek, M., McBurney, P., Luck, M.: Crossing the agent technology chasm: experiences and challenges in commercial applications of agents. Knowl. Eng. Rev. **21**(4), 345–392 (2006)
27. Rao, A.S., Georgeff, M.P.: An abstract architecture for rational agents. In: Rich, C., Swartout, W., Nebel, B. (eds.) Proceedings of the Third International Conference on Principles of Knowledge Representation and Reasoning, pp. 439–449. Morgan Kaufmann Publishers, San Mateo (1992)
28. van Riemsdijk, M.B., Jonker, C.M., Lesser, V.R.: Creating socially adaptive electronic partners: Interaction, reasoning and ethical challenges. In: Weiss, G., Yolum, P., Bordini, R.H., Elkind, E. (eds.) Conference on Autonomous Agents and Multiagent Systems (AAMAS), pp. 1201–1206. ACM (2015). https://dl.acm.org/citation.cfm?id=2773303
29. Robinette, P., Li, W., Allen, R., Howard, A.M., Wagner, A.R.: Overtrust of robots in emergency evacuation scenarios. In: Bartneck, C., Nagai, Y., Paiva, A., Sabanovic, S. (eds.) The Eleventh ACM/IEEE International Conference on Human Robot Interation, HRI 2016, Christchurch, New Zealand, 7–10 March 2016, pp. 101–108. IEEE/ACM (2016). https://doi.org/10.1109/HRI.2016.7451740
30. Schwartz, S.: An overview of the Schwartz theory of basic values. Online Read. Psychol. Cult. **2**(1), 11 (2012). https://doi.org/10.9707/2307-0919.1116
31. Sklar, E.I., Azhar, M.Q.: Explanation through argumentation. In: Imai, M., Norman, T., Sklar, E., Komatsu, T. (eds.) Proceedings of the 6th International Conference on Human-Agent Interaction, HAI 2018, Southampton, United Kingdom, 15–18 December 2018, pp. 277–285. ACM (2018). https://doi.org/10.1145/3284432.3284470
32. The IEEE Global Initiative for Ethical Considerations in Artificial Intelligence and Autonomous Systems: Ethically Aligned Design: A Vision For Prioritizing Wellbeing With Artificial Intelligence And Autonomous Systems, Version 1. IEEE (2016). https://standards.ieee.org/develop/indconn/ec/autonomous_systems.html
33. Thellman, S., Silvervarg, A., Ziemke, T.: Folk-psychological interpretation of human vs. humanoid robot behavior: exploring the intentional stance toward robots. Front. Psychol. **8**, 1–14 (2017). https://doi.org/10.3389/fpsyg.2017.01962
34. Verhagen, R.S., Neerincx, M.A., Tielman, M.L.: A two-dimensional explanation framework to classify AI as incomprehensible, interpretable, or understandable. In: Calvaresi, D., Najjar, A., Winikoff, M., Främling, K. (eds.) EXTRAAMAS 2021. LNCS (LNAI), vol. 12688, pp. 119–138. Springer, Cham (2021). https://doi.org/10.1007/978-3-030-82017-6_8
35. Winikoff, M.: Debugging agent programs with "Why?" questions. In: Proceedings of the 16th Conference on Autonomous Agents and Multiagent Systems, pp. 251–259 (2017)
36. Winikoff, M.: Towards trusting autonomous systems. In: El Fallah-Seghrouchni, A., Ricci, A., Son, T.C. (eds.) EMAS 2017. LNCS (LNAI), vol. 10738, pp. 3–20. Springer, Cham (2018). https://doi.org/10.1007/978-3-319-91899-0_1
37. Winikoff, M., Dignum, V., Dignum, F.: Why bad coffee? explaining agent plans with valuings. In: Gallina, B., Skavhaug, A., Schoitsch, E., Bitsch, F. (eds.) SAFECOMP 2018. LNCS, vol. 11094, pp. 521–534. Springer, Cham (2018). https://doi.org/10.1007/978-3-319-99229-7_47
38. Winikoff, M., Sidorenko, G., Dignum, V., Dignum, F.: Why bad coffee? explaining BDI agent behaviour with valuings. Artif. Intell. **300**, 103554 (2021). https://doi.org/10.1016/j.artint.2021.103554

A General-Purpose Protocol
for Multi-agent Based Explanations

Giovanni Ciatto[1]([✉])(iD), Matteo Magnini[1](iD), Berk Buzcu[2](iD),
Reyhan Aydoğan[2,3](iD), and Andrea Omicini[1](iD)

[1] Department of Computer Science and Engineering (DISI), Alma Mater Studiorum
– Università di Bologna, via dell'Università 50, 47522 Cesena, FC, Italy
{giovanni.ciatto,matteo.magnini,andrea.omicini}@unibo.it
[2] Department of Computer Science, Özyeğin University, Nisantepe Mah. Orman Sok.
No: 34-36 Alemdağ, Çekmeköy, 34794 Istanbul, Turkey
berk.buzcu@ozu.edu.tr
[3] Interactive Intelligence, Delft University of Technology, Mekelweg 4, 2628 CD
Delft, The Netherlands
reyhan.aydogan@ozyegin.edu.tr

Abstract. Building on prior works on explanation negotiation protocols, this paper proposes a general-purpose protocol for multi-agent systems where recommender agents may need to provide explanations for
their recommendations. The protocol specifies the roles and responsibilities of the explainee and the explainer agent and the types of information
that should be exchanged between them to ensure a clear and effective
explanation. However, it does not prescribe any particular sort of recommendation or explanation, hence remaining *agnostic* w.r.t. such notions.
Novelty lays in the extended support for both ordinary and contrastive
explanations, as well as for the situation where no explanation is needed
as none is requested by the explainee.

Accordingly, we formally present and analyse the protocol, motivating
its design and discussing its generality. We also discuss the reification of
the protocol into a re-usable software library, namely PYXMAS, which is
meant to support developers willing to build explainable MAS leveraging
our protocol. Finally, we discuss how custom notions of recommendation
and explanation can be easily plugged into PYXMAS.

Keywords: XAI · recommender systems · multi-agent systems ·
explanation protocols · SPADE · PYXMAS

1 Introduction

Explainable AI (XAI) is an area of research aimed at developing AI systems
that can provide understandable explanations of their decisions or behaviours
to humans [11]. The need for XAI arises from the fact that many modern AI
systems, particularly those based on deep learning and other forms of machine
learning, are often seen as "black boxes" that are difficult to interpret or explain

© The Author(s), under exclusive license to Springer Nature Switzerland AG 2023
D. Calvaresi et al. (Eds.): EXTRAAMAS 2023, LNAI 14127, pp. 38–58, 2023.
https://doi.org/10.1007/978-3-031-40878-6_3

[13]. This lack of transparency and interpretability can create significant challenges, particularly in applications such as healthcare, finance, and criminal justice, where decisions can have profound consequences for human lives [4].

The current focus of XAI research is on developing techniques for "opening up" these black boxes and providing insights about how an intelligent system reached a particular decision or prediction [10]. This involves developing methods for visualising the internal workings of the system, such as feature importance scores, attention maps, or decision trees. In all such cases, the goal is to support the AI expert willing to figure out how the intelligent system works, rather than the non-expert user who wants to understand why the system is behaving in a particular way. Furthermore, and more importantly, all such methods are based on the assumption that software tools should aid humans' interpretation of the system. However, the expectations of the XAI community go beyond merely opening up black boxes. Ideally, XAI systems should be able to *automatically* provide explanations that surpass mere descriptions of how a system works [7]. Instead, they should offer insights into why the system is – or is not – behaving in a particular way, possibly, by autonomously interacting with the explainee.

To achieve this goal, XAI researchers are increasingly focusing on the automation and interactivity of the explanation process [8]. This involves developing AI systems that can generate explanations on the fly and adapt their explanations to the needs and knowledge level of the explainee [5]. Along this line, multi-agent systems (MAS) are likely the most adequate metaphor for intelligent explainable systems. There, interaction and autonomy are first-class citizens. Hence, explanation can be smoothly modelled as a multi-agent interaction, where the explainee and the explainer agent (either a human or a software agent [18]) interact to achieve a common goal, namely, providing a clear and effective explanation.

Accordingly, in this paper we focus on the general problem enabling the interaction between explainee and explainer agents. To address this problem, this paper proposes a general-purpose protocol for multi-agent based recommendation and explanations. The protocol specifies the roles and responsibilities of the explainee and the explainer agent and the types of information that should be exchanged between them to ensure a clear and effective explanation. Notably, our protocol builds on top of prior attempts to model explanations as multi-agent interactions, such as the work by [3]. The key features of our proposal are the *(i)* the separation of recommendations from explanations, and *(ii)* the support for contrastive explanations.

As a side contribution, the paper also describes the design of a Spade-based Python library implementing the proposed protocol—namely, PyXMas. It supports the plugging of different sorts of explanation strategies, and representations. This library can be used as a starting point for building intelligent explainable systems where both recommendation and explanation behaviours are delegated to individual agents. Overall, this paper represents an important step towards developing XAI systems that can provide automatic and interactive explanations.

2   Background and Related Works

This section briefly overviews the literature on recommender systems, with an emphasis on food recommender system and interactive/explainable recommendations.

2.1   Interactive Recommendation Systems

In the past, *interactive* recommender systems have received significant attention from the recommender systems researchers due to their ability to provide *personalised* recommendations to users dynamically based on their feedback and interactions [12]. The key point behind interactivity is getting feedback from the user during the recommendation session for the next recommendation. The authors of [6], for instance, follow a one-shot recommendation where a few questions learned offline from past observations (i.e., from previous sessions) are asked prior to the recommendation. Answers to these questions let the recommender system personalise and improve future recommendations.

Building interactivity, researchers has also started to incorporate explainability into recommender systems [3,21] in order to increase the transparency of a recommender system. They do so through a repeated recommendation session where the system also provides explanations (other than recommendations), and it gets feedback given the positive effect observed in more transparent recommendations [17]. For instance, in [15], the authors implement visual explanations to users for music recommendation using grouped bar charts in a live comparison of the user's specified preferences for six categories and the song's matching percentage of that category. Similarly, the authors of [20] propose a recommendation system mimicking a human salesman: the system applies conversational explanations to convince users to buy more fitting alternatives.

2.2   Prior Work on Explanation Protocols

This paper proposes an extension of the protocol introduced by [3], which is tailored on food recommendations and explanations. There, the user starts the interaction by providing their constraints, which include ingredients that they are allergic to (such as milk or peanuts), preferred or disliked ingredients (such as certain meats or vegetables), and the desired cuisine type (like Middle Eastern or Mexican). The agent responds by suggesting a recipe and providing an explanation. The user then can accept the recommendation, decline it, or provide feedback on either the recipe, the explanation, or both at once. Accordingly, the agent may generate a new recipe or provide further explanation, or both at once. This interaction continues in a turn-taking fashion until the user accepts, leaves the session early, or reaches a time limit. Figure 1 illustrates how agents interact in line with this protocol.

Summarising, the key contribution of [3] is a framework for creating a nutrition-related personalised recommender system that simultaneously produces recommendations and explanations. On the one hand, presenting recommendations and explanations altogether establishes a transparent interaction

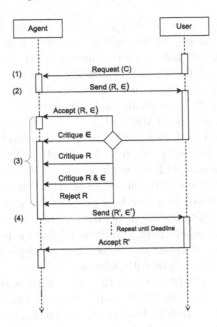

Fig. 1. FIPA representation of the negotiation protocol presented in [3]

with the user and may lead the user to accept such recommendations. On the other hand, unrequested explanations could be perceived as redundant and create an additional cognitive load on the user. The latter point has been studied, for instance, by Mualla *et al.* in [16]. There, *parsimony* has been outlined as one of the key features allowing successful human-agent interaction. In particular, parsimonious explanation are defined as the least complex that describes the situation adequately.

Accordingly, this work aims to revise the design to generate explanations based on users' requests, letting the user decide *when* to receive an explanation rather than providing a one-for-all solution. Furthermore, our revised protocol supports "zooming" explanations, where further explanatory details are only presented if and when the explainee is asking for them. In this way, the protocol lets users dynamically decide what degree of parsimony is fine for them

2.3 SPADE: Multi-agent Programming in Python

SPADE[1] is an open-source multi-agent system platform developed in Python. It provides a programming library for developing and simulating intelligent agents in various environments. The library is designed to be highly modular and extensible, allowing developers to easily create agents that can interact with each other and their environment.

---

[1] https://spade-mas.readthedocs.io.

At the modelling level, SPADE design and architecture are very close to the ones of JADE [2]. Accordingly, SPADE systems are distributed systems composed by agents which may or may not lay on the same network node. Agents' activities are governed by a set of behaviours, which are executed concurrently by the agent. Both agents and behaviours are implemented as abstract Python classes, which developers may extend to create their custom agents and behaviours.

Notable differences among SPADE and JADE mostly lay at the technological level. While JADE is a Java-based platform, SPADE is implemented in Python. This allows SPADE to be easily integrated with other Python libraries, there including the many ML and AI framework which are nowadays available for the Python platform. Furthermore, SPADE assumes agent interactions are mediated by an XMPP service, which is a standard protocol for instant messaging. This makes SPADE's agent communication facilities quite robust, interoperable, and scalable—as opposed to other agent platforms relying on proprietary or ad-hoc protocols. It also makes it easier to realise blended applications where agents interact with humans, other than with software agents.

Along this line, the SPADE framework comes with a range of features for developing intelligent agents, including communication protocols, message passing, and event handling. Notably, it supports the implementation of interaction protocols via finite-state machine behaviours—similarly to what JADE does.

Overall, the Spade library provides a powerful and flexible platform for developing intelligent agents, making it a popular choice for researchers and developers working in the field of multi-agent systems.

3 Explanation-Based Recommendation Protocol

The word "explanation" derives from the Latin word "explicare", which means "to unfold". There, the idea is that an explanation is a process of unfolding the meaning of a concept. Such a process is typically interactive, as it involves the interaction between some explainee and some explainer. In this sense, explanations is an inherently social protocol.

As far as interactions among human beings are concerned, the protocol is typically informal and unstructured. However, it typically involves an *explainee*, asking for help in understanding a given matter to some – allegedly, more knowledgeable – *explainer*. Explainers will then try their best to provide clear and effective explanations, possibly by trial-and-error, exploiting different explanation strategies or levels of detail. As interaction proceeds, explainers would also try to adapt their explanations to the needs and knowledge level of the explainee. Furthermore, explanations are commonly provided upon request, possibly in response to some prior information provided by either the explainee or the explainer, or someone else.

Modern intelligent systems are supposed to support decision-making by providing *recommendations*—possibly relying on artificial intelligence. Hence, when it comes to *explainable* intelligent systems, users commonly play the role of explainee, whereas software systems play both the role of the recommender and the explainer. By adopting a multi-agent perspective, we can model both recommendation and explanation as a single interaction protocol among two agents—where one of the two (commonly, the explainee) is a human being [8]. Hence, we may interchangeably use the terms "explainee" and "user" (resp. "explainer" and "agent"). Accordingly, in this section, we propose a general-purpose interaction protocol for multi-agent based recommendation and explanation.

Our protocol assumes that the user is in charge of initiating the interaction. Hence, the agent waits for the user to trigger a query. When receiving a query, the agent should respond by producing a recommendation.

While computing the recommendation, the agent may leverage on any information available to it at that moment, including the user's profile, the history of previous interactions, and – possibly – aggregated information about other users. Furthermore, it may take advantage of both symbolic AI reasoning facilities, and machine learning predictors.

In response to a recommendation, the user may either simply accept/discard the recommendation, or ask for explanations.

The explanation phase may involve several rounds of interaction, where the user may either ask for further details or request comparisons; and the agent attempts to provide all such kinds of information. Eventually, enlightened by the explanation process, the user may either accept or reject the recommendation. In both cases, the agent may consider the acceptance/rejection of its recommendation – as well as the amount of explanatory information provided required by the user to reach a decision – as feedback for future recommendations. In the particular case of a rejection, the agent may also be interested in the reason for the rejection, so as to improve its recommendation and explanation strategy.

Notably, explanations are always (i) provided upon request, (ii) related to the recommendation, and (iii) directed towards the user. Furthermore, explanations may be of two broad types, namely:

Ordinary explanations, which aim at answering the question "why did you recommend me this?";

Contrastive explanations, which aim at answering the question "why did you not recommend me that instead?".

Accordingly, our protocol supports both types of explanations, and it lets the user decide which type of explanation to request. Of course, the exchanged messages may be different depending on the type of explanation requested.

3.1 Abstract Formulation of the Protocol

Here, we propose an abstract formulation of the protocol which is agnostic w.r.t. the particular way in which the recommendation and explanation are represented and computed. In other words, we only focus on the messages exchanges among explainers and explainees, what information they should carry, and in which order they should be exchanged. Accordingly, the protocols relies on 13 types of messages, which may carry data fields of 5 different types, to be exchanged among agents playing 2 possible roles.

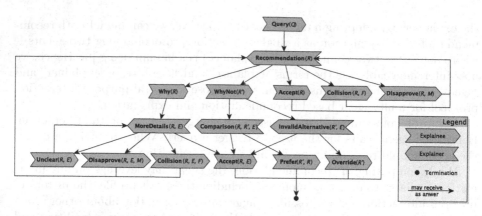

Fig. 2. Message communication diagram between an explainer agent (blue boxes) and an explainee (green boxes). Each box represents a message. Each message is connected to the ones it can receive as reply. (Color figure online)

Roles are of course explainee (a.k.a. user) and explainer (a.k.a. agent). The explainee initiates the protocol, while the explainer waits for the protocol to be started by the explainee.

We also identify 5 data types which represent the potential payload that agents may exchange during the protocol. As far as the *abstract* formulation of our protocol is concerned, we do not constrain the shape / structure of these types, but we simply assume they exist. In this way, implementers of the protocol will be free to define their own specification for these types, tailoring them on their particular application domain. In particular, the data types are:

Queries (denoted by Q), i.e. recommendation requests concerning a given topic, issued by the explainee when initiating the protocol;

Recommendations (denoted by R, R'), i.e. responses to queries, issued by the explainer;

Explanations (denoted by E, E'), i.e. chunks of explanatory information issued by the explainer to clarify their recommendation;

Features (denoted by F), i.e. aspects of the user which are relevant which justify some recommendation rejection, which the explainer should memorise and take into account in future interactions;

Motivations (denoted by M), i.e. reasons for the rejection of a recommendation, which may affect how the agent reacts to a rejection.

Finally, we identify 13 types of messages, which are exchanged among agents playing the explainee and explainer roles. We denote messages as named records of the form: Name(*Payload*), where Name represents the type of the message and *Payload* represents the data carried by the message—which consists of instances of the aforementioned data types. Payloads consist of ordered tuples of data types, where items suffixed by a question mark are *optional*. A summary of message types and their admissible payloads in Fig. 2. Accordingly, message types are (description follows a breadth-first traversal the diagram in the figure):

1. Query(Q) is the message issued by the explainee to initiate the protocol: it carries a recommendation request Q;
2. Recommendation(Q, R) is the message issued by the explainer in response to a query: it carries the query Q and the corresponding recommendation R computed by the explainer;
3. Why(Q, R) is the message issued by the explainee to request an explanation of a recommendation: it carries the original query Q and the recommendation R;
4. WhyNot(Q, R, R') is the message issued by the explainee to request a *contrastive* explanation of a recommendation: it carries the original query Q, the recommendation R, and a second recommendation R', which the explainee wants the explainer to contrast with R;
5. Accept($Q, R, E?$) is the message issued by the explainee to accept a recommendation: it carries the original query Q, the recommendation R, and optionally the explanation E provided by the explainer;
6. Collision($Q, R, F, E?$) is the message issued by the explainee to notify the explainer that the provided recommendation is colliding with some personal feature/preference of theirs: it carries the original query Q, the recommendation R, a description of the feature F, and optionally the explanation E provided by the explainer;
7. Disapprove($Q, R, M, E?$) is the message issued by the explainee to notify the explainer that the provided recommendation is not acceptable for some reason: it carries the original query Q, the recommendation R, a description of the reason M, and optionally the explanation E provided by the explainer;
8. Details(Q, R, E) is the message issued by the explainer to provide more details about a recommendation: it carries the original query Q, the recommendation R, and the explanation E;
9. Comparison(Q, R, R', E) is the message issued by the explainer to provide a contrastive explanation of a recommendation, in the case the one recommendation proposed by the explainee is admissible as well: it carries the original query Q, the recommendation R computed by the explainer and the one R' proposed by the explainee, and an explanation E comparing the two;
10. Invalid(Q, R', E) is the message issued by the explainer to notify the explainee that the proposed recommendation is invalid: it carries the original query Q, the proposed (and invalid) recommendation R', and an explanation E motivating the invalidity;
11. Unclear(Q, R, E) is the message issued by the explainee to notify the explainer that the provided explanation is unclear: it carries the original query Q, the recommendation R, and the provided (and unclear) explanation E;
12. Prefer(Q, R, R') is the message issued by the explainee to notify the explainer that they prefer a different recommendation: it carries the original query Q, the recommendation R proposed by the explainer, and the preferred recommendation R' proposed by the explainee;

13. Override(Q, R, R') is the message issued by the explainee to notify the explainer that want to force the decision to some recommendation which is considered invalid by the explainer: it carries the original query Q, the recommendation R proposed by the explainer, and the forced recommendation R' proposed by the explainee.

Notably, messages are designed by keeping the representational sate transfer (ReST, [9]) architectural style into account. Hence, each message type is designed to carry all the information necessary for any involved party to decide which action to take next. This is the reason why all/most messages carry the original query Q and the recommendation R (or R') which they are referring to.

The message communication diagram from Fig. 2 depicts not only the messages exchanged by the explainee and explainer, but also the admissbile request–response patterns which the protocol allows. There, a more detailed view of the message *flow* is provided, which we briefly summarise in the following. The explanation-based recommendation protocol consists in the following phases (depth-first traversal of Fig. 2):

1. the explainee initiates the protocol, by issuing a message Query(Q);
2. the explainer provides a message Recommendation(Q, R) in return;
3. the explainee may now:
   - 3.1 accept the recommendation, by answering Accept(Q, R), hence terminating the protocol;
   - 3.2 reject
     the recommendation because of M, by answering Disapprove(Q, R, M); or signal it as colliding with F, by answering Collision(Q, R, F). In this case, the explainer should propose another recommendation (go to 2.);
   - 3.3 ask for *ordinary* explanations, by answering Why(Q, R). In this case, the explainer should propose an explanation, by answering Details(R, E). The explainee may now:
     - 3.3.1. accept, reject, or signal R in light of E, by answering Accept(Q, R, E), Disapprove(Q, R, M, E), or Collision(Q, R, F, E), respectively, with outcomes similar to cases 3.1. and 3.2;
     - 3.3.2. ask for a better explanation via Unclear(Q, R, E) (go to 3.3.).
   - 3.4 ask for *contrastive* explanations motivating why not R', by answering WhyNot(Q, R, R'). The explainer may now:
     - 3.4.1. explain the difference E among R and R', if R' is admissible w.r.t. its current knowledge base, by answering Comparison(Q, R, R', E). Now, the explainee may either (in both cases, the protocol terminates):
       - 3.4.1.1. accept R, via Accept(Q, R, E), or
       - 3.4.1.2. state that they prefer R', via Prefer(Q, R, R').
     - 3.4.2 explain that R' is not an admissible recommendation because of E, by answering Invalid($()Q, R', E$). At this point, the explainee may either (in both cases, the protocol terminates):
       - 3.4.2.1. accept R, via Accept(Q, R, E), or
       - 3.4.2.2. override the explainer's decision, by stating that they prefer R', via Override(Q, R, R')—hence forcing the explainer to update their own knowledge base accordingly.

3.2   Relevant Scenarios and Protocol Analysis

The protocol is general enough to cover multiple relevant situations, corresponding to different needs/desires of the users. For instance, users may: *(i)* simply
want a recommendation; *(ii)* want the recommendation to be explained; *(iii)*
want more details for a given explanation; *(iv)* want to simulate other possible
recommendations; *(v)* provide positive or negative feedback about recommendations or explanations.

All such situations correspond to relevant usage scenarios of the protocol.
These are briefly summarised in Fig. 3, and discussed below.

Quick Accept. This is the scenario depicted in Fig. 3a. There is no need for
explanations, and the user simply accepts the recommendation provided by the
agent.

For instance, the user asks for a restaurant recommendation, and the agent
proposes a restaurant, and the user is fine with it.

Quick Retry. This is the scenario depicted in Fig. 3b. There is no need for explanations, and the user simply rejects the recommendation provided by the agent,
by either disapproving it or stating that it is in conflict with their own preferences. In both cases, the agent shall produce a new recommendation.

For instance, the user asks for a restaurant recommendation, and the agent
proposes a steakhouse, but the user does not like it because: *(i)* they are vegetarian or they do not like steak, or *(ii)* they do not want to eat meat that day. In
the former case, the user shall signal a collision among the recommendation and
its preference—which the agent is expected to learn and to take into account for
future recommendations. In the latter case, the user shall simply disapprove the
recommendation—but the agent is not supposed to memorise such an event.

Ordinary Explanation Loop. This is the scenario depicted in Fig. 3c. The user
asks for a recommendation, and the agent provides one. The user is not satisfied
with the recommendation, and asks for an explanation. The agent provides an
explanation, and the user is not satisfied with it. The user asks for further
details, and the agent provides them. The loop may be repeated several times.
Eventually, the user accepts the recommendation, or asks for a new one, similarly
to the quick accept/retry scenarios.

The interesting part here is the explanation loop. It is a flexible mechanism,
supporting zooming in/out explanations: the agent change the granularity of
the explanations, by providing more or less details. For instance, the agent may
provide *local* explanations first – i.e., explanations describing how the recommendation was produced – and then *global* explanations—i.e., explanations describing how recommendations are computed in general. The agent may also change
the representation means of the explanations, by providing textual explanations
first, and then visual explanations—or vice versa.

As an example, consider the situation where the user asks for a restaurant
recommendation. The agent recommends some Asian restaurant in the users'

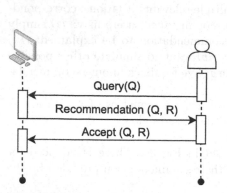

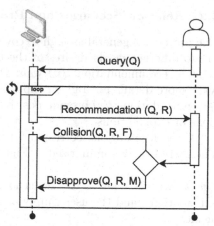

(a) Quick accept: the user accepts the recommendation without asking for explanations.

(b) Quick retry: the user rejects the recommendation without asking for explanations. Another recommendation is proposed, accordingly.

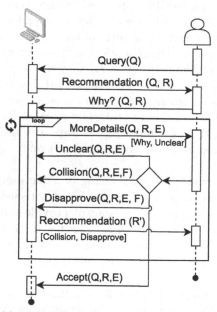

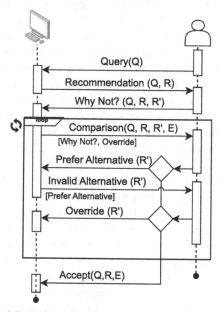

(c) Ordinary explanation loop: the user asks 'why' after a recommendation, and then agent answers with further details. The request for details may be repeated several times.

(d) Contrastive explanation loop: the user asks 'why not' another recommendation. The agent may then explain why the other recommendation is acceptable or invalid. The user may either accept the original recommendation or prefer their own.

Fig. 3. Sequence diagrams describing most common scenarios of the protocol.

surroundings, having 4.3 stars (out of 5) on ACME-Advisor.com. The user is curious in understanding the reason why the agent proposed that restaurant, and asks for an explanation. The agent provides an explanation, stating that the restaurant is close to the user, and that it has a good rating. Furthermore, the agent reminds to the user that – to the best of its knowledge – they like Sushi. The user is still not satisfied, and asks for further details. The agent provides a textual explanation, stating that it commonly recommends the highest ranked restaurant matching the user's tastes, and having a distance which is not higher than 1 km. Eventually, the user may be satisfied with the explanation, and accept the recommendation; or they may reject the recommendation and possibly request a new one.

Contrastive Explanation Loop. This is the scenario depicted in Fig. 3d. The user asks for a recommendation, and the agent provides one (R). The user was not expecting that recommendation, but rather another one (R'), and they ask for a *contrastive* explanation. If the users' recommendation is acceptable as well, the agent provides a comparison between the two recommendations, arguing one of the two is better than the other. Otherwise, if the users' recommendation is not acceptable, the agent provides an explanation for why it is not acceptable. In both cases, the user may either accept the original recommendation or prefer their own—possibly overriding the agent's recommendation. In the case of an override, the agent should learn from the user's preferences, and possibly update its recommendation policy. In any case, the interaction ends.

The key points here are the possibility, for the user, to *(i)* simulate alternative recommendations, and *(ii)* contradict the recommender agent in order to let it learn.

Consider for instance the aforementioned restaurant recommendation case. The user may not be satisfied with the agent's recommendation concerning an Asian restaurant, and propose the local steakhouse instead. The agent may then either consider the proposal acceptable or not, depending on the dietary goals and physiological condition of the user. If the agent considers the proposal acceptable, it may provide a comparison between the two recommendations, stating that the Asian restaurant is closer. At this point, the user may either accept the original recommendation (Asian restaurant) or prefer their own (steakhouse). Otherwise, if the agent considers the proposal unacceptable, it may provide an explanation for why it is not acceptable—e.g. steak is violating the user's dietary goals. In this case, the user may either accept the original recommendation (Asian restaurant) or override it (steakhouse).

3.3   Which Sorts of Explanations and Recommendation?

The explanation protocol is agnostic w.r.t. the particular way in which explanations and recommendations are represented. Indeed, it is implementers' responsibility to define the representation means of explanations and recommendations— other than deciding how they should be computed in practice. The protocol simply dictates *when* explanations and recommendations should be computed.

Accordingly, in this subsection we provide a few insights about the possible design choices for explanations and recommendations.

Recommendations are commonly supported by means of one or more ML predictors, trained on users' data. Whether predictors' training is a responsibility of the recommender agent, or simply the agent is endowed with pre-trained predictors at deployment time, is an implementation detail. In either cases, the recommender agent is supposed to know (or be able to access or acquire) profile-related information about the user. Such information may come, for instance, from some initial configuration phase, as well as be inferred by the agent itself, from the accepted/rejected recommendations. To support the latter case, the agent should be endowed with some learning algorithm, making it able to (re)train the predictors when new user data is avaliable. Under this perspective, as far as recommendations are concerned, the explainer agent is simply proxying the ML predictor(s).

Explanations, on the other hand, are not necessarily supported by ML predictors. In this case, the XAI literature is full of possible approaches, including both visual, textual, and numeric explanations. The interested reader may refer to high impact surveys such as [1, 10] for a comprehensive overview of the state of the art. The key point here is that the explainer agent should not only wrap the ML predictor(s), but also encapsulate the logic for representing and computing explanations.

Along this line, one critical situation is the one where recommendations and explanations come with different representation means—e.g. textual and visual. In this case, the explainer agent should be able to bridge the gap between the two, by providing a unified representation of the recommendation and its explanation.

To mitigate this issue, designers may consider adopting computational logic as the reference framework for both recommendations and explanations. In computational logic, both knowledge bases, and queries, are represented as logic formulæ. Logic formulæ, in turn, can be exploited to represent both recommendations and explanations. In fact, logic queries may be used to represent recommendation requests, and logic solutions may be used to represent recommendations, whereas proof trees may be exploited to compute explanations.

For instance, a recommendation query may consist of the logic goal *should_eat* (*Food*, lunch), where *Food* is a logic variable, i.e., a placeholder for unknown values. Recommendations $R, R', R'', \ldots$ may be logic solutions, i.e., assignments of logic variables (e.g., *Food* = paella). Explanations $E, E', E'', \ldots$ may be of many sorts:

- local explanation, e.g. the path in the proof tree computed by the explainer agent to provide the recommendation;
- global explanation, e.g. the logic program used by the explainer agent to provide the recommendation;
- contrastive explanation, e.g. quality metrics comparing two or more recommendations, or the violated constraints making some recommendation unacceptable;
- any combination of the above.

User features $F, F', F'', \ldots$ may be raw facts describing the user (e.g., $age(\mathtt{31})$, $goal(\mathtt{lose\_weight})$, $category(\mathtt{vegetarian})$). Disapprove motivations may be predefined facts such as $\mathtt{dislike}$ – the user does not like a recommendation and the agent should learn that – or $\mathtt{not\_now}$—the user simply does not want that recommendation now, may they like it in general (so the agent should not memorise that).

4 From Theory to Practice with PyXMas

In this section, we describe how our protocol can be reified into actually usable agent-oriented software. Accordingly, we discuss the design of PyXMas[2], i.e., our Python library for explainable multi-agent systems.

PyXMas is an agent-oriented software library based on SPADE. It comes with predefined – yet parametric – implementation of the protocol described in Sect. 3, in the form of reusable agent behaviours. In this way, researchers and developers can easily take advantage of our protocol to build explainable MAS, without wasting time in re-implementing the protocol. Rather, they can focus on the design of the actual recommender and explainer agents, as well as on the representation means of recommendations and explanations. In particular, PyXMas requires designers to define which particular notion of recommendation and explanation they want their agents to support, and how should agents compute or react to them.

4.1 PyXMas Architecture

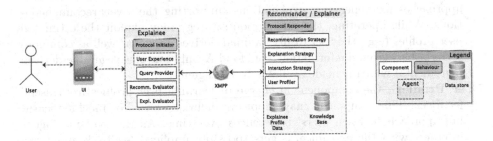

Fig. 4. Architecture of PyXMas. Behaviours are provided by the library, and they are parametric w.r.t. *components*. PyXMas users are responsible for implementing and plugging their own components, in order to tailor PyXMas to their specific needs. Optionally, if the explainee agent is human, users may also need to implement a UX component supporting interaction with the explainee—possibly via some device.

Figure 4 summarises the modular architecture of PyXMas. Leveraging on SPADE facilities, PyXMas is implemented two behaviours, which can be easily

---

[2] https://github.com/pikalab-unibo/pyxmas.

plugged into any agent—namely, the protocol *initiator* and the *responder*. On the one hand, the initiator behaviour is responsible for sending recommendation queries to the explainer agent, and for receiving and processing the corresponding recommendations and explanations. Consequently, the initiator behaviour is meant to be plugged into the explainee agent. On the other hand, the responder behaviour is responsible for receiving requests from the initiator, and for computing and sending recommendations and explanations. Consequently, the responder behaviour is meant to be plugged into the explainer agent.

Figure 4 also shows that PYXMAS is designed to be highly parametric. In particular, the initiator and responder behaviours are parametric with respect to a number of *components* which dictate the actual behaviour of the agents. In this way, users of PYXMAS can easily tailor the library to their specific needs, by implementing and plugging their own components.

Explainer Agent. As far as the explainer agent is concerned, the responder behaviour requires the following components provided by developers:

Recommendation Strategy— the component in charge of computing recommendations for any given query. In addition to users' requests and feedback, the recommendation strategy should consider profile information about users (e.g., their goals/interests, such 'losing weight'), as well as their preferences and interests (e.g., vegetarian users do not eat meat). Agents may have limited information about their users' preferences but could learn more over-time— also thanks to our protocol. The learned preferences and constraints could be exploited to generate well-targeted recommendations.

Explanation Strategy—the component in charge of computing explanations for any given recommendation. This is where designers can develop different approaches for generating explanations supporting the given recommendations. While operating, the explanation strategy may exploit the estimated user profiles (e.g., the user dislikes animal-derived food), as well as common-sense or background information (e.g., food X only contains vegetable-derived ingredients) which is made available to the agent.

User Profiler—the component in charge of learning user profiles from users' feedback. This component may adopt any heuristic-based or machine learning approach to learn users' preferences over time. As the explainer agent interacts with the explainee, it gets (possibly implicit) feedback about the recommendations it provides. It this way, it may infer valuable information about their preferences and interests. In this way, the agent can update the explainee's profile information—in order to eventually provide better explanations or recommendations.

Interaction Strategy—the component in charge of which recommendation and explanation strategies to exploit, and how to present recommendations and explanations to the explainee. This component is responsible for processing the content of the exchanged messages and transferring them to the related components. Note that not only the content of the messages but also how these messages are expressed/represented plays a crucial role in interactive intelligent systems. Consider for instance the case where the explainer agent is a humanoid robot. In the case, the interaction component is in charge of selecting the best gestures or facial expressions supporting the action taken (e.g., a surprising facial expression when it discovers unexpected knowledge about the user). The interaction strategy component may also operate the other way around. For instance, if the robot can sense people facial emotion, tones, or gestures, it may adjust other components behaviour accordingly.

It is worth mentioning that, to operate correctly, the explainer agent is supposed to collect and store two sorts of information, namely: *(i)* profile data about the explainee, and *(ii)* common-sense/background knowledge about the domain of interest. For an architectural perspective (cf. Fig. 4), information of these sorts are store in to *ad-hoc* data stores. In particular, profile data is stored in the *user profile* data store, while the common-sense/background knowledge is stored in the *knowledge base* data store. These data stores are local w.r.t. the explainer agent, and act as its memory/belief base. They are subject to reads/updates by the different components of the explainer agent.

Explainee Agent. As far as the explainee agent is concerned, the initiator behaviour requires the following components provided by developers:

Query Provider—the component in charge of generating queries for the explainer agent, depending on the current goals of the explainee.

Recommendation Evaluator—the component in charge of evaluating the recommendations provided by the explainer agent and deciding whether to accept or reject them.

Explanation Evaluator—the component in charge of evaluating the explanations provided by the explainer agent, and affecting the recommendation evaluator accordingly.

In the particular case where the explainee agent is a human user, the explainer agent should be implemented as a simple proxy agent, which acts on behalf of the human and mediates their interaction with the recommender agent. In that case, the proxy agent is responsible for human-computer interaction, possibly via some user interface (UI) presented to the human on top of some device (e.g., a smartphone). This is the situation depicted in Fig. 4. When this is the case, the proxy agent is supposed to include one further component, namely the **User Experience** (UX) one.

When present, the UX component is in charge of governing the UI, hence grasping humans' inputs and presenting recommendations and explanations to them. In this case, the other components are simply in charge of processing the humans' inputs and generating the appropriate messages to be sent to the explainer agent.

4.2   PYXMAS Design

PYXMAS consists of a Python library providing:

- abstract classes defining the (de)serialisation message payloads exchanged between the explainer and explainee agents (cf. Sect. 3.1),
- abstract classes defining the initiator and responder behaviours.

In both cases, we exploit abstract classes as we leave room for costumisation. In fact, developers may want to extend the provided abstract classes and override specific methods to plug their own components.

Accordingly, in this subsection, we describe the abstract classes available in PYXMAS and the way they are supposed to be extended to so ars to build some actual explainable MAS.

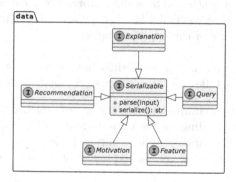

Fig. 5. Abstract classes for message payloads in PYXMAS.

Data Types for Message Payloads. As shown in Fig. 5, PYXMAS provides 5 abstract classes for the as-many data types defined in Sect. 3.1. These classes simply force developers to make these data types *serialisable*—i.e., to support their conversion into/from strings. This is necessary for the explainer and explainee agents to exchange messages over the network.

How to actually represent queries, recommendations, explanations, and so on, is left to developers. In fact, the only constraint is that the serialised version of these data types should be both machine- and human-interpretable.

When it comes to design some actual explainable MAS, developers may plug their custom notion of query, recommendation, explanation, and so on, by extending the provided abstract classes, and by implementing their (de)serialization-related methods.

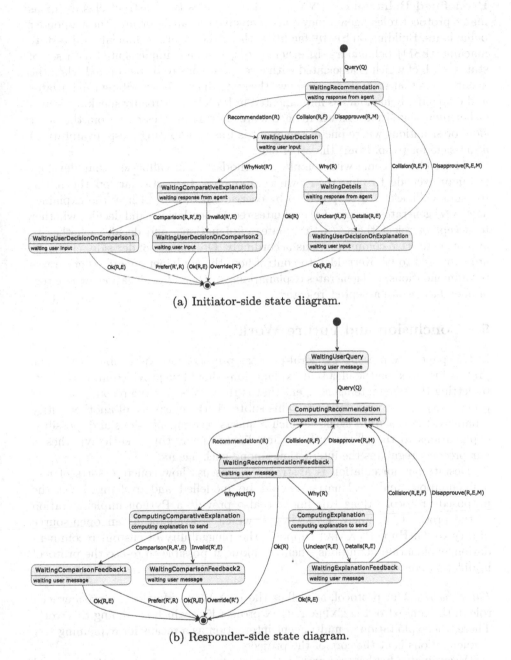

(a) Initiator-side state diagram.

(b) Responder-side state diagram.

Fig. 6. State diagrams describing the initiator and responder behaviours as implemented in PyXMas.

Predefined Behaviours. PyXMas also provides two abstract classes for as many protocol roles agents may play, namely: the initiator and the responder behaviours. Building on SPADE facilities, these classes are technically finite-state machine (FSM) behaviours. In other words, they are implemented as a set of states, each of which is associated with a set of actions to be performed when the agent enters that state. Figure 6 shows the state diagrams describing the initiator and responder behaviours as implemented in PyXMas. Broadly speaking, states either represent situations where agents are waiting for messeges from the other side, or situations where one of the agent is busy computing (resp. evalutation) a message for (resp. from) the other side.

These classes come with template methods (a.k.a. callbacks) that developers may override to plug their own components. In particular, on the initiator side, callbacks are supposed to be overridden to control how the explainee agent: *(i)* generates queries, *(ii)* evaluates recommendations and decides whether to accept or reject them, *(iii)* evaluates explanations, and decides whether to accept or reject recommendations accordingly. On the responder side, callbacks are supposed to be overridden to control how the explainer agent: *(i)* generates recommendations, *(ii)* generates explanations, *(iii)* handles situations where recommendations are accepted/rejected.

5   Conclusion and Future Work

In this paper, we present a general-purpose protocol for explainable MAS. The protocol is based on the idea that explanations should be provided upon request, by letting the same intelligent agent that is responsible for the recommendation process explain its own decisions. This subtends the existence of another entity – namely, the *explainee* agent – which requests recommendations and, possibly, explanations to the aforementioned intelligent agent. Under such hypotheses, our protocol regulates the interaction among such agents.

Despite our formulation is abstract, we discuss how concrete sorts of recommendations and explanations could be modelled and exchanged via the proposed protocol. Along this line, we also provide a Python implementation of the protocol – namely, PyXMas –, which is available as an open-source library on GitHub. PyXMas supports the pluggability of custom recommendation/explanation definitions—hence making it possible to re-use the protocol in different contexts.

Future works. Our protocol, as well as the PyXMas technology, plays a crucial role in the context of the EXPECTATION project [5]—which is funding this work. There, the exploitation of multi-agent interaction as a means for explaining recommendations is at the core of the project.

Accordingly, further research is needed to investigate how the protocol impacts human-user interaction as a means for XAI. Along this way, we are planning both theoretical extensions of the protocol and technical improvements of the PyXMas technology—possibly enabling empirical studies on the impact on explainability.

In particular, concerning the protocol, we are planning to extend the formulation to support meta-data describing the emotional state of the explainee agent—hence studying of such meta-information may affect the recommendation/explanation process.

Concerning the PYXMAS technology, we are planning to support the exploitation of symbolic knowledge extraction [19] and injection [14] as a means for explaining recommendations. This would imply leveraging on symbolic AI techniques to represent explanations and recommendations. Finally, we plan to provide better support towards human-computer interaction based on PYXMAS. In this regard, our intention is to develop a Web- or Telegram-based graphical user interface for letting humans interact with PYXMAS agents.

Acknowledgements. This work has been supported by the CHIST-ERA IV project "EXPECTATION", the Italian Ministry for Universities and Research (G.A. CHIST-ERA-19-XAI-005), and by the Scientific and Research Council of Turkey (TÜBİTAK, G.A. 120N680).

References

1. Barredo Arrieta, A., et al.: Explainable explainable artificial intelligence (XAI): concepts, taxonomies, opportunities and challenges toward responsible AI. Inf. Fusion **58**, 82–115 (2020). https://doi.org/10.1016/j.inffus.2019.12.012
2. Bellifemine, F.L., Caire, G., Greenwood, D.: Developing Multi-Agent Systems with JADE. Wiley, Hoboken (2007). http://eu.wiley.com/WileyCDA/WileyTitle/productCd-0470057475.html
3. Buzcu, B., Varadhajaran, V., Tchappi, I., Najjar, A., Calvaresi, D., Aydogan, R.: Explanation-based negotiation protocol for nutrition virtual coaching. In: Aydogan, R., Criado, N., Lang, J., Sánchez-Anguix, V., Serramia, M. (eds.) PRIMA 2022: Principles and Practice of Multi-Agent Systems - 24th International Conference, Valencia, Spain, 16–18 November 2022, Proceedings. Lecture Notes in Computer Science, vol. 13753, pp. 20–36. Springer, Heidelberg (2022). https://doi.org/10.1007/978-3-031-21203-1_2
4. Calegari, R., Ciatto, G., Omicini, A.: On the integration of symbolic and sub-symbolic techniques for XAI: a survey. Intelligenza Artificiale **14**(1), 7–32 (2020). https://doi.org/10.3233/IA-190036
5. Calvaresi, D., et al.: EXPECTATION: personalized explainable artificial intelligence for decentralized agents with heterogeneous knowledge. In: Calvaresi, D., Najjar, A., Winikoff, M., Främling, K. (eds.) EXTRAAMAS 2021. LNCS (LNAI), vol. 12688, pp. 331–343. Springer, Cham (2021). https://doi.org/10.1007/978-3-030-82017-6_20
6. Christakopoulou, K., Radlinski, F., Hofmann, K.: Towards conversational recommender systems. In: KDD 2016: Proceedings of the 22nd ACM SIGKDD International Conference on Knowledge Discovery and Data Mining, pp. 815–824 (2016). https://doi.org/10.1145/2939672.2939746
7. Ciatto, G., Calegari, R., Omicini, A., Calvaresi, D.: Towards XMAS: eXplainability through multi-agent systems. In: Savaglio, C., Fortino, G., Ciatto, G., Omicini, A. (eds.) AI&IoT 2019 - Artificial Intelligence and Internet of Things 2019, CEUR Workshop Proceedings, vol. 2502, pp. 40–53. Sun SITE Central Europe, RWTH Aachen University (2019). http://ceur-ws.org/Vol-2502/paper3.pdf

8. Ciatto, G., Schumacher, M.I., Omicini, A., Calvaresi, D.: Agent-based explanations in AI: towards an abstract framework. In: Calvaresi, D., Najjar, A., Winikoff, M., Främling, K. (eds.) EXTRAAMAS 2020. LNCS (LNAI), vol. 12175, pp. 3–20. Springer, Cham (2020). https://doi.org/10.1007/978-3-030-51924-7_1

9. Fielding, R.T., Taylor, R.N.: Principled design of the modern Web architecture. ACM Trans. Internet Technol. **2**(2), 115–150 (2002). https://doi.org/10.1145/514183.514185

10. Guidotti, R., Monreale, A., Ruggieri, S., Turini, F., Giannotti, F., Pedreschi, D.: A survey of methods for explaining black box models. ACM Comput. Surv. **51**(5), 93:1–93:42 (2018). https://doi.org/10.1145/3236009

11. Gunning, D.: Explainable artificial intelligence (XAI). Funding Program DARPA-BAA-16-53, DARPA (2016). http://www.darpa.mil/program/explainable-artificial-intelligence

12. Knijnenburg, B.P., Willemsen, M.C., Hirtbach, S.: Receiving recommendations and providing feedback: the user-experience of a recommender system. In: Buccafurri, F., Semeraro, G. (eds.) EC-Web 2010. LNBIP, vol. 61, pp. 207–216. Springer, Heidelberg (2010). https://doi.org/10.1007/978-3-642-15208-5_19

13. Lipton, Z.C.: The mythos of model interpretability. Commun. ACM **61**(10), 36–43 (2018). https://doi.org/10.1145/3233231

14. Magnini, M., Ciatto, G., Omicini, A.: On the design of PSyKI: a platform for symbolic knowledge injection into sub-symbolic predictors. In: Calvaresi, D., Najjar, A., Winikoff, M., Främling, K. (eds.) Explainable and Transparent AI and Multi-Agent Systems, 4th International Workshop, EXTRAAMAS 2022, Virtual Event, Revised Selected Papers, Lecture Notes in Computer Science, 9–10 May 2022, vol. 13283, chap. 6, pp. 90–108. Springer, Heidelberg (2022). https://doi.org/10.1007/978-3-031-15565-9_6

15. Millecamp, M., Htun, N.N., Conati, C., Verbert, K.: To explain or not to explain: The effects of personal characteristics when explaining music recommendations. In: IUI 2019: Proceedings of the 24th International Conference on Intelligent User Interfaces, pp. 397–407. Association for Computing Machinery, New York (2019). https://doi.org/10.1145/3301275.3302313

16. Mualla, Y., et al.: The quest of parsimonious XAI: a human-agent architecture for explanation formulation. Artif. Intell. **302**, 103573 (2022). https://doi.org/10.1016/j.artint.2021.103573

17. O'Donovan, J., Smyth, B., Gretarsson, B., Bostandjiev, S., Höllerer, T.: Peer-Chooser: visual interactive recommendation. In: CHI 2008: Proceedings of the SIGCHI Conference on Human Factors in Computing Systems, pp. 1085–1088 (2008). https://doi.org/10.1145/1357054.1357222

18. Omicini, A.: Not just for humans: explanation for agent-to-agent communication. In: Vizzari, G., Palmonari, M., Orlandini, A. (eds.) AIxIA 2020 DP – AIxIA 2020 Discussion Papers Workshop. AI*IA Series, vol. 2776, pp. 1–11. Sun SITE Central Europe, RWTH Aachen University, Aachen (2020). http://ceur-ws.org/Vol-2776/paper-1.pdf

19. Sabbatini, F., Ciatto, G., Calegari, R., Omicini, A.: Symbolic knowledge extraction from opaque ML predictors in PSyKE: platform design & experiments. Intelligenza Artificiale **16**(1), 27–48 (2022). https://doi.org/10.3233/IA-210120

20. Shimazu, H.: ExpertClerk: a conversational case-based reasoning tool for developing salesclerk agents in e-commerce webshops. Artif. Intell. Rev. **18**, 223–244 (2002). https://doi.org/10.1023/A:1020757023711

21. Zhang, Y., Chen, X.: Explainable recommendation: a survey and new perspectives. Found. Trends Inf. Retr. **17**(1), 1–101 (2020). https://doi.org/10.1561/1500000066

Dialogue Explanations for Rule-Based AI Systems

Yifan Xu$^{(\boxtimes)}$ ⓘ, Joe Collenette ⓘ, Louise Dennis ⓘ, and Clare Dixon ⓘ

Department of Computer Science, The University of Manchester, Manchester, UK
{yifan.xu,joe.collenette,louise.dennis,clare.dixon}@manchester.ac.uk

Abstract. The need for AI systems to explain themselves is increasingly recognised as a priority, particularly in domains where incorrect decisions can result in harm and, in the worst cases, death. Explainable Artificial Intelligence (XAI) tries to produce human-understandable explanations for AI decisions. However, most XAI systems prioritize factors such as technical complexities and research-oriented goals over end-user needs, risking information overload. This research attempts to bridge a gap in current understanding and provide insights for assisting users in comprehending the rule-based system's reasoning through dialogue. The hypothesis is that employing *dialogue* as a mechanism can be effective in constructing explanations. A dialogue framework for rule-based AI systems is presented, allowing the system to explain its decisions by engaging in "Why?" and "Why not?" questions and answers. We establish formal properties of this framework and present a small user study with encouraging results that compares dialogue-based explanations with proof trees produced by the AI System.

1 Introduction

Reasoning, the process of synthesising facts and beliefs to make new decisions, is a fundamental component of humans' explanatory mechanisms [11]. Giving the current generation of AI systems human-like capabilities for explaining themselves is challenging because their data-driven nature makes it hard to identify reasoning-like processes. In contrast, in the early days of AI, explainability was regarded as an easy task since most systems were logic-based [26]. Such *Rule-based systems* (RBS) may be learned, and, in particular, there have been recent results in the extraction of decision trees (and rules) from neural networks for the purposes of improved explainability [21,31]. Even in this work, however, the assumption is that once converted to the RBS, the resulting system is inherently explainable. Because when the rule-chaining process of such a system becomes very complex, their explanations are difficult to follow [14].

As a starting point, we focus on explaining hand-crafted RBS, with the aim of extending our learned rules to RBS extracted from machine learning models in the future. The utility of an explanation depends upon the user's context – why they are seeking an explanation. Are they surprised by a recommendation and

© The Author(s), under exclusive license to Springer Nature Switzerland AG 2023
D. Calvaresi et al. (Eds.): EXTRAAMAS 2023, LNAI 14127, pp. 59–77, 2023.
https://doi.org/10.1007/978-3-031-40878-6_4

want to know more? Do they want to challenge a recommendation? In particular, we have focused on situations where the user's information is different from that possessed by the system and we've used the user's ability to discover this mismatch following the explanatory process as one of our metrics for assessing the utility of the explanation.

We propose a formal framework for dialogues involving two participants (presumed to be a RBS and a user) that specifies allowable utterances (in the form of questions or "one step" explanations) and how each participant's mental model of the other is updated given these utterances. We have implemented this framework together with a simple RBS based on rules around Covid-19 restrictions. To assess our explanation, consider Miller's [16] findings that a good explanation must be short, be selected, and be social, we compared the dialogue system with providing the RBS' deduction tree with encouraging results.

2 Related Work

Early rule-based expert system explanations [22] focused in particular on the explanation framework [5,19,25,29], and the human-computer interface (HCI) through which the explanation was supplied [13,24]. The most sophisticated approaches involved an "intelligent" conversation with the system user that was done in simple terms and using interactive methods [9]. Naturalness was recognised as a condition for a good explanation [12,17] so the social aspect of explanation was known. The user's inquiry is restricted to asking why this information is being requested by the system [5]. However, little progress was made in terms of enabling users to really guide an explanation to a desired outcome, also it becomes challenging to construct a coherent explanation when there are numerous chained rules involved [14].

To solve the issue mentioned above, several dialogue models for explanation have been proposed [23,27]. Walton's shift model for dialogue proposes an explanation and examination dialogue with three stages and two rules governed by the explainee to determine the success of an explanation [3,28]. These models, however, don't appear to have iterative aspects like cyclic dialogues and lack a data-based foundation or validation. Madumal introduced an interaction protocol for interactive explanations by analyzing transcripts from real explanation dialogue datasets [15].

Argumentation, as an important reasoning strategy, has also been incorporated into dialogue models to enhance the explainability of AI systems [2,7, 18,20,26]. Walton and Bex [3] utilize argumentation models and dialogue and enable the explainee to question and dispute the provided explanations which are modeled as arguments. This enables the explainee to query and interrogate the provided explanations in order to achieve better comprehension. Although the proposed framework offers a high-level structure for explanation-based conversations, it does not place a strong emphasis on explaining rule-based deductions or using arguments to fully comprehend the beliefs of the other person. Furthermore, there are very few actual human experiments that have been done to evaluate the efficacy of such arguments.

A dialogue framework has been developed to explain the behavior of a system programmed using the BDI (Beliefs-Desires-Intentions) paradigm which has many similarities to RBS [8]. It defines a turn-based system and allows users to ask questions about the reasons behind selecting plans of action within the system, but does not provide a way to explain deductive reasoning (which is our focus). Building upon the foundational works of Dennis and Oren [8], we aim to ensure that the user gains a genuine comprehension of the explanation without overwhelming them with excessive information.

Miller highlights the importance of concise, carefully chosen, and socially relevant explanations [16]. He emphasizes that explanations serve as answers to "why" questions. Similarly, Winikoff also emphasizes the significance of addressing "why" questions when providing explanations [30]. Our dialogue explanations also prioritize addressing both "why" and "why not" questions to generate collaborative explanations.

3   Framework

Our starting point is two "players" (assumed to be some RBS and a user). Each possesses a set of facts (F) and a set of rules (R) and uses these to deduce whether some conclusion (C) is true or false. Deductions are represented as trees. When the players disagree they engage in a dialogue. Each player can ask *why* a particular node in a tree is believed in which case they are informed that it was either an initial fact, or it was deduced from its parent nodes using a rule. A player can also ask *why not* questions. In this case, the other player turns this around and asks the other player why they believe that something does hold. Note we assume that both players reason correctly.

3.1   Proof Trees

We assume:

- A language of terms, $\mathcal{L}$, defined in the standard way (See [10], p. 99).
- A set of labels L which include two special labels: *initial* and *unprovable*.
- A set of initial facts, F (positive literals in $\mathcal{L}$).
- A set of rules, R. A rule is a clause consisting of a non-empty set of literals in $\mathcal{L}$ (the antecedants, A), a consequent, a positive literal $C \in \mathcal{L}$, and a label $l \in L \backslash \{initial, unprovable\}$, written as $l : A \to C$. We assume that labels in R are unique and that rules that are identical up to the renaming of variables have the same label[1].

We use the notation $pos(A)$ for the set of terms that appear positively in some set of literals, A, and $neg(A)$ for the set of terms that appear negatively in some set of literals A (i.e. if $t \in neg(A)$ then $\neg t \in A$)

---

[1] We don't need to label rules for our system to work, but labels are a useful convenience when referring to rules.

Definition 1. Proof Tree

A proof tree is a directed rooted tree written $\langle N, E \rangle$, *where* N *is a set of nodes of the form* (t, l) *where* $t \in \mathcal{L}$ *is a ground positive literal and* $l \in L$ *is a label.* $E \subseteq N \times N$ *is the set of edges. An edge between two nodes* n_1 *and* n_2 *is written as* $n_1 \mapsto n_2$.

We use standard terminology so the **root** of a proof tree is the single node, n such that there is no edge $n' \mapsto n$. The **parent nodes** of a node n are the set of nodes n' such that there exists an edge $n \mapsto n'$. The **parent trees** of a node n are the set of sub-trees with a parent of n as their root.

If (t, l) is the root node of a tree, then we refer to t as the **root term** of the tree.

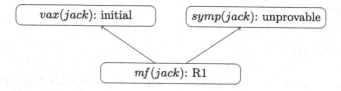

Fig. 1. A Proof Tree showing why Jack can meet his friends using $R1$: $\{vax(X), \neg symp(X)\} \rightarrow mf(X)$. R1: You can meet friends if you have been vaccinated and display no symptoms, and the initial fact set $\{vax(jack)\}$ means Jack is vaccinated.

Definition 2. Provable, Unprovable and Undecided in T

If $\langle N, E \rangle = T$ *is a proof tree and* t *is a ground positive literal in* $\mathcal{L}$. *We say:* t *is* **provable** *in* T *iff there exists a node* $(t, l) \in N$ *such that* $l \neq unprovable$; t *is* **unprovable** *in* T *iff* $(t, unprovable) \in N$; t *is* **undecided** *in* T *iff there is no node* $(t, l) \in N$.

Therefore, in Fig. 1, if our proof tree is T, then $vax(jack)$ and $mf(jack)$ are both provable in T, $symp(jack)$ is unprovable in T and any other term (e.g., $fever(jack)$) is undecided in T.

Definition 3. Proof Tree for F and R

A Proof Tree, T, for a set of facts, F, and rules, R is defined recursively as follows:

- $\langle \{(t, initial)\}, \emptyset \rangle$ is a proof tree for F and R iff $t \in F$
- $\langle \{(t, unprovable)\}, \emptyset \rangle$, is a proof tree for F and R iff no proof tree, T', for F and R exists such that t is provable in T'
- If $E \neq \emptyset$ then a proof tree $T = \langle N, E \rangle$ with root node (t, l) is a proof tree for F and R iff:
  - The parent trees of (t, l) are all proof trees for F and R
  - There exists a rule, $l : A \rightarrow C \in R$ and a substitution, θ for the free variables in A and C such that $C\theta = t$ and $t \notin F$, and

* if (t', l') is a parent of (t, l) in T then either
  · $\exists t_i \in pos(A).\, t_i\theta = t'$ and $l' \neq unprovable$ or,
  · $\exists t_i \in neg(A).\, t_i\theta = t'$ and $l' = unprovable$; and
* $\forall t' \in A\theta$ there exists a unique label, l' such that (t', l') is a parent node of (t, l) in T.

A proof tree with some statement t at its root (either as a provable or unprovable statement) can be constructed from F and R by standard backward reasoning with negation as a failure as used in logic programming languages such as Prolog [6]. From this point, we will stop referring to substitutions, θ, etc. for reasons of readability and present our theory only for the case where rules contain no free variables. Our proofs can be adapted straightforwardly to the more general case.

Note that our proof trees are essentially SLDNF-trees (Selective Linear Definite Clause with Negation as Failure) from logic programming [1] extended with rule labels. We assume that our facts and rules are such that SLDNF-resolution is complete – for instance that they represent an acyclic program [4].

4   Dialogues

We formalise the idea of a disagreement between two RBSs as a difference in their initial facts or rules. The purpose of a dialogue will be to identify at least one such difference from a starting point where one RBS has deduced some fact to be the case and the other has deduced that it is not the case.

Definition 4. Deduction

We formalise a deduction as a tuple $\mathcal{D}(F, R, T)$ where F is a set of initial facts, R a set of rules and T is a set of proof trees for F and R. We will refer to T as the deduction trees.

Our problem is: given two deductions $\mathcal{D}(F_1, R_1, T_1) \neq \mathcal{D}(F_2, R_2, T_2)$ which disagree about some deduced fact can we identify the disagreement in terms of their initial facts or rules? More formally if there exists a $T_1 \in T_1$ (resp. $T_2 \in T_2$) which has some provable root term t that is unprovable in at least one $T_2 \in T_2$ (resp. $T_1 \in T_1$), can we identify at least one fact, t' such that $t' \in F_1$ and $t' \notin F_2$ (or vice versa) or at least one rule r such that $r \in R_1$ and $r \notin R_2$ (or vice versa).

We can trivially identify the differences if we have full access to F_1, F_2, R_1, R_2, etc., so we assume that this is not the case but take the viewpoint of one of the parties making the deduction – so either we have access to F_1 and R_1 but not F_2 and R_2 or vice versa. We do assume that rules with the same label in R_1 and R_2 are identical up to the renaming of variables – i.e., if $l : A_1 \rightarrow C_1 \in R_1$ and $l : A_2 \rightarrow C_2 \in R_2$ then $A_1 = A_2$ and $C_1 = C_2$. This means we can use rule labels without loss of generality as proxies for the rules themselves rather than having to match antecedents and consequents.

Definition 5. Provable/Unprovable for Deductions

Given a deduction $D = \mathcal{D}(F, R, \mathcal{T})$ we say a term t is provable in D if t is provable in some $T \in \mathcal{T}$ and that t is unprovable in D if t is unprovable in some $T \in \mathcal{T}$.

To simplify our proofs we introduce a completeness property for deductions. This specifies that if some term is unprovable in the deduction then the deduction contains the evidence for why it is unprovable – in particular it contains proof trees for all the antecedents of any rule with the term as its consequent. These can then be inspected to understand why that rule did not apply.

Definition 6. Complete Deduction *We say that a deduction $D = \mathcal{D}(F, R, \mathcal{T})$ is complete if, for any t that is unprovable in D, if there is a rule $l : A \to t \in R$ then all terms $t' \in pos(A) \cup neg(A)$ are either provable or unprovable in D.*

In practice, we can generate necessary additional proof trees on the fly during a dialogue and add them to deductions in order to make them complete. But this process complicates the presentation here so we assume our dialogue starts out with all the proof information it needs to justify an agent's conclusions.

A dialogue is a sequence of moves taken by two players. P_1 knows all the information in $D_1 = \mathcal{D}(F_1, R_1, \mathcal{T}_1)$ while P_2 knows all the information in $D_2 = \mathcal{D}(F_2, R_2, \mathcal{T}_2)$.

We will extend our simple example from Fig. 1 into a scenario involving two players, P_1 and P_2, that will be used to illustrate our dialogue definition. There's already one rule ($R1$), and we introduce another rule:

$$R2 : \{\neg tns(X)\} \to symp(X)$$

(if X has lost their sense of taste and smell (tns) then they have symptoms).

Scenario:

- $F_1 = \{vax(jack), tns(jack)\}$ while $F_2 = \{vax(jack)\}$. So the difference between our two players is that one is aware that Jack retains his sense of taste and smell while the other is not.
- Both players have rules $R1$ and $R2$ in their rule set.
  - $R1 : \{vax(X), \neg symp(X)\} \to mf(X)$
  - $R2 : \{\neg tns(X)\} \to symp(X)$
- P_1 has deduced that Jack can meet his friends and P_2 has deduced he can not. We start the dialogue with complete deductions.
  - $D_1 = \mathcal{D}(F_1, R_1, \mathcal{T}_1)$ where $\mathcal{T}_1$ contains the proof tree shown in Fig. 1 and a proof tree consisting of a single node $(tns(jack), initial)$ (This is because, for the deduction to be complete, we need the antecedents of $R2$ to be either provable or unprovable in D_1).
  - $D_2 = \mathcal{D}(F_2, R_2, \mathcal{T}_2)$ and $\mathcal{T}_2$ contains a proof tree consisting of the single node $(mf(jack), unprovable)$. For deduction to be complete the antecedents for $R1$ must be provable or unprovable in D_2. Therefore $\mathcal{T}_2$ also contains the proof tree shown in Fig. 2 and a proof tree consisting of the single node $(vax(jack), initial)$.

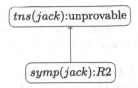

Fig. 2. P_2's proof tree for why Jack has symptoms.

Note that $mf(jack)$ is provable in D_1 and unprovable in D_2; $symp(jack)$ is provable in D_2 and unprovable in D_1; $tns(jack)$ is unprovable in D_2 and provable in D_1; and $vax(jack)$ is provable in both D_1 and D_2.

The two players gradually build up a *mental model* of how the other player has reasoned. This model consists of four sets OB_{ij}, OF_{ij}, OD_{ij} and OR_{ij}:

- OB_{ij} consists of terms t that P_i has established that P_j believes. We refer to OB_{ij} as the *opponent belief set*.
- OF_{ij} consists of terms t that P_i has established that P_j had as an initial fact. Note that $OF_{ij} \subseteq OB_{ij}$. We refer to OF_{ij} as the *opponent fact set*.
- OD_{ij} consists of terms t that P_i has established that P_j does not believe. We refer to OD_{ij} as the *opponent disbelief set*.
- OR_{ij} consists of labels l that P_i has established label one of P_j's rules. We refer to OR_{ij} as the *opponent rule set*.

There are seven possible statements that can be made in the course of a dialogue:

1. $df(t, i, j)$ (the two players have different initial facts) – $t \in F_i$ and $l \notin F_j$.
2. $dr(l : A \rightarrow C, i, j)$ (the two players have different rules) – $l : A \rightarrow C \in R_i$ and $l : A \rightarrow C \notin R_j$.
3. $initial(t)$ – t is an initial fact for the player.
4. $l : A \rightarrow t$ – the player deduced t from the terms in A using the rule labelled l
5. $why(t)$ - why do you believe t?
6. $whynot(t)$ – why don't you believe t?
7. $pass$ – the dialogue participant has no question to ask and skips its turn.

The first two statements terminate the dialogue.

Definition 7. Player State *The state of P_i at statement k in a dialogue with P_j is $S_k^i = \langle D_i, OB_{ij}, OF_{ij}, OD_{ij}, OR_{ij} \rangle$ where D_i is a deduction, and $OB_{ij}, OF_{ij}, OD_{ij}, OR_{ij}$ are P_i's opponent belief set, fact, set, disbelief set and rule set respectively.*

The initial state of the two players is one where the only thing they know is that they disagree on some term t. So their opponent's belief sets etc., are empty.

Definition 8. Initial Player State *The initial state of P_i is either $\langle \mathcal{D}(F_i, R_i,) T_i, \{t\}, \emptyset, \emptyset, \emptyset \rangle$ where t is unprovable in T_i or $\langle \mathcal{D}(F_i, R_i, T_i), \emptyset, \emptyset, \{t\}, \emptyset \rangle$ where t is the root term of some $T_i \in \mathcal{T}_i$.*

Definition 9. Dialogue State S_k *is the state of the dialogue after the utterance of the kth statement. It consists of the two-player states, the last dialogue statement, stmt, and whose turn it is,* P_i. $S_k = \langle S_k^1, S_k^2, stmt, P_i \rangle$

A dialogue is a sequence of dialogue states $S_0, \ldots, S_n$. The starting point for the dialogue is the disagreement over the term t in Definition 8. Without loss of generality, we assume this is provable in T_1 and unprovable in T_2. Therefore, $S_0 = \langle S_0^1, S_0^2, stmt_0, P_i \rangle$ where $S_0^1 = \langle D_1, \emptyset, \emptyset, \{t\}, \emptyset \rangle$, $S_0^2 = \langle D_2, \{t\}, \emptyset, \emptyset, \emptyset \rangle$, and either $P_i = P_1$ and $stmt_0 = why(t)$ (P_2 started the dialogue by asking P_1 why they believe t and it is now P_1's turn) or $P_i = P_2$ and $stmt_0 = whynot(t)$ (P_1 started the dialogue by asking P_2 why they don't believe t).

Suppose $S_k = \langle S_k^1, S_k^2, stmt_k, P_i \rangle$ is the state of a dialogue at utterance k and $S_{k+1} = \langle S_{k+1}^1, S_{k+1}^2, stmt_{k+1}, P_j \rangle$ is the next state. We define what it means for S_{k+1} to be a legal next state. S_{k+1}^i defines how each player has updated their mental model of the other in response to $stmt_k$ and $stmt_{k+1}$ is the next utterance.

First, we consider how the two players update their state. P_j ($j \neq i$) does not alter their state – they uttered the last statement and have not learned any new information. So $S_{k+1}^j = S_k^j$.

P_i, on the other hand has gained information from P_j's utterance and so their state changes. Before the utterance their state was $S_k^i = \langle D_i, OB_{ij}, OF_{ij}, OD_{ij}, OR_{ij} \rangle$. We provide four rules below that govern the state and can be updated.

Upd.1 If $stmt_k = initial(t)$ then $S_{k+1}^i = \langle D_i, OB_{ij} \cup \{t\}, OF_{ij} \cup \{t\}, OD_{ij}, OR_{ij} \rangle$ (P_i adds t to the things P_j believes and P_j's initial facts).
Upd.2 If $stmt_k = l$, $l : A \to C \in R_i$ then $S_{k+1}^i = \langle \mathcal{D}(F_i, R_i, T), OB_{ij} \cup pos(A), OF_{ij}, OD_{ij} \cup neg(A), OR_{ij} \cup \{l\} \rangle$ (P_i adds all the positive literals in A to OB_{ij} (these are things the other player believes) and all the negative literals in A to OD_{ij} (these are all the things the other player does not believe), and adds l to OR_{ij}).
Upd.3 If $stmt_k = why(t)$, $D_i = \mathcal{D}(F_i, R_i, T)$, and t is provable in T then $S_{k+1}^i = \langle \mathcal{D}(F_i, R_i, T), OB_{ij}, OF_{ij}, OD_{ij} \cup \{t\}, OR_{ij} \rangle$ (P_i adds t to OD_{ij} (the other player doesn't believe t)).
Upd.4 If $stmt_k = whynot(t)$, and t is unprovable in D_i then $S_{k+1}^i = \langle D_i, OB_{ij} \cup \{t\}, OF_{ij}, OD_{ij}, OR_{ij} \rangle$ (P_i adds t to OB_{ij} (the other player believes t)).

Note: The states where P_i is asked either why it believes something it does not, or why it does not believe something that it does should not occur in a legal dialogue and so these have been omitted. For the purposes of our theoretical results, we assume that if this does occur the dialogue terminates with an error and no next state is generated. We will prove that error states can not arise as corollaries to Lemmas 4 and 5.

We now consider the utterances P_i can make – possible values for $stmt_{k+1}$. In some dialogue states, there may be several possible utterances.

Utt.1 $stmt_{k+1} = initial(t)$ is legal iff $stmt_k = why(t)$ and $t \in F_i$

Utt.2 $stmt_{k+1} = l$ is legal iff: $l : A \to C \in R_i$; $stmt_k = why(t)$; $t \notin F_i$; and there exists a proof tree $\langle N, E \rangle \in \mathcal{T}_{i_k}$ such that $(t, l) \in N$.

Utt.3 $stmt_{k+1} = whynot(t)$ is legal iff: $\forall t'.stmt_k \neq why(t') \wedge stmt_k \neq whynot(t')$ (you can not answer a question by asking why not); $\forall l.l \leq k \to stmt_l \neq whynot(t)$ (this question has not been asked before); t is provable for D_i; and $t \in OD_{ij}$. P_i identifies a term t that it believes and it has established the other doesn't and asks why not.

Utt.4 $stmt_{k+1} = why(t)$ is legal iff either $stmt_k = whynot(t)$; or $\forall t'. stmt_k \neq why(t') \wedge stmt_k \neq whynot(t')$ (you can not answer a question by asking $why(t)$ unless that question was $whynot(t)$); $\forall l.l \leq k \to stmt_l \neq why(t)$ (this question has not been asked before); t is unprovable for D_i; and $t \in OB_{ij}$ P_i identifies a term t that it does not believe and it has established the other does and asks why.

Utt.5 $stmt_{k+1} = df(t, j, i)$ is legal iff $t \in OF_{ij}$ and $t \notin F_i$

Utt.6 $stmt_{k+1} = df(t, i, j)$ is legal iff $t \in OD_{ij}$ and $t \in F_i$

Utt.7 $stmt_{k+1} = dr(l, j, i)$ is legal iff $l \in OR_{ij}$ and there is no rule $l : A \to C \in R_i$

Utt.8 $stmt_{k+1} = pass$ is legal iff no other utterance is legal and $stmt_k \neq pass$.

Finally, the player whose turn it is switched.

Figure 3 shows an example dialogue for our scenario. We show the opponent's belief, fact, disbelief, and rule sets for each player as they are built up, as well as the statement uttered and whose turn it is next. We also comment on the changes with reference to the updates and utterances defined by the dialogue framework.

5   Theoretical Results

We demonstrate that error states in dialogues cannot arise, that opposing belief sets etc., are correct representations of the other player's deductions, and that the debate process ends when a discrepancy is discovered.

We establish via a set of lemmas that the assumptions made by the update process are correct (for instance in Lemma 1 that if one player has uttered $initial(t)$ then t is indeed an initial fact for that player).

Lemma 1 (Statements about initial facts are truthful). *If the current dialogue state is $\langle S_k^1, S_k^2, initial(t), P_i \rangle$, $i \neq j$ and $D_j = \mathcal{D}(F_j, R_j, \mathcal{T}_j)$ then $t \in F_j$ and is provable for D_j.*

Lemma 2 (Statements about the use of rules are truthful). *If the current dialogue state is $\langle S_k^1, S_k^2, l : A \to t, P_i \rangle$, $i \neq j$ and $D_j = \mathcal{D}(F_j, R_j, \mathcal{T}_j)$ is P_j's deduction then there exists a proof tree, $T_j \in \mathcal{T}_j$ such that (t, l) is a node in T_j; $l : A \to t \in R_j$, for all $t \in pos(A)$, t is provable in D_j; and for all $t \in neg(A)$, t is unprovable D_j.*

| k | P_i State | | | | $stmt_k$ | j |
	OB_{ij}	OF_{ij}	OD_{ij}	OR_{ij}		
0	$\emptyset$	$\emptyset$	$\{mf(jack)\}$	$\emptyset$	$whynot(mf(jack))$	2

P_1 has asked why P_2 thinks Jack can't meet friends

1	$\{mf(jack)\}$	$\emptyset$	$\emptyset$	$\emptyset$	$why(mf(jack))$	1

Upd.4 applies but P_2 already knows P_1 thinks Jack can meet friends as our initial condition; P_2 asks why P_1 thinks Jack can meet friends (**Utt.3**)

2	$\emptyset$	$\emptyset$	$\{mf(jack)\}$	$\emptyset$	R1	2

Upd.3 applies but makes no change; P_1 responds with the Rule it used (**Utt.2**)

3	$\{mf(jack), vax(jack)\}$	$\emptyset$	$\{symp(jack)\}$	$\{R1\}$	$whynot(symp(jack))$	1

Upd.2 applies and changes P_2's state. P_2 asks why P_1 does not believe Jack has symptoms – note they can't ask why P_1 believes Jack has been vaccinated because they could only ask this if they disagreed with this belief.

4	$\{symp(jack)\}$	$\emptyset$	$\{mf(jack)\}$	$\emptyset$	$why(symp(jack))$	2

Upd.4 applies and changes P_1's state; P_1 asks why P_2 believes Jack has symptoms (**Utt.3**)

5	$\{mf(jack), vax(jack)\}$	$\emptyset$	$\{symp(jack)\}$	$\{R1\}$	R2	1

Upd.3 applies but does not change P_2's state. P_2 responds with the rule it used.

6	$\{symp(jack)\}$	$\emptyset$	$\{mf(jack), tns(jack)\}$	$\{R2\}$	$whynot(tns(jack))$	2

Upd.2 applies and changes P_1's state. P_1 asks why P_2 does not believe Jack has a sense of taste and smell.

7	$\{mf(jack), vax(jack), tns(jack)\}$	$\emptyset$	$\{symp(jack)\}$	$\{R1\}$	$why(tns(jack))$	1

Upd.4 applies but makes no change to P_2's state; P_2 asks why P_1 believes Jack has a sense of taste and smell.

8	$\{symp(jack)\}$	$\emptyset$	$\{mf(jack), tns(jack)\}$	$\{R2\}$	$initial(tns(jack))$	2

Upd.3 applies but does not change P_1's state; P_1 replies that this is an initial fact.

9	$\{mf(jack), vax(jack), tns(jack)\}$	$\{tns(jack)\}$	$\{symp(jack)\}$	$\{R1\}$	$df(tns(jack),1,2)$	1

Upd.1 applies and changes P_2's state; P_2 replies announcing it has found a different fact and terminating the dialogue.

Fig. 3. Sample Dialogue for our scenario showing the current player's opponent belief, fact, disbelief and rule sets and the statement the player has uttered.

Lemma 3 (A player only asks the other "why not" about statements it believes to be true). *If the current dialogue state is $\langle S^1_{stmt}, S^2_{stmt}, whynot(t), P_i\rangle$, $i \neq j$ and D_j is P_j's deduction then t is provable in D_j.*

Lemmas 1, 2 and 3 follow trivially from the rules for legal utterances in dialogue. The equivalent to Lemma 3 for $why(t)$ is Lemma 6 but we need a few other results before we can prove this, in particular, we need to know that the dialogue participants' mental models of each other are correct.

Dialogue Mental Models are Correct. We establish that $t \in OF_{ij}$ iff $t \in F_j$ (i.e., P_i only decides P_j has t as an initial fact if P_j does indeed have t as an initial fact). The same for OB_{ij}, OD_{ij} etc. As a result, we can also show that the error states (where a participant is asked $why(t)$ for some term t they do not believe or $whynot(t)$ for some t they do believe) never occur.

Theorem 1 (The opponent fact set is correct). *Given two players P_i and P_j in a legal dialogue, if $\mathcal{D}(F_j, R_j, \mathcal{T}_j)$ is P_j's deduction and OF_{ij} is P_i's opponent fact set, $OF_{ij} \subseteq F_j$.*

Proof Sketch. The proof follows by induction on the size of OF_{ij} using Lemma 1

Theorem 2 (The opponent belief set is correct). *Given two players P_i and P_j in a legal dialogue where D_j is P_j's deduction and OB_{ij} is P_i's opponent belief set, then all terms $t \in OB_{ij}$ are provable in D_j.*

Proof Sketch. The proof follows by induction on the size of OB_{ij} using Lemmas 1, 2 and 3.

Lemma 4 (A player is only asked why about things it believes to be true). *If the dialogue state is $\langle S_k^1, S_k^2, why(t), P_i \rangle$ and D_i is P_i's deduction then t is provable in D_i.*

Proof. This holds in the initial state. Otherwise, $why(t)$ has been uttered because $t \in OB_{ji}$ (**Utt.4**) and this follows from Theorem 2 or $why(t)$ has been uttered in response to $whynot(t)$ (**Utt.5**) and this follows from Lemma 3.

Corollary. *If the dialogue state is $\langle S_k^1, S_k^2, why(t), P_i \rangle$ then the error state does not arise.*

Theorem 3 (The opponent disbelief set is correct). *Given two players P_i and P_j in a legal dialogue where D_j is P_j's deduction and OD_{ij} is P_i's opponent disbelief set, then all terms $t \in OD_{ij}$ are unprovable in D_j.*

Proof Sketch The proof follows by induction on the size of OD_{ij} using Lemmas 2 and 4.

Lemma 5 (A player is only asked why not about things it does not believe to be true). *If the dialogue state is $\langle S_k^1, S_k^2, whynot(t), P_i \rangle$ and D_i is P_i's deduction then t is unprovable for P_i.*

Proof. P_j can only ask $whynot(t)$ if $t \in OD_{ji}$ (**Utt.3**) so t is unprovable in D_i by Theorem 3.

Corollary. *If the dialogue state is* $\langle S_k^1, S_k^2, whynot(t), P_i \rangle$ *then the error state doesn't arise.*

Lemma 6 (A player only asks the other player why about things it believes are not the case). *If the dialogue state is* $\langle S_k^1, S_k^2, why(t), P_i \rangle$ *($i \neq j$) then t is unprovable for P_j.*

Proof. P_j can only ask $why(t)$ if either a) P_i asked $whynot(t)$ in which case t is unprovable for D_j by Lemma 5; or b) t is unprovable for D_j (**Utt.4**).

Theorem 4 (The opponent rule set is correct). *Given P_i and P_j in a legal dialogue where $\mathcal{D}(F_j, R_j, \mathcal{T}_j)$ is P_j's deduction and OR_{ij} is P_i's opponent rule set, then* $\forall l. l \in OR_{ij}. \exists A, C. l : A \rightarrow C \in R_j$

Proof Sketch The proof follows by induction on the size of OR_{ij} using Lemma 2 and the definition of proof trees.

5.1   Termination

Theorem 5. *Let $D_1 = \mathcal{D}(F_1, R_1, \mathcal{T}_1)$ and $D_2 = \mathcal{D}(F_2, R_2, \mathcal{T}_2)$ be two complete deductions. If $\mathcal{T}_1$ and $\mathcal{T}_2$ contain a finite number of finite proof trees then any dialogue starting from D_1 and D_2 terminates.*

Proof Sketch. By assumption, there are only a finite number of terms in $\mathcal{T}_1$ and $\mathcal{T}_2$. Therefore $why(t)$ can only be asked a finite number of times. The number of times all other utterances can be made depends upon how many times $why(t)$ is asked. Therefore the dialogue terminates.

Note this means that dialogues only terminate if complete deductions can be created from the attempt to prove or disprove some term t and this depends on the facts, rules, and t. However many sets of facts and rules have this property for given terms.

In order to show that when dialogues terminate a disagreement between the facts or rules of the two players has been found, we need to show that it is always possible for a player to ask questions about terms in OB_{ij} or OD_{ij} which requires these terms to be provable or unprovable in that player's deduction (because of the conditions on **Utt.3** and **Utt.4**). We establish this in two lemmas whose proofs rely on our completeness property for deductions.

Lemma 7. *Given two dialogue participants P_i and P_j where D_i is P_i's deduction and OB_{ij} is i's opponent belief set, then all terms $t \in OB_{ij}$ are either provable or unprovable in D_i.*

Lemma 8. *Given two dialogue participants P_i and P_j where D_i is P_i's deduction and OD_{ij} is i's opponent disbelief set, then all terms $t \in OD_{ij}$ are either provable or unprovable in D_i.*

Proof Sketch The proofs for both these lemmas proceed by induction on the size of OB_{ij} (resp. OD_{ij}), noting that the property holds at the start of the dialogue and exploiting Theorems 2 and 3 and the completeness of deductions together with Lemma 6 in the step case.

Having established this we then introduce the concept of a *disagreement tree* in order to prove that all dialogues terminate with a statement that either facts or rules are different.

Definition 10. *A* disagreement tree *is a tree that reveals the inference processes behind the disagreements between two dialogue participants. Every node in the tree is a tuple* $\langle t, i, lbl \rangle$ *where t is a term that is provable for one dialogue participant and unprovable for the other; i is the participant for which the term is provable, and lbl is either initial (meaning* $t \in F_i$*), l^- (meaning t was deduced by i using rule l and rule l is not in the rule set for the other participant) or l^+ (meaning t was deduced by i using rule l and rule l is in the rule set for the other participant). Nodes labeled initial or l^- are leaf nodes. Nodes labeled l^+ have child nodes consisting of all terms in pos(a) which are provable for i and not for j and all terms in neg(a) which are provable for j and not for i.*

Note that all nodes l^+ must have at least one child node. Figure 4 shows the disagreement tree for our scenario. The two players disagree on the truth of $mf(jack)$ which P_1 has deduced using $R1$ but which P_2 could not deduce because P_1 and P_2 disagree on the truth of $symp(jack)$ and so on.

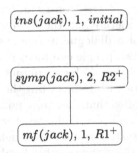

Fig. 4. The Disagreement Tree for Scenario 1.

Lemma 9. *Consider two players in a legal dialogue and a disagreement tree, DT, which has the initial disagreement term as its root. Let NT be the set of node terms closest to the root of DT (there may be several such terms since this is a tree) about which why(t) has not been asked. $\forall t \in NT$ the dialogue will continue deterministically until t is in the belief or disbelief set for at least one player.*

Proof Sketch. The proof observes that $why(t)$ will have been asked for the parents of each of these nodes and so a player either has or will, respond with **Utt.1** or **Utt.2** which has or will trigger an appropriate update in a player's state.

Theorem 6. *If the kth state in a legal sequence of dialogue states is* $\langle S_k^1, S_k^2, s, P_i \rangle$ *and* $s \neq df(t, i, j)$, $s \neq df(t, j, i)$ *and* $s \neq dr(l, i, j)$ *then there is a legal next dialogue state.*

Proof Sketch. We use Lemmas 7, 8 and 9 to show that one player exists who can ask $why(t)$ if it is their turn and they are not required by the framework to make some other utterance. If the current player is not capable of asking $why(t)$, then they can utter *pass*, and the other player will be able to respond.

Corollary. *If a dialogue terminates the last statement is:* $df(t, i, j)$, $df(t, j, i)$, $dr(l, i, j)$ *or* $dr(l, j, i)$.

6   Implementation

We applied our framework to an RBS that functions as a Covid Advice System (CAS) implemented in Prolog. This consists of a simple backward-chaining rule-based system with sets of example rules and facts based on Covid-19 restrictions paired with an implementation of the dialogue framework. The dialogue framework implementation tracks both participants' dialogue states and allows the human user to choose between legal next utterances. As a result, a theoretically legal dialogue can be generated, even where the human is not sure of the legal moves, or may not be reasoning correctly with their facts and rules.

Our implementation differs slightly from the theory in that dialogues did not start with complete deductions, instead, one participant starts with a deduction that contains a single proof tree consisting only of an unprovable node. Additional proof trees were generated on-the-fly during the dialogue as needed.

We present an example of a dialogue in our system. In this example both players have a rule that says two people can meet if a) they are both vaccinated, b) neither of them has been "pinged" by a contact tracing app and c) neither has symptoms. Harry and Sara wish to meet but the CAS is unaware that Harry has been vaccinated and so states that they may not. The user thinks Harry and Sara should be able to meet and so a dialogue starts with a why not question from the user. The dialogue system responds on behalf of the CAS with a why question using **Utt.4** and displays the possible legal user responses (Fig. 5).

Why do you believe Sara and Harry can meet? Please State your reason:

1. Because it's a user initial fact.
2. Because it is a new fact deduced by a rule.

Fig. 5. The computer asks why

The user selects 2 because they've used a rule. The dialogue system has stored the rules provided to the test participants so it offers a choice of these rules (Fig. 6). The user selects rule 1. The dialogue system updates the CAS

> **PLEASE SELECT A RULE NUMBER FROM YOUR RULES:**
> 1. If both X and Y are vaccinated, and none of them have been pinged (close contact with someone who has Covid-19), and none of them have symptoms, then X and Y can meet.
> 2. If X does not have taste or smell, then X has symptoms.
> 3. If X has a fever, then X has symptoms.
> 4. If X has a cough, then X has symptoms.

Fig. 6. The user is offered a choice of rules

> *Why do you believe Harry is vaccinated? Please state your reason:*
> 1. Because it's a user initial fact.
> 2. Because it is a new fact deduced by a rule.

Fig. 7. Why does the user believe Harry is vaccinated?

Table 1. Our results are broken up by scenario. Each participant marked their explanation on a scale of 0–4 for how easy it was to understand and how helpful they found it - we show the average mark for each explanation style. Additionally, we show what percentage of users correctly identified the difference between their facts and rules and that of the CAS.

	Ex1		Ex2		Ex3		Ex4		Ex5		Ex6	
Type	Tree	Dlog	Tree	Dlog	Tree	Dlog	Tree	Dlog	Tree	Dlog	Tree	Dlog
Ease	0.5	2.5	2.25	2.5	1.5	3	0.25	2.75	2.5	2.75	3	3
Helpful	0.25	2.5	2	2	1.75	2.5	0.25	2.25	2.5	2	2.75	2.25
Correct	0	100%	100%	100%	75%	75%	0	75%	100%	75%	100%	100%

mental model of the user and consults the system's proof trees for a mismatch. In this case, it identifies that *vaccinated(harry)* is unprovable for the CAS. The dialogue system asks why the user believes this (Fig. 7). The user selects 1. The dialogue system then terminates announcing that a difference has been found.

7 User Evaluation

The purpose of our user evaluation was to test our hypothesis that dialogue is a useful mechanism for building explanations. The proof trees generated by the deductive process were used as an alternative explanation for comparison. We created six scenarios in which the CAS and the User were given different sets of facts and rules (differing either by one fact or by one rule) and the CAS presented a conclusion which the user should not be able to derive (if the user reasoned correctly). The user was then either shown the proof tree generated by the CAS or allowed to participate in a dialogue. Our expectation was that dialogue explanations would have an advantage firstly in situations where the CAS deduced something was unprovable (and so produced a proof tree consisting of a single unprovable node) and secondly, as proof trees grew beyond a certain size.

Our study comprised 24 volunteers from the Department of Computer Science. Each participant was presented with two scenarios (one where they viewed a proof tree and one where they could use the dialogue system). Each scenario was completed by the same number of participants, and followed by a short questionnaire. We summarise the features of the six scenarios in Table 2 – as can be seen, two of the examples feature trees consisting of a single unprovable node, while the others have trees of varying, though modest, sizes.

Table 2. Our examples, showing how many nodes the initial proof tree contains, how many of those nodes are unprovable and the cause of the disagreement between user and CAS

Ex.	Nodes	Unprovable Nodes	Cause
1	1	1	CAS missing Fact
2	14	4	User missing Fact
3	18	4	User missing Rule
4	1	1	CAS missing Rule
5	7	1	CAS missing Fact
6	6	3	CAS and User have different Rules

Out of 24 responses, 20 preferred the dialogue explanation, and 18 found the dialogue explanation easy. Table 1 shows a breakdown of our results by example. As can be seen, the dialogue explanations have a clear advantage where no meaningful tree was provided (Scenarios 1 and 4) while there is not much to tell between the two explanation styles in most other cases. Scenario 3, with the largest number of nodes, suggests that the dialogue explanation was beginning to outperform the tree in terms of ease of use and perceived helpfulness, but the sample size is too small (4 people) to draw strong conclusions. Classifying whether a user had correctly identified the difference proved more challenging than we expected. We allowed freeform answers to the question "What do you think the difference was between your information and the computer information?" and in some cases these answers were very minimal (e.g., "There was a different rule") and in some cases, it is difficult to decide whether or not they should count as correct (e.g., in scenario 2 one respondent correctly identified that they did not possess a rule, but stated the rule's antecedents incorrectly). We allowed minimal but correct answers to count as correct but did not allow other mistakes to count as correct. For Scenario 6 we counted as correct both answers which noted that they had slightly different rules for deducing whether someone was required to get a Covid test and answers which noted that the user did not have the rule the computer was using. Future study will define the rules in a way that is easier for general users to understand because in this experiment the rules provided to the user require a high cognitive load and consider the user with different background.

8   Discussion

We have proposed a dialogue approach to explain the reasoning in systems where derivations are represented as trees, typical of rule-based AI systems. A dialogue system assumes that an explanation is a collaborative process in which the system determines what information it is that the user wants. We have established some theoretical properties of the dialogue framework and performed a small user study.

The study shows a clear advantage for the dialogue process where no meaningful proof tree can be presented. There is some evidence that, as the amount of information in the proof tree increases, the dialogue explanation becomes more useful, and in further work, we intend to extend our study with larger scenarios. We also intend to examine how our explanations could be adapted to explore "what-if" scenarios which would allow a dialogue to progress beyond identifying a source of disagreement to exploring whether eliminating that disagreement would change the system's conclusion, and to evaluate whether dialogue is a useful explanatory process when applied to RBS extracted from statistical models as in [21]. In the future, we'll take into account the agreed-upon situation in which the user can request further information without causing a disagreement.

Acknowledgement. This work is supported by EPSRC, through EP/W01081X (*Computational Agent Responsibility*).

Data Access Statement. The code and data supporting the findings reported in this paper are available for open access at https://github.com/xuLily9/RBS_TheoryI (Code) and https://doi.org/10.6084/m9.figshare.22220494.v3 (User Evaluation).

Ethical Approval. We performed a light-touch ethical review for the user evaluation, using a tool provided by our university. This tool advised that since the only personal data gathered was names on consent forms and these were stored in a locked cabinet separate from the rest of the gathered data, further ethical approval was not required.

References

1. Apt, K.R., Van Emden, M.H.: Contributions to the theory of logic programming. J. ACM (JACM) **29**(3), 841–862 (1982)
2. Arioua, A., Tamani, N., Croitoru, M.: Query answering explanation in inconsistent datalog+/− knowledge bases. In: Chen, Q., Hameurlain, A., Toumani, F., Wagner, R., Decker, H. (eds.) DEXA 2015. LNCS, vol. 9261, pp. 203–219. Springer, Cham (2015). https://doi.org/10.1007/978-3-319-22849-5_15
3. Bex, F., Walton, D.: Combining explanation and argumentation in dialogue. Argument Comput. **7**(1), 55–68 (2016)

4. Cavedon, L., Lloyd, J.: A completeness theorem for SLDNF resolution. J. Logic Program. **7**(3), 177–191 (1989). https://www.sciencedirect.com/science/article/pii/0743106689900204
5. Clancey, W.J.: The epistemology of a rule-based expert system-a framework for explanation. Artif. Intell. **20**(3), 215–251 (1983)
6. Clocksin, W.F., Mellish, C.S.: Programming in Prolog, 5 edn. Springer, Heidelberg (2003). https://doi.org/10.1007/978-3-642-55481-0
7. Cocarascu, O., Stylianou, A., Čyras, K., Toni, F.: Data-empowered argumentation for dialectically explainable predictions. In: ECAI 2020, pp. 2449–2456. IOS Press (2020)
8. Dennis, L.A., Oren, N.: Explaining BDI agent behaviour through dialogue. In: Proceedings of the 20th International Conference on Autonomous Agents and Multiagent Systems (AAMAS 2021). International Foundation for Autonomous Agents and Multiagent Systems (IFAAMAS) (2021)
9. Fiedler, A.: Dialog-driven adaptation of explanations of proofs. In: International Joint Conference on Artificial Intelligence, vol. 17, pp. 1295–1300. Citeseer (2001)
10. Huth, M., Ryan, M.: Logic in Computer Science: Modelling and Reasoning about Systems. Cambridge University Press, Cambridge (2004)
11. Johnson-Laird, P.N.: Mental models in cognitive science. Cogn. Sci. **4**(1), 71–115 (1980)
12. Kass, R., Finin, T., et al.: The need for user models in generating expert system explanations. Int. J. Expert Syst. **1**(4) (1988)
13. Lacave, C., Díez, F.J.: A review of explanation methods for Bayesian networks. Knowl. Eng. Rev. **17**(2), 107–127 (2002)
14. Lacave, C., Diez, F.J.: A review of explanation methods for heuristic expert systems. Knowl. Eng. Rev. **19**(2), 133–146 (2004)
15. Madumal, P., Miller, T., Sonenberg, L., Vetere, F.: A grounded interaction protocol for explainable artificial intelligence. In: Proceedings of the 18th International Conference on Autonomous Agents and MultiAgent Systems, pp. 1033–1041 (2019)
16. Miller, T., Howe, P., Sonenberg, L.: Explainable AI: beware of inmates running the asylum or: how i learnt to stop worrying and love the social and behavioural sciences. arXiv preprint arXiv:1712.00547 (2017)
17. Moore, J.D., Paris, C.L.: Requirements for an expert system explanation facility. Comput. Intell. **7**(4), 367–370 (1991)
18. Oren, N., van Deemter, K., Vasconcelos, W.W.: Argument-based plan explanation. In: Vallati, M., Kitchin, D. (eds.) Knowledge Engineering Tools and Techniques for AI Planning. LNCS, pp. 173–188. Springer, Cham (2020). https://doi.org/10.1007/978-3-030-38561-3_9
19. Reggia, J.A., Perricone, B.T.: Answer justification in medical decision support systems based on Bayesian classification. Comput. Biol. Med. **15**(4), 161–167 (1985)
20. Sendi, N., Abchiche-Mimouni, N., Zehraoui, F.: A new transparent ensemble method based on deep learning. Procedia Comput. Sci. **159**, 271–280 (2019)
21. Shams, Z., et al.: REM: an integrative rule extraction methodology for explainable data analysis in healthcare (2021)
22. Shortliffe, E.H., Axline, S.G., Buchanan, B.G., Merigan, T.C., Cohen, S.N.: An artificial intelligence program to advise physicians regarding antimicrobial therapy. Comput. Biomed. Res. **6**(6), 544–560 (1973)
23. Singh, R., Miller, T., Newn, J., Sonenberg, L., Velloso, E., Vetere, F.: Combining planning with gaze for online human intention recognition. In: Proceedings of the 17th International Conference on Autonomous Agents and Multiagent Systems, pp. 488–496 (2018)

24. Studer, R., Benjamins, V.R., Fensel, D.: Knowledge engineering: principles and methods. Data Knowl. Eng. **25**(1–2), 161–197 (1998)
25. Swartout, W.R.: XPLAIN: a system for creating and explaining expert consulting programs. Artif. Intell. **21**(3), 285–325 (1983)
26. Vassiliades, A., Bassiliades, N., Patkos, T.: Argumentation and explainable artificial intelligence: a survey. Knowl. Eng. Review **36**, e5 (2021)
27. Walton, D.: A dialogue system specification for explanation. Synthese **182**, 349–374 (2011)
28. Walton, D.: A Dialogue System for Evaluating Explanations, pp. 69–116. Springer, Cham (2016). https://doi.org/10.1007/978-3-319-19626-8_3
29. Wick, M.R., Thompson, W.B.: Reconstructive expert system explanation. Artif. Intell. **54**(1–2), 33–70 (1992)
30. Winikoff, M., Sidorenko, G., Dignum, V., Dignum, F.: Why bad coffee? Explaining BDI agent behaviour with valuings. Artif. Intell. **300**, 103554 (2021)
31. Zarlenga, M.E., Shams, Z., Jamnik, M.: Efficient decompositional rule extraction for deep neural networks. arXiv preprint arXiv:2111.12628 (2021)

Estimating Causal Responsibility for Explaining Autonomous Behavior

Saaduddin Mahmud[1(✉)], Samer B. Nashed[1], Claudia V. Goldman[2], and Shlomo Zilberstein[1]

[1] University of Massachusetts, Amherst, MA, USA
{smahmud,snashed,shlomo}@cs.umass.edu
[2] General Motors Research, Tel Aviv, Israel
claudia.goldman@gm.com

Abstract. There has been growing interest in causal explanations of stochastic, sequential decision-making systems. Structural causal models and causal reasoning offer several theoretical benefits when exact inference can be applied. Furthermore, users overwhelmingly prefer the resulting causal explanations over other state-of-the-art systems. In this work, we focus on one such method, MEANRESP, and its approximate versions that drastically reduce compute load and assign a responsibility score to each variable, which helps identify smaller sets of causes to be used as explanations. However, this method, and its approximate versions in particular, lack deeper theoretical analysis and broader empirical tests. To address these shortcomings, we provide three primary contributions. First, we offer several theoretical insights on the sample complexity and error rate of approximate MEANRESP. Second, we discuss several automated metrics for comparing explanations generated from approximate methods to those generated via exact methods. While we recognize the significance of user studies as the gold standard for evaluating explanations, our aim is to leverage the proposed metrics to systematically compare explanation-generation methods along important quantitative dimensions. Finally, we provide a more detailed discussion of MEANRESP and how its output under different definitions of responsibility compares to existing widely adopted methods that use Shapley values.

Keywords: Causal Inference · Explainable AI · MDPs

1 Introduction

Researchers from many fields have shown that developing trust in AI systems is required for their timely adoption and proficient use [18,35,39]. It is also widely accepted that autonomous agents with the ability to explain their decisions increase user trust [7,13,22]. However, there are many challenges in generating explanations. Consider, for example, an agent managing load on a power grid by setting electricity prices and engaging other physical resources within the grid.

S. Mahmud and S. B. Nashed—Authors contributed equally.

© The Author(s), under exclusive license to Springer Nature Switzerland AG 2023
D. Calvaresi et al. (Eds.): EXTRAAMAS 2023, LNAI 14127, pp. 78–94, 2023.
https://doi.org/10.1007/978-3-031-40878-6_5

Generating suitable explanations of such a system is hard due to the complexity of planning, which may involve large state spaces, stochastic actions, imperfect observations, and complicated objectives. Furthermore, useful explanations must somehow reduce the internal reasoning process to a form understandable by a user who likely does not know all of the algorithmic details. One significant class of autonomous decision-making models for which there is a desire to generate explanations is the Markov decision process (MDP) and its derivatives.

In our previous work [27], we developed a framework, based on *structural causal models* (SCMs) [11], for applying causal analysis to sequential decision-making agents. This framework creates an SCM representing the computation needed to derive a policy for an MDP and applies causal inference to identify variables that cause certain agent behavior. Explanations are then generated using these variables, for example by completing natural language templates. This framework is both theoretically sound, based on formalisms from the causality literature, as well as flexible, allowing multiple semantically different types of explanans.

This method, known as MEANRESP, has many different approximate versions and is compatible with several definitions of responsibility [8]. The theoretical characteristics of approximate MEANRESP, as well as its performance compared to the exact version, are yet to be explored in detail. Since in practice, the approximate versions are the most likely to be deployed, we see this as a critical gap in our current understanding of how to explain MDP agent behaviors. Moreover, as MEANRESP may produce many causes related to a decision, it is often necessary to reduce the size of this set to make the explanations more concise and therefore easier to understand. MEANRESP supports this type of 'top k' analysis natively, but little work has been done on understanding how to compare different outputs, either against each other or against the output from the exact version of the algorithm. To this end, we also propose several metrics which may be used to compare MEANRESP outputs at different levels of approximation. These metrics capture diverse types of differences and underscore the difficulty of devising a single metric for evaluating objects as complicated, nuanced, and context-dependent as explanations.

Our results include theoretical analyses regarding the correctness and sampling error rates for causal and responsibility determination for approximate MEANRESP, discuss several potential metrics for comparing explanations, empirical analyses of sampling error convergence rates and explanation dissimilarity between different versions of MEANRESP and Shapley-value based methods. Overall, these results establish several key facts about approximate MEANRESP as well as open the door for a variety of avenues of continued research.

2 Related Work

While this paper focuses specifically on deepening analysis related to one particular algorithm for automatic explanation generation, MEANRESP, the body of work on explainable machine learning (XML)—a focus area that aims to explain the decisions of black-box machine learning algorithms [15,19,26]—and explainable planning (XAIP)—a focus area that aims to explain the outputs of planning algorithms or modify planning algorithms so that they produce plans that are inherently more explainable—is large and growing rapidly. In this section, we aim to provide some context to the existing literature to highlight the importance of MEANRESP.

In XAIP literature, one common method for explaining complex planners is via policy summarization, where either A) the original reasoning problem is made simpler and then the solution is explained exactly, or B) the original problem is not reduced, but the solution (e.g., policy) is simplified post-hoc and the simplified policy is explained. For example, Pouget et al. [30] identify key state-action pairs via spectrum-based fault localization, and Russell et al. [31] use decision trees to approximate a given policy and analyze the decision nodes to determine which state factors are most influential for immediate reward. Panigutti et al. [29] used similar methods to explain classifiers. These methods are appealing in that they parallel our intuitions about simplification in a number of other settings, such as analogizing during an explanation [9], science communication [32], and even other AI tools, like automated text simplification [28,34] or summarization [1]. However, these methods are driven primarily by heuristics and may be difficult to generalize to the many different forms of planners and models.

Research on explanations of stochastic planners specifically, such as MDPs, is relatively sparse. However, there are several notable existing efforts. Most present heuristics that are specifically designed for MDPs, such as generating counterfactual states and then identifying important state factors by analyzing how the value function changes given perturbations to different state factors [10]. Wang et al. [38] try to explain policies of partially observable MDPs by communicating the relative likelihoods of different events or levels of belief. However, research clearly indicates that humans are not good at using this kind of numerical information [23].

A more common heuristic approach is to analyze (and produce explanations that reference) the reward function. Khan et al. [16] first presented a technique to explain policies for factored MDPs by analyzing the expected occupancy frequency of states with extreme reward values. Later, Sukkerd et al. [36] proposed explaining factored MDPs by annotating them with "quality attributes" (QAs) related to independent, measurable cost functions. Explanations describe the QA objectives, expected consequences QA values given a policy, and how those values contribute to the policy's expected cost. The system also explains whether the policy achieves the best possible QA values simultaneously, or if there are competing objectives that required reconciliation and proposes counterfactual alternatives. Thus, it explains entire policies, not individual actions, using custom graphics and natural language templates, the latter of which has become the de facto standard for automatic explanations. Instead of looking at how the policy is affected by the reward function overall, Juozapaitis et al. [14] analyze how extreme reward values impact action selection in decomposed-reward RL agents, and Bertram and Peng [4] look at reward sources in deterministic MDPs.

While these approaches are computationally cheap and easy to implement, they have limited scope in the explanations they provide, and do not have many theoretical advantages, if any. Thus, recently, some research has investigated the application of causal modeling and causal analysis to the automatic generation of explanations for planners, including MDPs. One particularly compelling framework for doing so, which we study in this paper, is a method called MEAN-RESP [27]. MEANRESP is based on a responsibility attribution method called RESP, introduced in [3] to explain classification outcomes, which has its roots in prior work on formal definitions of causality and responsibility [8,11,12]. In this paper, we examine several choices related to the definitions of responsibility for use within MEANRESP.

The most similar work to this paper is other research that has proposed using SCMs for explaining MDPs and their variants in both planning and learning scenarios. Madumal et al. [21] use SCMs to encode the influence of particular actions available to the agent in a model-free, reinforcement learning, where it requires several strong assumptions including the prior availability of a graph representing causal direction between variables, discrete actions, and the existence of sink states.

Finally, our approach to estimating causal responsibility can be viewed as a form of feature attribution, which is a common approach in explainable Machine Learning (XML) for feature ranking [25], most often via Shapley values and their approximations [20,33]. In this paper, we conduct a quantitative comparison between Shapley values and different versions of MEANRESP. Specifically, we analyze the approximation error between a prominent Shapley value-based feature attribution method [37] and various versions of approximate MEAN-RESP, considering the number of samples. Additionally, we assess the dissimilarity between explanations generated by these two attribution methods. The purpose of this comparison is to investigate whether there exists a significant disparity in the content of the explanations produced by these methods, potentially motivating future research on the relative advantages of each method.

3   Background

Here, we review some concepts and notations relevant to the three main ideas this paper builds upon: structural causal models (SCMs), our working definition of cause, and Markov decision processes (MDPs).

3.1   Structural Causal Models, Actual Causes, and Responsibility

SCMs model scenarios $\mathcal{S} = \langle U, V, \mathcal{M} \rangle$, which break causality or attribution problems down into three components:

1. A set of variables U, known as the context, which are required to define the scenario, but which should not be identified as causal. These variables are considered fixed for a given scenario. The choice of which variables belong in the context is a design choice, and the main function of the context is to bound the size of the total problem.

2. A set of variables V, known as the endogenous variables, which we may want to identify as causal or highlight in an explanation. All variables in a scenario must be in $U \cup V$.
3. A set of equations, $\mathcal{M}$, which model how variables in V are calculated as functions of variables in U or other variables in V.

Nashed et al. [27] define several SCM representations of an MDP with different choices of the context U. For the purpose of analysis, throughout the rest of this paper, we will consider one of the most natural of those choices and describe its mathematical definition and interpretation in the following subsection. We now review our working definition of cause from [12].

Definition 1. *Let $X \subseteq V$ be a subset of the endogenous variables, and let x be a specific assignment of values for those variables. Given an event ϕ, defined as a logical expression, for instance $\phi = (\neg a \wedge b)$, a weak cause of ϕ satisfies the following conditions:*

1. *Given the context $U = u$ and $X = x$, ϕ holds.*
2. *Some $W \subseteq (V \setminus X)$ and some $\bar{x}$ and w exist such that:*
   A) using these values produces $\neg\phi$.
   B) for all $W' \subseteq W$, $Z \subseteq V \setminus (X \cup W)$, where
      $w' = w|W'$ and $z = Z$ given $U = u$, ϕ holds when $X = x$.

Here, condition 2B) is saying that given context $U = u$, $X = x$ alone is sufficient to cause ϕ, independent of some other variables W. This and similar definitions of cause are often called "but-for" definitions. There is a related, slightly older definition due to [11] in which condition 2B) is replaced with the following, simpler statement: for all $Z \subseteq V \setminus (X \cup W)$, where $w = W$ and $z = Z$ given $U = u$, ϕ holds when $X = x$.

Actual causes are defined as minimal weak causes. That is, an actual cause is a weak cause C_W for which no set $C'_W \subset C_W$ is also a weak cause. Note that in this paper, we only consider $|C_W| = 1$, and therefore the above definition also defines actual causes. Table 1 provides a reference for the common related notation used throughout the paper.

3.2   Markov Decision Processes

A Markov decision process (MDP) is a model for reasoning in fully observable, stochastic environments [2], defined as a tuple $\langle S, A, T, R, d \rangle$. S is a finite set of states, where $s \in S$ may be expressed in terms of a set of *state factors*, $\langle f_1, f_2, \ldots, f_N \rangle$, such that s indexes a unique assignment of values to the factors f; A is a finite set of actions; $T : S \times A \times S \rightarrow [0, 1]$ represents the probability of reaching a state $s' \in S$ after performing an action $a \in A$ in a state $s \in S$; $R : S \times A \times S \rightarrow \mathbb{R}$ represents the expected immediate reward of reaching a state $s' \in S$ after performing an action $a \in A$ in a state $s \in S$; and $d : S \rightarrow [0, 1]$ represents the probability of starting in a state $s \in S$. A solution to an MDP is a policy $\pi : S \rightarrow A$ indicating that an action $\pi(s) \in A$ should be performed in

Table 1. Important notations, summarized from [11].

Notation	Meaning
X	A set of decision variables, $X = \{X_1, X_2, X_3\}$
x	An assignment of values to the set X, $\{X_1 = x_1, X_2 = x_2, X_3 = x_3\}$
$\mathcal{P}(X)$	Power set of X
$\mathcal{D}(X_1)$	Domain of the joint assignments of all $x \in X$
$x' \leftarrow x|X'$	x' is the restriction of x to X', e.g., if $X' = \{X_1\}$ and $x = \{X_1 = x_1, X_2 = x_2, X_3 = x_3\}$, then $x' = \{X_1 = x_1\}$
$x \leftarrow [x\langle x'\rangle]$	Replace values of x with values from x', e.g., if $x = \{X_1 = x_1, X_2 = x_2\}$ and $x' = \{X_1 = b\}$, then $x = \{X_1 = b, X_2 = x_2\}$

a state $s \in S$. A policy π induces a value function $V^\pi : S \to \mathbb{R}$ representing the expected discounted cumulative reward $V^\pi(s) \in \mathbb{R}$ for each state $s \in S$ given a discount factor $0 \leq \gamma < 1$. An optimal policy π^* maximizes the expected discounted cumulative reward for every state $s \in S$ by satisfying the Bellman optimality equation $V^*(s) = \max_{a \in A} \sum_{s' \in S} T(s, a, s')[R(s, a, s') + \gamma V^*(s')]$.

One of the most natural ways to represent an MDP as an SCM is to let U consist of all variables related to the reward function R, transition function T, start distribution d, and discount factor γ. Then, V can be defined as $F \cup \Pi$, where F is the set of variables representing state factors, $F = \{f_1, f_2, \ldots, f_N\}$, and Π is the set of variables representing the optimal policy, $\Pi = \{\pi_{s_1 a_1}, \pi_{s_1 a_2}, \ldots, \pi_{s_{|S|} a_{|A|}}\}$. Here, π_{sa} is a variable that is true when action a may be taken in state s. Thus, an obvious choice for an event ϕ is a subset of Π and their assignment. For example, if action a is taken in state s instead of a', we have

$$\phi = \langle [\pi(s) = a], [\pi(s) = a'] \rangle = \langle \text{TRUE}, \text{FALSE} \rangle.$$

Under this modeling setup, counterfactual settings to F do not result in new MDP policies as they would be variables from R or T to be used in V. Instead, this setup represents a fixed world model and a fixed model of agent capability, where counterfactual inputs represent different situations, or states, that the agent may encounter. Although MEANRESP may be applied to other components of the MDP using different definitions of U and V, we focus on this particular setup as it is computationally less demanding for empirical analysis. We should note that none of our theoretical analysis relies on this particular definition of U and V, or even that MEANRESP is used to analyze an MDP instead of a classifier.

4   MeanRESP

Chockler and Halpern [8] defined the *responsibility* (RESP), of an actual cause X' with contingency set W as $\frac{1}{1+|W|}$. Based on that, we define the MEANRESP score, ρ, of an actual cause X', to be the expected number of different ways X'

satisfies the definition of actual cause weighted by a responsibility share. Hence, the MEANRESP score equates the strength of the causal effect with the number of different scenarios under which X' can be considered a cause for the event.

There are several plausible versions of MEANRESP, all of which detect sets of variables that satisfy the definition of actual cause given above. To facilitate understanding throughout the rest of the paper, we now provide a novel, high-level description of a generalized version of MEANRESP and its relation to different definitions of cause. Moreover, we would like the MEANRESP score to behave in a manner summarized by the following properties:

1. **Property 1:** A set of variables $X' \subseteq X$ that is not a cause of the event ϕ should have $\rho = 0$. A set of variables $X' \subseteq X$ that is a cause of the event ϕ should have $\rho > 0$.
2. **Property 2:** As the cause allows a set of witness variables, ρ should divide the causal responsibility among the cause and witness in a principled manner.
3. **Property 3:** A relatively higher value of ρ for a cause $X' \subseteq X$ should indicate the event ϕ is relatively more affected by the assignment $x|X'$ of X'.

Responsibility scores are important in practice since they allow both users and developers to differentiate between causes that are highly relevant to the given scenario and those which may have less explanatory value. Here, we present a generalized version of MEANRESP (Algorithm 1) that has all three of these properties. The algorithm considers witness sets of size up to $|W|_{max} = |X| - 1$ (Line 4). After fixing a witness $W = w$ (lines 6–7), it calculates the RESP score (line 9) using either RESP-UC (Algorithm 2) if we use weak cause Definition 1 or RESP-OC (Algorithm 3) if we use the original weak cause definition. In RESP-UC, if 2B holds from Definition 1 (lines 4–9), then we check for 2A (lines 10–12). Notice that RESP-UC will return a value greater than 0 whenever both 2A and 2B hold. This ensures property 1. Note that the condition in definition 1 always holds for a deterministic policy or classifier, and therefore is not explicitly checked. Additionally, in both RESP-UC and RESP-OC, accumulating the RESP scores in lines 13 and 10, respectively, provides property 2. Intuitively, the RESP score scales with the number of different ways X' satisfies the definition of actual cause weighted by a responsibility share. Hence, the RESP score equates the strength of the causal effect with the number of different scenarios under which X' can be considered a cause for the event. This gives MEANRESP property 3. We use the following notation to denote whether event ϕ occurred.

$$\phi(x_a) = \begin{cases} \text{TRUE} & \text{if } \Pi(x) = \Pi(x_a) \\ \text{FALSE} & \text{if } \Pi(x) \neq \Pi(x_a) \end{cases} \tag{1}$$

Overall, there are several design choices one can make regarding how exactly to compute the mean RESP scores, generating a family of closely related algorithms. First, either RESP-UC or RESP-OC may be used, depending on the desired definition of cause. Not only will this affect the resultant RESP scores, but most importantly, it will change what is identified as a cause; some sets of variables will have RESP scores of zero under one definition but not the other.

Algorithm 1. MEANRESP

1: **Input:** All Causal Variables X, Variable of Interest X', Inference Model Π,
   Variable assignment x, Responsibility function $RESP$
2: **Output:** Mean Responsibility Scores ρ.
3: $MEANRESP \leftarrow 0$
4: **for all** $\beta = 0...|W|_{max}$ **do**
5:    $\sigma, T \leftarrow 0$
6:    **for all** $W \in \mathcal{P}(X \setminus X')$ such that $|W| = \beta$ **do**
7:      **for all** $w \in Dom(W)$ **do**
8:       $T \leftarrow T + 1$
9:       $\sigma \leftarrow \sigma + RESP(\Pi, X, X', x, d \sim Dom(X'), W, w)$
10:   $MEANRESP \leftarrow MEANRESP + \frac{\sigma}{T}$
11: **return** $\frac{MEANRESP}{|W|_{max}+1}$

Algorithm 2. RESP-UC

1: **Input:** Π, X, x, d, W, w
2: **Output:** score, σ.
3: $D_1, D_2 \leftarrow 1$
4: **for all** $W' \in \mathcal{P}(W)$ **do**
5:   $w' \leftarrow w|W'$
6:   $x_p \leftarrow [x\langle w'\rangle]$
7:   **if** $\neg\phi(x_p)$ **then**
8:     $D_1 \leftarrow 0$
9:     **break**
10: $x_m \leftarrow [x\langle(d \cup w)\rangle]$
11: **if** $\phi(x_m)$ **then**
12:   $D_2 \leftarrow 0$
13: **return** $\frac{D_1}{1+|W|}D_2$

Second, the mean RESP score can be calculated in two ways. It may be tallied over only the witness sets of size β_{min}, where β_{min} is the smallest β for which there exists a satisfying witness set (as in [27]). Or, it may be tallied overall witness sets, regardless of β, as in Algorithm 1. Actual causes with at least some small witness sets will receive lower RESP scores under the latter design.

Third, as responsibility incrementally accrues with respect to an actual causal set, these increments can either be counted equally or can be normalized by the size of the domain of the actual cause. We refer to this as the option to perform domain normalization, and the theory behind it is that with a larger domain the chance that some assignment $X = \bar{x}$ will meet the conditions of Definition 1 increases, and thus the responsibility should correspondingly decrease.

None of these choices interfere with properties 1–3, but they may subtly alter relative responsibility assigned to different actual causes. As there is no clear reason based on first principles as to the correct choice, these decisions involve tradeoffs. For example, short-circuiting after finding a single witness set of size β that satisfies Definition 1 will save compute time, but may give a slightly higher

---

Algorithm 3. RESP-OC

1: **Input:** Π, X, x, d, W, w
2: **Output:** score, σ.
3: $D_1, D_2 \leftarrow 1$
4: $x_p \leftarrow [x\langle w\rangle]$
5: **if** $\neg\phi(x_p)$ **then**
6:     $D_1 \leftarrow 0$
7: $x_m \leftarrow [x\langle (d \cup w)\rangle]$
8: **if** $\phi(x_m)$ **then**
9:     $D_2 \leftarrow 0$
10: **return** $\frac{D_1}{1+|W|} D_2$

Algorithm 4. SAMPLED MEANRESP

1: **Input:** All Causal Variable X, Variable of Interest X', Inference Model Π,
    Variable assignment x, Responsibility function $RESP$, Sample Size, T
2: **Output:** Mean Responsibility Scores ρ.
3: $\sigma \leftarrow 0$
4: **for all** $t = 0...T$ **do**
5:     $W \sim \mathcal{P}(X \setminus X')$
6:     $w \sim Dom(W)$
7:     $\sigma \leftarrow \sigma + RESP(\Pi, X, X', x, d \sim Dom(X'), W, w)$
8: **return** $\frac{\sigma}{T}$

---

or lower ρ score depending on whether the variables of interest are important under many counterfactual scenarios or only a few.

4.1 Approximating MEANRESP

Algorithm 1 is an exact algorithm that iterates over all possible scenarios to count where X' satisfies the definition of cause. When the state space is very large, due to either continuous variables or large discrete domains, we can use essentially the same algorithm adapted to sample witness set assignments using Monte Carlo sampling. Algorithm 4 approximates exact MEANRESP, and reproduces the exact algorithm in the limit. Sampling may be constrained along several dimensions independently, depending on the most expensive features of the problem. Here, we present in detail a novel sample-based algorithm to calculate responsibility scores. We then discuss its connection to the popular Shapley value-based attribution score. In subsequent sections, we will theoretically and empirically analyze this algorithm.

The main difference is that instead of going through all possible scenarios (i.e. $W \in \mathcal{P}(X \setminus X'), w \in \mathcal{D}(W), d \in \mathcal{D}(X')$) we sample different scenarios uniformly. The expression being estimated can be written as the following equation for RESP-UC:

$$E_{W \sim \mathcal{P}(X \setminus X'), w \sim \mathcal{D}(W), d \sim \mathcal{D}(X')}[\frac{D1}{1+|W|}(\phi(x) - \phi(x_m))] \tag{2}$$

For RESP-OC it can be written as:

$$E_{W \sim \mathcal{P}(X \setminus X'), w \sim \mathcal{D}(W), d \sim \mathcal{D}(X')}[\frac{\phi(x_p)}{1 + |W|}(\phi(x) - \phi(x_m))] \tag{3}$$

It can be verified that this expression is the same as the following:

$$E_{W \sim \mathcal{P}(X \setminus X'), w \sim \mathcal{D}(W), d \sim \mathcal{D}(X')}[\frac{\phi(x_p)}{1 + |W|}(\phi(x_p) - \phi(x_m))] \tag{4}$$

This rewrite provides us with insight into the connection between Shapley value and RESP. In particular, the Monte Carlo approximation of the expected Shapely value can be written as:

$$E_{W \sim \mathcal{P}(X \setminus X'), w \sim \mathcal{D}(W), d \sim \mathcal{D}(X')}[(\phi(x_p) - \phi(x_m))] \tag{5}$$

Intuitively, from Eqs. 4, and 5 MEANRESP can be thought of as distanced weighted Shapely Value. Here, $1+|W|$ captures the difference in the original input x and x_m. $\phi(x_p)$ captures the difference in original output $\Pi(x)$ and $\Pi(x_p)$.

Finally, when the action space is continuous the definition of ϕ might not work as well due to the floating point values. Therefore, we can consider a softer version of ϕ as below:

$$\phi_{soft}(x_a) = e^{-\beta(|\Pi(x_a) - \Pi(x)|)} \tag{6}$$

Here, $\beta \to \infty : \phi_{soft}(x_a) \to \phi(x_a)$. However, note that for ϕ_{soft} Eqs. 2 and 3 are no longer equivalent.

5   Theoretical Analysis

Proposition 1. MEANRESP *Score* $\rho > 0$ *using the* RESP-UC *function implies actual cause according to Definition 1.*

Proof Sketch: MEANRESP Score $\rho > 0$ iff there exists at least one contingency (W, w) for which all the in Definition 1 is satisfied.

Proposition 2. *The false positive rate of sampled MeanResp is 0.*

Proof Sketch: $\rho > 0$ iff all the constraint of the Definition 1 is satisfied at least once.

Proposition 3. *The false negative rate of sampled Mean Resp with n sample is at most* $(1 - (\rho^*(|W|_{max} + 1)))^n$.

Proof Sketch The probability of not classifying X' as a weak cause i.e. false negative rate using 1 sample will be at most $1 - (\rho^*(|W|_{max} + 1))$. If samples are drawn independently then the false negative rate using n samples is at most $(1 - (\rho^*(|W|_{max} + 1)))^n$.

Proposition 4. $P(|\rho - \rho^*| \geq \sqrt{\epsilon \rho^*}) \leq 2e^{-\epsilon n/3k}; k = W_{max} + 1, n = number$ *of samples. In words, probability that the estimated ρ deviate by $\sqrt{\frac{\epsilon}{\rho^*}}$ is at most* $2e^{-\epsilon n/3k}$.

Proof Sketch According to Chernoff Bound:

$$P(|\frac{\rho n}{k} - \frac{\rho^* n}{k}| \geq \delta \frac{\rho^* n}{k}) \leq 2e^{-\delta^2 \rho^* n/3k} \tag{7}$$

Setting $\delta = \sqrt{\frac{\epsilon}{\rho*}}$:

$$P(|\frac{\rho n}{k} - \frac{\rho^* n}{k}| \geq \frac{\sqrt{\epsilon \rho* n}}{k}) \leq 2e^{-\epsilon n/3k} \tag{8}$$

This is same as:

$$P(|\rho - \rho^*| \geq \sqrt{\epsilon \rho*}) \leq 2e^{-\epsilon n/3k} \tag{9}$$

Proposition 5. MEANRESP *score is upper-bounded by Shapley Value.*

Proof Sketch Since in Eq. 3, $0 \leq \frac{\phi(x_p)}{1+|W|} \leq 1$, MEANRESP score will always be upper-bounded by shapely value when we consider Eq. 4.

6   Empirical Analysis

In this section, we will first discuss several metrics for comparing feature rankings generated using exact methods to those generated using different forms of approximation. Then, we will use some of these metrics to empirically evaluate the sampling error of the two proposed approximate MEANRESP methods (UC and OC) in conjunction with the Shapley value [37]. Finally, we will examine the disagreement in feature rankings among these three methods. This disagreement will help us evaluate whether MEANRESP differs significantly from the widely adopted Shapley value, potentially warranting a human-subject study.

6.1   Environment Details

To conduct experiments in this section, we used three open-source environments designed for sequential decision-making. The initial two environments were obtained from the OpenAI Gym library [6]: Blackjack and Taxi. The Blackjack environment consists of 704 states and 2 actions, with each state being represented by 3 features. As for the Taxi environment, it comprises 500 states and 6 actions, with each state being represented by 4 features. Additionally, we employed the highway-fast-v0 environment from the Highway-env library [17] (referred to as "Highway" hereafter). This environment encompasses 20 features (we exclude the features indicating *presence* from our experiment) within its feature space, each with continuous values. To facilitate our analysis, we discretized the feature domain into 20 equidistant points. Consequently, the total number of states in this environment amounted to 20^{20}. In this environment, the agent can select from 5 different discrete actions. Note that due to such a large state space, it is computationally infeasible to estimate exact MEANRESP. For the empirical analysis, we used a very large sample size of 10^5 to emulate exact MEANRESP For the Blackjack and Taxi environments, we employed value iteration [5] to compute the optimal policy, while for Highway, we utilized Deep Q-learning [24] to approximate an optimal policy.

Table 2. Example of the top $k = 5$ features identified as causes by the exact (ground truth) MEANRESP and approximate MEANRESP-OC methods after different numbers of samples ($N = 2000, 1000, 500, 50$). We observe that: a) After 50 samples, MEANRESP identifies most of the causal variables (4 out of 5). b) By 500 samples, the first 3 rankings match exactly with the exact method. c) After 2000 samples, the ranking completely matches the exact method. While the score estimates fluctuate with additional samples, highly influential variables (ranks 1, 2, and 3) are relatively easy to identify. Other weakly influential variables appear frequently but may not always be ranked correctly. We observe this trend of influential variables stabilizing early throughout our experiment.

Ground Truth	N = 2000	N = 1000	N = 500	N = 50
1. Vehicle-2_X	1. Vehicle-2_X	1. Vehicle-1_Y	1. Vehicle-1_Y	1. Vehicle-1_Y
2. Vehicle-1_Y	2. Vehicle-1_Y	2. Vehicle-2_X	2. Vehicle-2_X	2. Vehicle-3_X
3. Vehicle-3_Y	3. Vehicle-3_Y	3. Vehicle-3_Y	3. Vehicle-3_Y	3. Vehicle-Ego_Y
4. Vehicle-Ego_Y	4. Vehicle-Ego_Y	4. Vehicle-3_X	4. Vehicle-3_X	4. Vehicle-2_Y
5. Vehicle-3_X	5. Vehicle-3_X	5. Vehicle-2_Y	5. Vehicle-2_Y	5. Vehicle-2_X

6.2 Metrics

In this subsection, we will discuss several existing metrics used to compare feature rankings generated by exact methods with those generated using various forms of approximation. Some of these metrics rely solely on the contents of the explanations, others rely only on the relative rankings of feature sets, and some require ranking scores.

To evaluate the effectiveness of these metrics, we employed the Highway environment. In Table 2, we present snapshots of the top k most responsible variables for a given action outcome, illustrating examples of both the exact MEANRESP-OC method and various stages during the sampling process within the approximate MEANRESP-OC approach. The behavior of these metrics throughout the sampling process is summarized in Figs. 1.

Ranking Only. We can calculate the ranking of the features by sorting the ρ-values in descending order. Using Kendall's τ and Spearman's ρ, we can calculate a rank correlation coefficient to compare the ranks of features. This coefficient should increase towards 1 as we increase the number of samples used to estimate. Finally, one simple approach is to check if the two rankings are identical.

Responsibility. Having access to the raw responsibility scores provides an opportunity for additional nuance in our metrics. Here, we present several options.

First, let us treat each feature in X as a 2D point. The x-value will be the true ρ-value, as given by the exact method, ρ^* for that set. The y-value will be the estimated ρ-value. As the number of samples increases, the slope of the least squares fit line on these 2D points should approach 1.

Second, we can use Pearson's r correlation factor to calculate the correlation between ρ^* and ρ. As the number of samples increases the correlation factor should reach 1.

Third, we take the top k features from both exact and approximate methods, sum the ρ^* values associated with the exact results, and call this ρ_k^*. Then, sum the ρ^*-values for approximate results and call this ρ_k^{approx}. The fraction $\frac{\rho_k^{approx}}{\rho_k^*}$ should approach 1 as number of samples increases.

Finally, we can calculate the Euclidean distance between the vectors representing ρ^*- and ρ-values for every potential feature set. As the number of samples increases the distance should reach 0.

Feature Set Contents. The previous metrics concern the relative importance assigned to different causes identified by MEANRESP or a similar algorithm. Here, we consider the presence or absence of information represented within the causal sets, since in many cases, the user will ultimately see only the causes and not their relative importance. If we create a set C_k^* that is the union of the top k features from the exact algorithm, and similarly define a set C_k that is the union of the top k features from the approximate algorithm, we then have a basis to understand what has been erroneously included or omitted. In this case, we propose simply finding the number of insertions and deletions required to make the sets identical or the edit distance. This number should approach 0 as the number of samples increases.

Discussion. In the context of explanation generation, we argue that feature-set content is a more interpretable metric as it tells us exactly how many different factors will be communicated to the user. For sample error estimation, while all the metrics under responsibility are good candidates, we found no clear winner. We opted for using Euclidean distance for our experiments.

6.3   Sampling Error

In Fig. 2 we show estimation error versus the number of samples used to estimate the score. We use the following estimation error:

$$\frac{1}{|S|} \sum_{s \in S} \sqrt{\sum_{i \in [1,|s|]} (\rho_i^*(s) - \rho_i(s))^2} \tag{10}$$

Here, $\rho_i^*(s)$ and $\rho_i(s)$ are the ground truth value and approximated value respectively for the i−th feature of state s. Note that this is equivalent to the average Euclidean distance. We use the average of 30 different evaluations of Eq. 10 to create Fig. 2. In both environments, we see MEANRESP-UC and MEANRESP-OC perform similarly. However, for the same amount of samples, we see 10%-70% more error in the estimation for the Shapley Value compared to both MEANRESP.

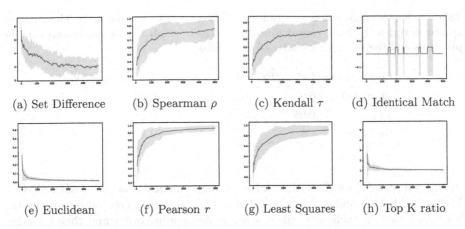

(a) Set Difference   (b) Spearman ρ (c) Kendall τ (d) Identical Match

(e) Euclidean   (f) Pearson r (g) Least Squares   (h) Top K ratio

Fig. 1. Traces of all 8 metrics over time as they compare exact and approximate responsibility estimates. The solid blue line represents the mean (50 runs using 50 different states total), and the blue-shaded regions represent one standard deviation. Clearly, some metrics are more sensitive than others. Moreover, some shifts appear to be detected universally, for example, near 200 samples, while at other points some metrics respond to updated estimates while others do not. Notably, different is the boolean metric in (d) that checks whether the top $k = 5$ items in the set are identically ranked. The trace also shows us when the first time results become identical to the exact methods. Due to the Monte Carlo sampling, we see some oscillation. In addition, (a) is a version of edit distance, measuring how many insertions or deletions need to be made before the sets are identical. Here, both absolute and relative responsibility scores are irrelevant; only inclusion somewhere in the top k is captured. (Color figure online)

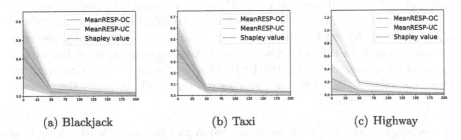

(a) Blackjack   (b) Taxi   (c) Highway

Fig. 2. Convergence rates for different attribution methods. Horizontal axes represent the number of samples taken, and vertical axes represent the absolute error in attribution value with respect to the exact solution. Color gradients represent one standard deviation.

6.4   Explanation Dissimilarity

Table 3. The average Feature-set Difference among attribution methods. The numbers appear surprisingly high, considering the relatively small number of features in the problems and the relative similarity of the methods.

Environment	OC vs. UC	OC vs. Shapley	UC vs. Shapley
BlackJack	0.05	0.15	0.13
Taxi	0.05	0.11	0.14
Highway	1.8	2.5	2.4

In this subsection, we show the ranking disagreement among MEANRESP-OC, MEANRESP-UC, and Shapley in Table 3. The ranking is created by sorting the features based on their attribution score in descending order and then selecting the top 33% of the features (for Taxi, the top feature, for Blackjack, the top 2 features, and for Highway, the top 7 features). We then calculated pairwise explanation dissimilarity using the Feature-set Difference metric discussed previously. We generated explanations for 100 sampled states in each environment and reported the average Feature-set Difference. In all cases, we see that the disagreement between MEANRESP-OC and MEANRESP-UC is smaller than Shapley. Also, MEANRESP-OC is more similar to Shapley than MEANRESP-UC, in two out of three environments. These results suggest that there is a significant amount of difference in the explanations created by these methods, especially in larger environments. This motivates a potential future human subject study of explanation preference.

7   Conclusion

In summary, this study provides a comprehensive examination of MEANRESP, a framework for causal analysis of MDPs using structural causal models. The theoretical and empirical analyses shed light on crucial properties of approximate MEANRESP, including the convergence of error rates. Additionally, we introduce various metrics that contribute to a deeper understanding of the ranking generated by approximate MEANRESP. Moving forward, future research will involve conducting user preference studies to empirically evaluate the effectiveness of these methods.

Acknowledgments. This work was supported in part by the National Science Foundation grant number IIS-1954782.

References

1. Allahyari, M., et al.: Text summarization techniques: a brief survey. arXiv preprint arXiv:1707.02268 (2017)
2. Bellman, R.: On the theory of dynamic programming. Natl. Acad. Sci. United States Am. **38**(8), 716 (1952)
3. Bertossi, L., Li, J., Schleich, M., Suciu, D., Vagena, Z.: Causality-based explanation of classification outcomes (2020). arXiv preprint arXiv:2003.06868
4. Bertram, J., Wei, P.: Explainable deterministic MDPs (2018). arXiv preprint arXiv:1806.03492
5. Bertsekas, D.P.: Dynamic programming and optimal control (1995)
6. Brockman, G., et al.: OpenAI Gym (2016). https://arxiv.org/abs/1606.01540
7. Chen, J.Y., Lakhmani, S.G., Stowers, K., Selkowitz, A.R., Wright, J.L., Barnes, M.: Situation awareness-based agent transparency and human-autonomy teaming effectiveness. Theor. Issues Ergon. Sci. **19**(3), 259–282 (2018)
8. Chockler, H., Halpern, J.Y.: Responsibility and blame: a structural-model approach. J. Artif. Intell. Res. **22**, 93–115 (2004)
9. David Wong, E.: Understanding the generative capacity of analogies as a tool for explanation. J. Res. Sci. Teach. **30**(10), 1259–1272 (1993)
10. Elizalde, F., Sucar, E., Noguez, J., Reyes, A.: Generating explanations based on markov decision processes. In: Aguirre, A.H., Borja, R.M., Garciá, C.A.R. (eds.) MICAI 2009. LNCS (LNAI), vol. 5845, pp. 51–62. Springer, Heidelberg (2009). https://doi.org/10.1007/978-3-642-05258-3_5
11. Halpern, J.Y., Pearl, J.: Causes and explanations: a structural-model approach. Part I: Causes. Brit. J. Phil. Sci. **52**(3), 613–622 (2005)
12. Halpern, J.Y., Pearl, J.: Causes and explanations: a structural-model approach. Part II: explanations. Brit. J. Phil. Sci. **56**(4), 889–911 (2005)
13. Hayes, B., Shah, J.A.: Improving robot controller transparency through autonomous policy explanation. In: ACM/IEEE International Conference on Human-Robot Interaction (HRI), pp. 303–312 (2017)
14. Juozapaitis, Z., Koul, A., Fern, A., Erwig, M., Doshi-Velez, F.: Explainable reinforcement learning via reward decomposition. In: IJCAI/ECAI Workshop on Explainable Artificial Intelligence (2019)
15. Karimi, A.H., Schölkopf, B., Valera, I.: Algorithmic recourse: from counterfactual explanations to interventions. In: Proceedings of the ACM Conference on Fairness, Accountability, and Transparency, pp. 353–362 (2021)
16. Khan, O., Poupart, P., Black, J.: Minimal sufficient explanations for factored Markov decision processes. In: International Conference on Automated Planning and Scheduling (ICAPS), vol. 19 (2009)
17. Leurent, E.: An environment for autonomous driving decision-making (2018)
18. Linegang, M.P., et al.: Human-automation collaboration in dynamic mission planning: a challenge requiring an ecological approach. Proc. Human Fact. Ergon. Soc. Annual Meet. **50**(23), 2482–2486 (2006)
19. Lucic, A., Haned, H., de Rijke, M.: Why does my model fail? contrastive local explanations for retail forecasting. In: Proceedings of the ACM Conference on Fairness, Accountability, and Transparency, pp. 90–98 (2020)
20. Lundberg, S.M., Lee, S.I.: A unified approach to interpreting model predictions. Adv. Neural Inf. Process. Syst. **30** (2017)
21. Madumal, P., Miller, T., Sonenberg, L., Vetere, F.: Explainable reinforcement learning through a causal lens. In: Proceedings of the AAAI Conference on Artificial Intelligence, vol. 34, pp. 2493–2500 (2020)

22. Mercado, J.E., Rupp, M.A., Chen, J.Y., Barnes, M.J., Barber, D., Procci, K.: Intelligent agent transparency in human-agent teaming for Multi-UxV management. Hum. Fact. **58**(3), 401–415 (2016)
23. Miller, T.: Explanation in artificial intelligence: insights from the social sciences. Artif. Intell. **267**, 1–38 (2019)
24. Mnih, V., et al.: Playing Atari with deep reinforcement learning (2013). https://arxiv.org/abs/1312.5602
25. Molnar, C.: Interpretable Machine Learning, 2 edn. (2022). https://christophm.github.io/interpretable-ml-book
26. Mothilal, R.K., Sharma, A., Tan, C.: Explaining machine learning classifiers through diverse counterfactual explanations. In: Proceedings of the ACM Conference on Fairness, Accountability, and Transparency, pp. 607–617 (2020)
27. Nashed, S.B., Mahmud, S., Goldman, C.V., Zilberstein, S.: Causal explanations for sequential decision making under uncertainty (2022)
28. Nisioi, S., Štajner, S., Ponzetto, S.P., Dinu, L.P.: Exploring neural text simplification models. In: Proceedings of the 55th Annual Meeting of the Association for Computational Linguistics, pp. 85–91 (2017)
29. Panigutti, C., Perotti, A., Pedreschi, D.: Doctor XAI: an ontology-based approach to black-box sequential data classification explanations. In: Proceedings of the ACM Conference on Fairness, Accountability, and Transparency, pp. 629–639 (2020)
30. Pouget, H., Chockler, H., Sun, Y., Kroening, D.: Ranking policy decisions (2020). arXiv preprint arXiv:2008.13607
31. Russell, J., Santos, E.: Explaining reward functions in Markov decision processes. In: Thirty-Second International FLAIRS Conference (2019)
32. Scharrer, L., Bromme, R., Britt, M.A., Stadtler, M.: The seduction of easiness: how science depictions influence laypeople's reliance on their own evaluation of scientific information. Learn. Inst. **22**(3), 231–243 (2012)
33. Shapley, L.S., et al.: A value for n-person games (1953)
34. Srikanth, N., Li, J.J.: Elaborative simplification: content addition and explanation generation in text simplification (2020). arXiv preprint arXiv:2010.10035
35. Stubbs, K., Hinds, P.J., Wettergreen, D.: Autonomy and common ground in human-robot interaction: a field study. IEEE Intell. Syst. **22**(2), 42–50 (2007)
36. Sukkerd, R., Simmons, R., Garlan, D.: Tradeoff-focused contrastive explanation for mdp planning. In: 2020 29th IEEE International Conference on Robot and Human Interactive Communication (RO-MAN), pp. 1041–1048. IEEE (2020)
37. Štrumbelj, E., Kononenko, I.: Explaining prediction models and individual predictions with feature contributions. Knowl. Inf. Syst. **41**, 647–665 (2014)
38. Wang, N., Pynadath, D.V., Hill, S.G.: The impact of POMDP-generated explanations on trust and performance in human-robot teams. In: International Conference on Autonomous Agents and Multiagent Systems (AAMAS), pp. 997–1005 (2016)
39. Zhang, Y., Liao, Q.V., Bellamy, R.K.: Effect of confidence and explanation on accuracy and trust calibration in AI-assisted decision making. In: Proceedings of the ACM Conference on Fairness, Accountability, and Transparency, pp. 295–305 (2020)

Explainable Machine Learning

The Quarrel of Local Post-hoc Explainers for Moral Values Classification in Natural Language Processing

Andrea Agiollo[1]([✉]) [iD], Luciano Cavalcante Siebert[2] [iD],
Pradeep Kumar Murukannaiah[2] [iD], and Andrea Omicini[1] [iD]

[1] Dipartimento di Informatica – Scienza e Ingegneria (DISI), Alma Mater
Studiorum—Università di Bologna, Cesena, Italy
{andrea.agiollo,andrea.omicini}@unibo.it
[2] Delft University of Technology, Delft, The Netherlands
{L.CavalcanteSiebert,P.K.Murukannaiah}@tudelft.nl

Abstract. Although popular and effective, *large language models*
(LLM) are characterised by a performance vs. transparency trade-off
that hinders their applicability to sensitive scenarios. This is the main
reason behind many approaches focusing on local post-hoc explanations
recently proposed by the XAI community. However, to the best of our
knowledge, a thorough comparison among available explainability techniques is currently missing, mainly for the lack of a general metric to measure their benefits. We compare state-of-the-art local post-hoc explanation mechanisms for models trained over moral value classification tasks
based on a measure of correlation. By relying on a novel framework for
comparing global impact scores, our experiments show how most local
post-hoc explainers are loosely correlated, and highlight huge discrepancies in their results—their "quarrel" about explanations. Finally, we
compare the impact scores distribution obtained from each local post-
hoc explainer with human-made dictionaries, and point out that there
is no correlation between explanation outputs and the concepts humans
consider as salient.

Keywords: Natural Language Processing · Moral Values
Classification · eXplainable Artificial Intelligence · Local Post-hoc
Explanations

1 Introduction

Large Language Models (LLMs) represent the de-facto solution for dealing with
complex Natural Language Processing (NLP) tasks such as sentiment analysis [45], question [19], and many others [34]. The ever-increasing popularity of
such data-driven approaches is widely caused by their performance improvements
against their counterparts. Indeed, Neural Network (NN) based approaches have
shown uncanny performance over different NLP tasks such as grammar acceptability of a sentence [43] and text translation [40]. However, following the quest

© The Author(s), under exclusive license to Springer Nature Switzerland AG 2023
D. Calvaresi et al. (Eds.): EXTRAAMAS 2023, LNAI 14127, pp. 97–115, 2023.
https://doi.org/10.1007/978-3-031-40878-6_6

for higher performance, research efforts gave birth to ever more complex NN architectures such as BERT [15], GPT [12], and T5 [35].

Although being powerful and, empirically, reliable, LLM suffer from a performance vs. transparency trade-off [11,47]. Indeed, LLM are *black-box* models, as they rely on the optimisation of their numerical sub-symbolical components, which are mostly unreadable by humans. The *black-box* nature of LLMs hinder their applicability to some scenarios where transparency represents a fundamental requirement, e.g., NLP for medical analysis [29,39], etc. Therefore, there exists the need to identify relevant mechanisms capable of opening such LLM black-boxes and diagnose their reasoning process, and presenting it in a human-understandable fashion. Towards this aim, a few different explainability approaches, focusing mostly on *Local Post-hoc explainer (LPE)* mechanisms, have been recently proposed. An LPE represents a popular solution to explain the reasoning process by highlighting how different portions of the input sample impact differently the produced output, by assigning a relevance score to each input component. These approaches apply to single instances of input sample, thus being *local*, and to optimised LLM—thus being *post-hoc*.

Despite a broad variety of LPE approaches, the state of the art lacks a fair comparison among them. A common trend for proposals of novel explanation mechanisms is to highlight its advantages through a set of tailored experiments. This hinders comparison fairness, making it very difficult to identify the best approach for obtaining explanations of NLP models or even to know if such a best approach exists. This is why we present a framework for comparing several well-known LPE mechanisms for text classification in NLP. Aiming at obtaining comparison fairness, we rely on aggregating the local explanations obtained by each local post-hoc explainer into a set of global impact scores. Such scores identify the set of concepts that best describe the underlying NLP model from the perspective of each LPE. These concepts, along with their aggregated impact scores, are then compared for each LPE against other LPE counterparts. The comparison between the aggregated global impact scores rather than the single explanations is justified by the locality of LPE approaches. Indeed, it is reasonable for local explanations of different LPEs to differ somehow, depending on the approach design, therefore making it complex to compare the quality of two LPEs over the same sample. However, it is also expected for the aggregated global impacts to be aligned between different LPE as they are applied to the same NN, which leverages the same set of relevant concepts for its inference. Therefore, when comparing the aggregated impact scores of different LPE, we expect them to be correlated—at least up to a certain extent.

We perform our comparison between LPE explanations across the social domains available in the Moral Foundation Twitter Corpus (MFTC) [20]. MFTC represents an example of a challenging task, as it is proposed to tackle moral values classification. Moral values are inherently subjective to human readers, therefore introducing possible disagreement inside annotations and making the overall optimisation pipeline sensitive to small changes. Moreover, identifying moral values represents a sensitive task, as it requires a deep and safe understanding of complex concepts such as harm and fairness. Consequently, we believe MFTC to represent a suitable option for analysing the behaviour of LPEs over different scenarios. Moreover, relying on MFTC enables a comparison between extracted relevant concepts and a set of humanly tailored impact scores, namely Moral Foundations Dictionary (MFD) [21]. Therefore, allowing us to study the extent of correlation between LPE-extracted concepts and humanly salient concepts. Surprisingly, our experiments show how there are setups where the explanations of different LPEs are far from being correlated, highlighting how explanation quality is highly dependent on the chosen eXplainable Artificial Intelligence (xAI) approach and the respective scenario at hand. There are huge discrepancies in the results of different state-of-the-art local explainers, each of which identifies a set of relevant concepts that largely differs from the others—at least in terms of relative impact scores. Therefore, we stress the need for identifying a robust approach to compare the quality of explanations and the approaches for their extraction. Moreover, the comparison between the distribution of LPEs' impact scores and the set of human-tailored impact scores shows how there exists almost always no correlation between salient concepts extracted from the NN model and concepts relevant for humans. The obtained results highlight the fragility of xAI approaches for NLP, caused mainly by the complexity of large NN models, their inclination to the extreme fitting of data—with no regard for concept meaning— and the lack of sound techniques for comparing xAI mechanisms.

2   Background

2.1   Explanation Mechanisms in NLP

The set of explanations extraction mechanisms available in the xAI community are often categorized along two main aspects [2,17]: *(i) local* against *global* explanations, and *(ii) self-explaining* against *post-hoc* approaches. In the former context, *local* identifies the set of explainability approaches that given a single input, i.e., sample or sentence, produce an explanation of the reasoning process followed by the NN model to output its prediction for the given input [32]. In contrast, *global* explanations aim at expressing the reasoning process of the NN model as a whole [18,22]. Given the complexity of the NN models leveraged for tackling most NLP tasks, it is worth noticing how there is a significant lack of *global* explainability systems, whereas a variety of *local* xAI approaches are available [31,37].

About the latter aspect, we define *post-hoc* as those set of explainability approaches which apply to an already optimized black-box model for which it is required to obtain some sort of insight [33]. Therefore, a *post-hoc* approach requires additional operations to be performed after that the model outputs its predictions [14]. Conversely, inherently explainable, i.e., *self-explaining*, mechanisms aim at building a predictor having a transparent reasoning process by design, e.g., CART [30]. Therefore, a self-explaining approach can be seen as generating the explanation along with its prediction, using the information emitted by the model as a result of the process of making that prediction [14].

In this paper, we focus on *local post-hoc* explanation approaches applied to NLP. Here, it represents a popular solution to explain the reasoning process by highlighting how different portions, i.e., words, of the input sample impact differently the produced output, by assigning a relevance score to each input component. The relevance score is then highlighted using some saliency map to ease the visualisation of the obtained explanation. Therefore, it is also common for local post-hoc explanations to be referred to as *saliency* approaches, as they aim at highlighting salient components.

2.2   Moral Foundation Twitter Corpus Dataset

In our experiments, we select the MFTC dataset as the target classification task. The MFTC dataset is composed of 35,108 tweets – sentences –, which can be considered as a collection of seven different datasets. Each split of MFTC corresponds to a different context. Here, tweets corresponding to the dataset samples are collected following a certain event or target. As an example, tweets belonging to the Black Lives Matter (BLM) split were collected during the period of Black Lives Matter protests in the US. The list of all MFTC subjects is the following: *(i)* All Lives Matter (ALM), *(ii)* BLM, *(iii)* Baltimore protests (BLT), *(iv)* hate speech and offensive language (DAV), *(v)* presidential election (ELE), *(vi)* MeToo movement (MT), *(vii)* hurricane Sandy (SND). In our experiments we also considered training and testing the NN model over the totality of MFTC tweets. This was done to analyse the LPEs behaviour over an unbiased task, as the average morality of each MFTC split is influenced by the corresponding collection event.

Each tweet in MFTC is labelled, following the same moral theory, with one or more of the following 11 moral values: *(i)* care/harm, *(ii)* fairness/cheating, *(iii)* loyalty/betrayal, *(iv)* authority/subversion, *(v)* purity/degradation, *(vi)* non-moral. Ten of the 11 available moral values are obtained as a moral concept and its opposite expression—e.g., fairness refers to the act of supporting fairness and equality, while cheating refers to the act of refraining from cheating or exploiting others. Given morality subjectivity, each tweet is labelled by multiple annotators, and the final moral labels are obtained via majority voting.

Finally, similar to previous works [28,36], we preprocess the tweets before using them as input samples for our LLM training. We preprocess the tweets by removing URLs, emails, usernames and mentions, as well as correcting common spelling mistakes and converting emojis to their respective lemmas using the Ekphrasis package[1] and the Python Emoji package[2], respectively.

3   Methodology

In this section, we present our methodology for comparing LPE mechanisms. We first propose an overview of the proposed approach in Sect. 3.1. Subsequently, the set of LPE mechanisms adopted in our experiments are presented in Sect. 3.2,

---

[1] https://github.com/cbaziotis/ekphrasis.
[2] https://pypi.org/project/emoji/.

and the aggregation approaches leveraged to obtain global impact scores from LPE outputs are described in Sect. 3.3. Finally, in Sect. 3.4 we present the metrics used to identify the correlation between LPEs.

3.1   Overview

Given the complexity of measuring different LPE approaches over single local explanations, we here consider measuring how much LPEs correlate with each other over a set of fixed samples. The underlying assumption of our framework is that various LPE techniques aim at explaining the same NN model used for prediction. Therefore, while explanations may differ over local samples, it is reasonable to assume that reliable LPEs when applied over a vast set of samples— sentences or set of sentences—should converge to similar (correlated) results. Indeed, the underlying LLM considers being relevant for its inference always the same set of concepts—lemmas. A lack of correlation between different LPE mechanisms would hint that there exists a conflict between the set of concepts that each explanation mechanisms consider as relevant for the LLM, thus making at least one, if not all, of the explanations unreliable.

Being interested in analysing the correlation between a set of LPEs over the same pool of samples, we first define ϵ_{NN} as a LPE technique applied to a NN model at hand. Being local, ϵ_{NN} is applied to the single input sample $\mathbf{x}_i$, producing as output one impact score for each component (token) of the input sample l_k. Throughout the remainder of the paper, we consider l_k to be the lemmas corresponding to the input components. Mathematically, we define the output impact score for a single token or its corresponding lemma as $j\left(l_k, \epsilon_{NN}(\mathbf{x}_i)\right)$. Depending on the given ϵ_{NN}, the corresponding impact score j may be associated with a single label – i.e., moral value –, making j a scalar value, or with a set of labels, making j a vector—one scalar value for each label. To enable comparing different LPE, we define the aggregated impact scores of a LPE mechanism over a NN model and a set of samples $\mathcal{S}$ as $\epsilon_{NN}(\mathcal{S})$. In our framework we obtain $\epsilon_{NN}(\mathcal{S})$ aggregating $\epsilon_{NN}(\mathbf{x}_i)$ for each $\mathbf{x}_i \in \mathcal{S}$ using an aggregation operation $\mathcal{A}$, mathematically:

$$\epsilon_{NN}\left(\mathcal{S}\right) = \mathcal{A}\left(\{\epsilon_{NN}(\mathbf{x}_i) \text{ for each } \mathbf{x}_i \in \mathcal{S}\}\right). \tag{1}$$

Defining a correlation metric $\mathcal{C}$, we obtain from Eq. (1) the following for describing the correlation between two LPE techniques:

$$\mathcal{C}\left(\epsilon_{NN}\left(\mathcal{S}\right), \epsilon'_{NN}\left(\mathcal{S}\right)\right) = \mathcal{C}\big(\mathcal{A}\left(\{\epsilon_{NN}(\mathbf{x}_i) \text{ for each } \mathbf{x}_i \in \mathcal{S}\}\right), \\ \mathcal{A}\left(\{\epsilon'_{NN}(\mathbf{x}_i) \text{ for each } \mathbf{x}_i \in \mathcal{S}\}\right)\big) \tag{2}$$

where ϵ_{NN} and ϵ'_{NN} are two LPE techniques applied to the same NN model.

3.2   Local Post-hoc Explanations

In our framework, we consider seven different LPE approaches for extracting local explanations $j\left(l_k, \epsilon_{NN}(\mathbf{x}_i)\right)$ from an input sentence $\mathbf{x}_i$ and the trained

LLM—identified as NN. The seven LPEs are selected in order to represent as faithfully as possible the state-of-the-art of xAI approaches in NLP. Subsequently, we briefly describe each of the seven selected LPE. However, a detailed analysis of these LPEs is out of the scope of this paper and we refer interested readers to [14,32,38].

Gradient Sensitivity Analysis. The Gradient Sensitivity analysis (GS) probably represents the simplest approach for assigning relevance scores to input components. GS relies on computing gradients over inputs components as $\dfrac{\delta f_c(\mathbf{x}_i)}{\delta \mathbf{x}_{i,k}}$, which represents the derivative of the output with respect to the the kth component of $\mathbf{x}_i$. Following this approach local impact scores of an input component can be thus defined as:

$$j\left(l_k, \epsilon_{NN}(\mathbf{x}_i)\right) = \frac{\delta f_{\tau_m}(\mathbf{x}_i)}{\delta \mathbf{x}_{i,k}}, \tag{3}$$

where $f_{\tau_m}(\mathbf{x}_i)$ represents the predicted probability distribution of an input sequence $\mathbf{x}_i$ over a target class τ_m. While simple, GS has been shown to be an effective approach for understanding approximate input components relevance. However, this approach suffers from a variety of drawbacks, mainly linked with its inability to define negative contributions of input components for a specific prediction—i.e., negative impact scores.

Gradient × Input Aiming at addressing few of the limitations affecting GS, the Gradient × Input (GI) approach defines the relevance scores assignment as GS multiplied – element-wise – with $\mathbf{x}_{i,k}$ [25]. Therefore, mathematically speaking, GI impact scores are defined as:

$$j\left(l_k, \epsilon_{NN}(\mathbf{x}_i)\right) = \mathbf{x}_{i,k} \cdot \frac{\delta f_{\tau_m}(\mathbf{x}_i)}{\delta \mathbf{x}_{i,k}}, \tag{4}$$

where notation follows the one of Eq. (3). Being very similar to GS, GI inherits most of its limitations.

Layer-Wise Relevance Propagation. Building on top of gradient-based relevance scores mechanisms – such as GS and GI –, Layer-wise Relevance Propagation (LRP) proposes a novel mechanism relying on conservation of relevance scores across the layers of the NN at hand. Indeed, LRP relies on the following assumptions: *(i)* NN can be decomposed into several layers of computation; *(ii)* there exists a relevance score $R_d^{(l)}$ for each dimension $\mathbf{z}_d^{(l)}$ of the vector $\mathbf{z}^{(l)}$ obtained as the output of the lth layer of the NN; and *(iii)* the total relevance scores across dimensions should propagate through all layers of the NN model, mathematically:

$$f(\mathbf{x}) = \sum_{d \in L} R_d^{(L)} = \sum_{d \in L-1} R_d^{(L-1)} = \cdots = \sum_{d \in 1} R_d^{(1)}, \tag{5}$$

where, $f(\mathbf{x})$ represents the predicted probability distribution of an input sequence $\mathbf{x}$, and L the number of layers of the NN at hand. Moreover, LRP defines a propagation rule for obtaining $R_d^{(l)}$ from $R^{(l+1)}$. However, the derivation of such propagation rule is out of the scope of this paper and thus we refer interested readers to [8,10]. In our experiments, we consider as impact scores the relevance scores of the input layer, namely $j\left(l_k, \epsilon_{NN}(\mathbf{x}_i)\right) = R_d^{(1)}$.

Layer-Wise Attention Tracing. Since LLMs rely heavily on self-attention mechanisms [42], recent efforts propose to identify input components relevance scores analysing solely the relevance scores of attentions heads of LLM models, introducing Layer-wise Attention Tracing (LAT) [1,44]. Building on top of LRP, LAT propose to redistribute the inner relevance scores $R^{(l)}$ across dimensions using solely self-attention weights. Therefore, LAT defines a custom redistribution rule as:

$$R_i^{(l)} = \sum_{k \text{ s.t. } i \text{ is input for neuron } k} \sum_h \mathbf{a}^{(h)} R_{k,h}^{(l+1)}, \qquad (6)$$

where, h corresponds to the attention head index, while $\mathbf{a}^{(h)}$ are the corresponding learnt weights of the attention head. Similarly to LRP, we here consider as impact scores the relevance scores of the input layer, namely $j\left(l_k, \epsilon_{NN}(\mathbf{x}_i)\right) = R^{(1)}$.

Integrated Gradient. Motivated by the shortcomings of previously proposed gradient-based relevance score attribution mechanisms – such as GS and GI –, Sundararajan et al. [41] propose a novel Integrated Gradient approach. The proposed approach aims at explaining the input sample components relevance by integrating the gradient along some trajectory of the input space, which links some baseline value $\mathbf{x}_i'$ to the sample under examination $\mathbf{x}_i$. Therefore, the relevance score of the input k^{th} component of the input sample $\mathbf{x}_i$ is obtained following

$$j\left(l_k, \epsilon_{NN}(\mathbf{x}_i)\right) = \left(\mathbf{x}_{i,k} - \mathbf{x}_{i,k}'\right) \cdot \int_{a=0}^1 \frac{\delta f(\mathbf{x}_i' + t \cdot (\mathbf{x}_i - \mathbf{x}_i'))}{\delta \mathbf{x}_{i,k}}\, dt, \qquad (7)$$

where $\mathbf{x}_{i,k}$ represents the kth component of the input sample $\mathbf{x}_i$. By integrating the gradient along an input space trajectory, the authors aim at addressing the locality issue of gradient information. In our experiments we refer to the Integrated Gradient approach as HESS, as for its implementation we rely on the integrated hessian library available for hugging face models[3].

SHAP. SHapley Additive exPlanations (SHAP) relies on Shapley values to identify the contribution of each component of the input sample toward the final prediction distribution. The Shapley value concept derives from game theory,

---

[3] https://github.com/suinleelab/path_explain.

where it represents a solution for a cooperative game, found assigning a distribution of a total surplus generated by the players coalition. SHAP computes the impact of an input component as its marginal contribution toward a label τ_m, computed deleting the component from the input and evaluating the output discrepancy. Firstly defined for explaining simple NN models [31], in our experiments we leverage the extension of SHAP supporting transformer models such as BERT [26], available in the SHAP python library[4].

LIME. Similarly to SHAP, Local Interpretable Model-agnostic Explanations (LIME) relies on input sample perturbation to identify its relevant components. Here, the predictions of the NN at hand are explained via learning an explainable surrogate model [37]. More in detail, to obtain its explanations LIME constructs a set of samples from the perturbation of the input observation under examination. The constructed samples are considered to be close to the observation to be explained from a geometric perspective, thus considering small perturbation of the input. The explainable surrogate model is then trained over the constructed set of samples, obtaining the corresponding local explanation. Given an input sentence, we here consider obtaining its perturbed version via words – or tokens – removal and words substitution. In our experiments, we rely on the already available LIME python library[5].

3.3 Aggregating Local Explanations

Once local explanations of the NN model are obtained for each input sentence – i.e., tweet –, we aggregate them to obtain a global list of concept impact scores. Before aggregating the local impact scores, we convert the words composing local explanations into their corresponding lemmas – i.e., concepts – to avoid issues when aggregating different words expressing the same concept—e.g., hate and hateful. As there exists no bullet-proof solution for aggregating different impact scores, we adopt four different approaches in our experiments, namely:

- *Sum.* A simple summation operation is leveraged to obtain the aggregated score for each lemma. While simple this aggregation approach is effective when dealing with additive impact scores such as SHAP values. However, it suffers from lemma frequency issues, as it tends to overestimate frequent lemmas having average low impact scores. Global impact scores are here defined as $J(l_k, \epsilon_{NN}) = \sum_{i=1}^{N} j\left(l_k, \epsilon_{NN}\left(\mathbf{x}_i\right)\right)$. Therefore, we here define $\mathcal{A}$ as

$$\mathcal{A}\left(\{\epsilon_{NN}(\mathbf{x}_i) \text{ for each } \mathbf{x}_i \in \mathcal{S}\}\right) = \left\{\sum_{i=1}^{N} j\left(l_k, \epsilon_{NN}\left(\mathbf{x}_i\right)\right) \text{ for each } l_k \in \mathcal{S}\right\}. \tag{8}$$

---

[4] https://github.com/slundberg/shap.
[5] https://github.com/marcotcr/lime.

– *Absolute sum.* We here consider summing the absolute values of the local impact scores – rather than their true values – to increase the awareness of global impact scores towards lemmas having both high positive and high negative impact over some sentences. Mathematically, we obtain aggregated scores as $J(l_k, \epsilon_{NN}) = \sum_{i=1}^{N} |j(l_k, \epsilon_{NN}(\mathbf{x}_i))|$.

$$\mathcal{A}(\{\epsilon_{NN}(\mathbf{x}_i) \text{ for each } \mathbf{x}_i \in \mathcal{S}\}) = \left\{\sum_{i=1}^{N} |j(l_k, \epsilon_{NN}(\mathbf{x}_i))| \text{ for each } l_k \in \mathcal{S}\right\}. \tag{9}$$

– *Average.* Similar to the sum operation, we here consider obtaining aggregated scores averaging local impact scores, thus avoiding possible overshooting issues arising when dealing with very frequent lemmas. Mathematically, we define $J(l_k, \epsilon_{NN}) = \frac{1}{N} \cdot \sum_{i=1}^{N} j(l_k, \epsilon_{NN}(\mathbf{x}_i))$.

$$\mathcal{A}(\{\epsilon_{NN}(\mathbf{x}_i) \text{ for each } \mathbf{x}_i \in \mathcal{S}\}) = \left\{\frac{1}{N} \cdot \sum_{i=1}^{N} j(l_k, \epsilon_{NN}(\mathbf{x}_i)) \text{ for each } l_k \in \mathcal{S}\right\}. \tag{10}$$

– *Absolute average.* Similarly to absolute sum, we here consider to average absolute values of local impact scores for better-managing lemmas having a skewed impact as well as tackling frequency issues. Global impact scores are here defined as $J(l_k, \epsilon_{NN}) = \frac{1}{N} \cdot \sum_{i=1}^{N} |j(l_k, \epsilon_{NN}(\mathbf{x}_i))|$.

$$\mathcal{A}(\{\epsilon_{NN}(\mathbf{x}_i) \text{ for each } \mathbf{x}_i \in \mathcal{S}\}) = \left\{\frac{1}{N} \cdot \sum_{i=1}^{N} |j(l_k, \epsilon_{NN}(\mathbf{x}_i))| \text{ for each } l_k \in \mathcal{S}\right\}. \tag{11}$$

Being aware that the selection of the aggregation mechanism may influence the correlation between different LPEs, in our experiments we analyse LPEs correlation over the same aggregation scheme. Moreover, we also consider analysing how aggregation impacts the impact scores correlation over the same LPE, highlighting how leveraging the absolute value of impact score is highly similar to adopting its true value—see Sect. 4.3.

3.4 Comparing Explanations

Each aggregated global explanation J depends on a corresponding label τ_m – i.e., moral value – since LPEs produce either a scalar impact value for a single τ_m or a vector of impact scores for each τ_m. Therefore, recalling Sect. 4.3, we can define the set of aggregated global scores depending on the label they refer to as following:

$$\mathcal{J}_{\tau_m}(\epsilon_{NN}, \mathcal{S}) = \{J(l_k, \epsilon_{NN}) | \tau_m \text{ for each } l_k \in \mathcal{S}\}. \tag{12}$$

$\mathcal{J}_{\tau_m}(\epsilon_{NN}, \mathcal{S})$ represents a distribution of impact scores over the set of lemmas – i.e., concepts – available in the samples set for a specific label. To compare the distributions of impact scores extracted using two LPEs – i.e., $\mathcal{J}_{\tau_m}(\epsilon_{NN}, \mathcal{S})$ and $\mathcal{J}_{\tau_m}(\epsilon'_{NN}, \mathcal{S})$ – we use Pearson correlation, which is defined as the ratio between the covariance of two variables and the product of their standard deviations, and it measures their level of linear correlation. The selected correlation metric is applied to the normalised impact scores. Indeed, different LPEs produce impact scores which may differ relevantly in terms of their magnitude. Normalising the impact scores, we map impact scores to a fixed interval, allowing for a direct comparison of $\mathcal{J}_{\tau_m}$ over different ϵ_{NN}. Mathematically, we refer to the normalised global impact scores as $\|\mathcal{J}_{\tau_m}\|$. Therefore, we define the correlation score between two sets of global impact scores for a single label as:

$$\rho\left(\|\mathcal{J}_{\tau_m}\left(\epsilon_{NN}, \mathcal{S}\right)\|, \|\mathcal{J}_{\tau_m}\left(\epsilon'_{NN}, \mathcal{S}\right)\|\right) = \rho\big(\|\{J\left(l_k, \epsilon_{NN}\right)|\tau_m \text{ for each } l_k \in \mathcal{S}\}\|,$$
$$\|\{J\left(l_k, \epsilon_{NN}\right)|\tau_m \text{ for each } l_k \in \mathcal{S}\}\|\big) \tag{13}$$

where ρ refers to the Pearson correlation used to compare couples of $\mathcal{J}_{\tau_m}\left(\epsilon_{NN}, \mathcal{S}\right)$. Throughout our analysis we experimented with similar correlation metrics, such as Spearman correlation and simple vector distance – similarly to [27] –, obtaining similar results. Therefore, to avoid redundancy we here show only the Pearson correlation results. Throughout our experiments, we consider a simple *min-max* normalisation process, scaling the scores to the range $[0, 1]$.

As our aim is to obtain a measure of similarity between LPEs applied over the same set of samples, we can average the correlation scores ρ obtained for each label τ_m over the set of labels $\mathcal{T}$. Therefore, we mathematically define the correlation score of two LPEs, putting together Eqs. (2), (12) and (13) as:

$$\mathcal{C}\left(\epsilon_{NN}\left(\mathcal{S}\right), \epsilon'_{NN}\left(\mathcal{S}\right)\right) = \frac{1}{M} \cdot \sum_{m=1}^{M} \rho\left(\|\mathcal{J}_{\tau_m}\left(\epsilon_{NN}, \mathcal{S}\right)\|, \|\mathcal{J}_{\tau_m}\left(\epsilon'_{NN}, \mathcal{S}\right)\|\right) \tag{14}$$

where M is the total number of labels, i.e., moral principles, belonging to $\mathcal{T}$.

4   Experiments

In this section, we present the setup and results of our experiments. We present the model training details and its obtained performance in Sect. 4.1. We then focus on the comparison between the available LPEs, showing the correlation between their explanations in Sect. 4.2. Section 4.3 analyses how correlation scores are affected by the selected aggregation mechanism $\mathcal{A}$. Finally, in Sect. 4.4 we analyse the extent to which LPEs explanations are aligned with human notions of moral values.

4.1   Model Training

We follow state-of-the-art approaches for dealing with morality classification task [9,24]. Thus, we treat the morality classification problem as a multi-class multi-label classification task, leveraging BERT as the LLM to be optimised [15]. We define one NN model for each MFTC split and optimise its parameters over the 70% of tweets, leaving the remaining 30% for testing purposes. However, conversely from recent approaches, we here do not rely on the *sequential training* paradigm, but rather train each model solely on the MFTC split at hand. Indeed, in our experiments, we do not aim at obtaining strong transferability between domains, but rather we focus on analysing LPEs behaviour.

We leverage the pre-trained *bert-base-uncased* model – available in the Hugging Face python library[6] – as the starting point of our training process. Each model is trained for 3 epochs using a standard binary cross entropy loss [46], a learning rate of 5×10^{-5}, a batch size of 16 and a maximum sequence length of 64. We keep track of the macro F1-score for each model to identify its performance over the test samples. Table 1 shows the performance of the trained BERT model.

Table 1. BERT performance over MFTC datasets.

	ALM	BLM	BLT	DAV	ELE	MT	SND	TOT
F_1 score	63.04%	82.59%	64.51%	88.12%	63.14%	52.16%	56.85%	69.10%

4.2 Are Local Post-hoc Explainers Aligned?

We analyse the extent to which different LPEs are aligned in their process of identifying impactful concepts for the underlying NN model. With this aim, we train a BERT model over a specific dataset (following the approach described in Sect. 4.1) and compute the pairwise correlation $\mathcal{C}\left(\epsilon_{NN}\left(\mathcal{S}\right), \epsilon'_{NN}\left(\mathcal{S}\right)\right)$ (as described in Sect. 3) for each pair of LPE in the selected set. To avoid issues caused by model overfitting over the training set, which would render explanations unreliable, we apply each ϵ_{NN} over the test set of the selected dataset.

Using the pairwise correlation values we construct the correlation matrices shown in Figs. 1 and 2, which highlight how there exist a very weak correlation score between most LPEs over different datasets. Here, it is interesting to notice how, there exists few specific couples or clusters of LPE which highly correlate with each other. For example, GS, GI and LRP show moderate to high correlation score, mainly due to their reliance on computing the gradient of the prediction to identify impactful concepts. However, this is not the case for all LPE couples relying on similar approaches. For example, GI and gradient integration – HESS in the matrices – show little to no correlation, although they both are gradient-based approach for producing local explanations. Similarly, SHAP and LIME show no correlation even if they both rely on input perturbation and are considered the state-of-the-art.

Figures 1 and 2 highlight how the vast majority of LPE pairs show very small to no correlation at all, exposing how there exists a disagreement between the selected approaches. This finding represents a fundamental result of our study, as it highlights how there is no accordance between LPE even when they are applied

---

[6] https://github.com/huggingface.

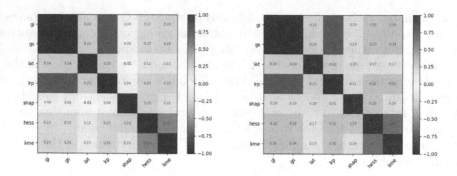

Fig. 1. $\mathcal{C}\left(\epsilon_{NN}\left(\mathcal{S}\right),\epsilon'_{NN}\left(\mathcal{S}\right)\right)$ using average aggregation (left) and absolute average aggregation (right) as $\mathcal{A}$ over the BLM dataset.

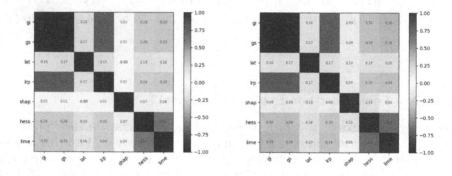

Fig. 2. $\mathcal{C}\left(\epsilon_{NN}\left(\mathcal{S}\right),\epsilon'_{NN}\left(\mathcal{S}\right)\right)$ using average aggregation (left) and absolute average aggregation (right) as $\mathcal{A}$ over the ELE dataset.

to the same model and dataset. The reason behind such large discrepancies among LPE might be various, but mostly bear down to the following:

- Few of the LPE considered in the literature do not represent reliable solutions for identifying the reasoning principles of LLMs.
- Each of the uncorrelated LPEs highlight a different set or subset of reasoning principles of the underlying model.

Therefore, our results show how it is also complex to identify a set of fair and reliable metrics to spot the best LPE or even reliable LPEs, as they seem to gather uncorrelated explanations. Similar results to the ones shown in Figs. 1 and 2 are obtained for all dataset splits and are made available at https://tinyurl.com/QU4RR3L.

4.3 How Does Impact Scores Aggregation Affect Correlation?

Since our LPE correlation metric is dependent on $\mathcal{A}$, we here analyse how the selection of different aggregation strategies impacts the correlation between

LPEs. To understand the impact of $\mathcal{A}$ on $\mathcal{C}$, we plot the correlation matrices for a single dataset, varying the aggregation approach, thus obtaining the four correlation matrices shown in Fig. 3.

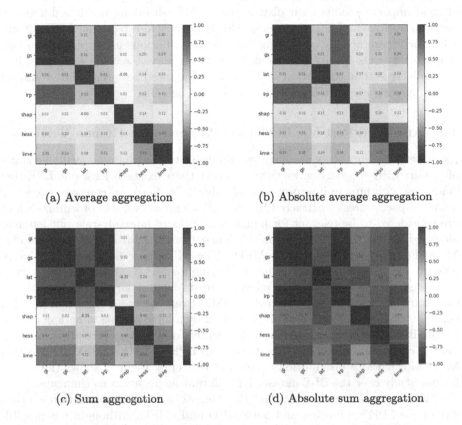

(a) Average aggregation   (b) Absolute average aggregation

(c) Sum aggregation   (d) Absolute sum aggregation

Fig. 3. $\mathcal{C}\left(\epsilon_{NN}\left(\mathcal{S}\right), \epsilon'_{NN}\left(\mathcal{S}\right)\right)$ using different aggregations over the ALM dataset.

From Figs. 3c and 3d it is possible to notice how there exists a strong correlation between different LPEs. This results seems to be in contrast with the results found in Sect. 4.2. However, the reason behind the strong correlation achieved when relying on summation aggregation is not caused by the actual correlation between explanations, but rather on the susceptibility of summation to tokens frequency. Indeed, since the summation aggregation approaches do not take into account the occurrence frequency of lemmas in $\mathcal{S}$, they tend to overestimate the relevance of popular concepts. Intuitively, using this aggregations, a rather impactless lemma appearing 5000 times would obtain a global impact higher than a very impactful lemma appearing only 10 times. These results highlight the importance of relying on average based aggregation approaches when considering to construct global explanations from the LPE outputs.

Figure 3 also highlights how leveraging the absolute value of LPEs incurs in higher correlation scores. The reason behind such a phenomenon is to be found in the impact scores distributions. Indeed, while true local impact scores are distributed over the set of real numbers $\mathbb{R}$, computing the absolute value of local impacts j shifts their distribution to $\mathbb{R}^+$, shrinking possible differences between positive and negative scores. Moreover, it is also true that LPE outputs rely much more heavily on scoring positive contributions using positive impact scores, and tends to give less focus to negative impact scores. Therefore, it is generally true that the output of LPEs is unbalanced towards positive impact scores, making negative impact scores mostly negligible.

4.4   Are Local Post-hoc Explainers Aligned with Human Values?

As our experiments show the huge variability in the response by available state-of-the-art LPE approaches, we check whether there exists at least one LPE that is aligned with human interpretation of values. To do so, we compare the set of global impact scores $\mathcal{J}$ extracted by each LPE against two sets of lemmas which are considered to be relevant for humans. The set of humanly-relevant lemmas, along with their impact scores are obtained from the MFD and the extended Moral Foundations Dictionary (eMFD). The MFD is a dictionary of relevant lemmas for the set of moral values belonging to MFTC. Such a dictionary is generated manually by picking relevant words from a large list of words for each foundation value [21]. Meanwhile, eMFD represents an extension of MFD constructed from text annotations generated by a large sample of human coders.

Similar to the comparison of Sect. 4.2, we rely on Pearson correlation, measuring the correlation coefficient $\mathcal{C}$ between each LPE and MFD or eMFD, treating MFD as if it was a distribution of relevant concepts. Figure 4 shows the results for our study over the BLT dataset for different aggregation mechanisms.

Alarmingly, the results show how there exists no positive correlation between any of the LPE approaches and both MFD and eMFD. Although it is possible that the trained model learns relevant concepts that are specific to the target domain – i.e., BLT in Fig. 4 – it is concerning how strongly uncorrelated LPE and human interpretation of values are. Indeed, while BERT may focus on a few specific concepts which are not human-like, it is assumed and proven to be effective in learning human-like concepts over the majority of NLP tasks. Especially if we consider our BERT model to be only fine-tuned on the target domain, it is very unreasonable to assume these results to be caused by BERT learning concepts that are not aligned with human values. Rather, it is fairly reasonable to deduce that the considered LPEs are far from being completely aligned to the real reasoning process of the underlying BERT model, thus incurring in such high discrepancy with human-labeled moral values.

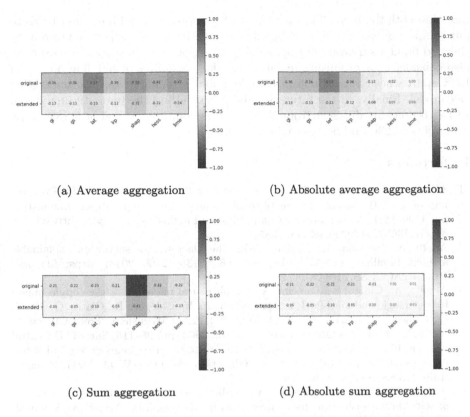

(a) Average aggregation   (b) Absolute average aggregation

(c) Sum aggregation   (d) Absolute sum aggregation

Fig. 4. $\mathcal{C}\left(\epsilon_{NN}\left(\mathcal{S}\right), MFD\right)$ using different aggregations over the BLT dataset.

5   Conclusion and Future Work

We propose a new approach for the comparison among state-of-the-art local post-hoc explanation mechanisms, aiming at identifying the extent to which their extracted explanations correlate. We rely on a novel framework for extracting and comparing global impact scores from local explanations obtained from LPEs, and apply such a framework over the MFTC dataset. Our experiments show how most LPEs explanations are far from being mutually correlated when LPEs are applied over a large set of input samples. These results highlight what we called the "quarrel" among state-of-the-art local explainers, apparently caused by each of them focusing on a different set or subset of relevant concepts, or imposing a different distribution on top of them. Further, we compare the impact scores distribution obtained from each LPEs with a set of human-made dictionaries. Our experiments alarmingly show how there exists no correlation between LPE outputs and the concepts considered to be salient by humans. Therefore, our experiments highlight the current fragility of xAI approaches for NLP.

Our proposal is a solid starting point for the exploration of the reliability and soundness of xAI approaches in NLP. In our future work, we aim at investigating more in-depth the issue of robustness of LPE approaches, adding novel LPEs to our comparison such as [16], and aiming at identifying if it is possible to rely on them to build a surrogate of the model from a global perspective. Moreover, we also consider as a promising research line the possibility of building on top of LPE approaches so as to obtain reliable global explanations of the underlying NN model. Finally, in the future we aim at extending the in-depth analysis of LPEs to domains different from NLP, such as computer vision [4,5,13], graph processing [6,23], and neuro-symbolic approaches [3,7].

References

1. Abnar, S., Zuidema, W.: Quantifying attention flow in transformers. In: Proceedings of the 58th Annual Meeting of the Association for Computational Linguistics, pp. 4190–4197. Association for Computational Linguistics, July 2020. https://doi.org/10.18653/v1/2020.acl-main.385
2. Adadi, A., Berrada, M.: Peeking inside the black-box: a survey on explainable artificial intelligence (XAI). IEEE Access 6, 52138–52160 (2018). https://doi.org/10.1109/ACCESS.2018.2870052
3. Agiollo, A., Ciatto, G., Omicini, A.: Graph neural networks as the copula mundi between logic and machine learning: a roadmap. In: Calegari, R., Ciatto, G., Denti, E., Omicini, A., Sartor, G. (eds.) WOA 2021–22nd Workshop "From Objects to Agents". CEUR Workshop Proceedings, vol. 2963, pp. 98–115. Sun SITE Central Europe, RWTH Aachen University, October 2021. http://ceur-ws.org/Vol-2963/paper18.pdf, 22nd Workshop "From Objects to Agents" (WOA 2021), Bologna, Italy, 1–3 September 2021. Proceedings
4. Agiollo, A., Ciatto, G., Omicini, A.: Shallow2Deep: restraining neural networks opacity through neural architecture search. In: Calvaresi, D., Najjar, A., Winikoff, M., Främling, K. (eds.) Explainable and Transparent AI and Multi-agent Systems. Third International Workshop, EXTRAAMAS 2021. LNCS, vol. 12688, pp. 63–82. Springer, Cham (2021). https://doi.org/10.1007/978-3-030-82017-6_5
5. Agiollo, A., Omicini, A.: Load classification: a case study for applying neural networks in hyper-constrained embedded devices. Appl. Sci. 11(24) (2021). https://doi.org/10.3390/app112411957, https://www.mdpi.com/2076-3417/11/24/11957, Special Issue "Artificial Intelligence and Data Engineering in Engineering Applications"
6. Agiollo, A., Omicini, A.: GNN2GNN: graph neural networks to generate neural networks. In: Cussens, J., Zhang, K. (eds.) Uncertainty in Artificial Intelligence. Proceedings of Machine Learning Research, vol. 180, pp. 32–42. ML Research Press, August 2022. https://proceedings.mlr.press/v180/agiollo22a.html, Proceedings of the Thirty-Eighth Conference on Uncertainty in Artificial Intelligence, UAI 2022, 1–5 August 2022, Eindhoven, The Netherlands
7. Agiollo, A., Rafanelli, A., Omicini, A.: Towards quality-of-service metrics for symbolic knowledge injection. In: Ferrando, A., Mascardi, V. (eds.) WOA 2022–23rd Workshop "From Objects to Agents", CEUR Workshop Proceedings, vol. 3261, pp. 30–47. Sun SITE Central Europe, RWTH Aachen University, November 2022. http://ceur-ws.org/Vol-3261/paper3.pdf
8. Ali, A., Schnake, T., Eberle, O., Montavon, G., Müller, K.R., Wolf, L.: XAI for transformers: better explanations through conservative propagation. In: International Conference on Machine Learning, pp. 435–451. PMLR (2022). https://proceedings.mlr.press/v162/ali22a.html

9. Alshomary, M., Baff, R.E., Gurcke, T., Wachsmuth, H.: The moral debater: a study on the computational generation of morally framed arguments. In: Proceedings of the 60th Annual Meeting of the Association for Computational Linguistics (Volume 1: Long Papers), pp. 8782–8797. Association for Computational Linguistics, Dublin, Ireland, May 2022. https://doi.org/10.18653/v1/2022.acl-long.601
10. Bach, S., Binder, A., Montavon, G., Klauschen, F., Müller, K.R., Samek, W.: On pixel-wise explanations for non-linear classifier decisions by layer-wise relevance propagation. PloS ONE **10**(7), e0130140 (2015). https://doi.org/10.1371/journal.pone.0130140
11. Bender, E.M., Gebru, T., McMillan-Major, A., Shmitchell, S.: On the dangers of stochastic parrots: can language models be too big? In: Proceedings of the 2021 ACM Conference on Fairness, Accountability, and Transparency, pp. 610–623 (2021). https://doi.org/10.1145/3442188.3445922
12. Brown, T., et al.: Language models are few-shot learners. In: Advances in Neural Information Processing Systems, vol. 33, pp. 1877–1901 (2020). https://dl.acm.org/doi/abs/10.5555/3495724.3495883
13. Buhrmester, V., Münch, D., Arens, M.: Analysis of explainers of black box deep neural networks for computer vision: a survey. Mach. Learn. Knowl. Extr. **3**(4), 966–989 (2021). https://doi.org/10.3390/make3040048
14. Danilevsky, M., Qian, K., Aharonov, R., Katsis, Y., Kawas, B., Sen, P.: A survey of the state of explainable AI for natural language processing. In: Proceedings of the 1st Conference of the Asia-Pacific Chapter of the Association for Computational Linguistics and the 10th International Joint Conference on Natural Language Processing, pp. 447–459. Association for Computational Linguistics, Suzhou, China, December 2020. https://aclanthology.org/2020.aacl-main.46
15. Devlin, J., Chang, M.W., Lee, K., Toutanova, K.: BERT: pre-training of deep bidirectional transformers for language understanding. In: Proceedings of the 2019 Conference of the North American Chapter of the Association for Computational Linguistics: Human Language Technologies, vol. 1 (Long and Short Papers), pp. 4171–4186. Association for Computational Linguistics, Minneapolis, MN, USA, June 2019. https://doi.org/10.18653/v1/N19-1423
16. Främling, K., Westberg, M., Jullum, M., Madhikermi, M., Malhi, A.: Comparison of contextual importance and utility with LIME and Shapley values. In: Calvaresi, D., Najjar, A., Winikoff, M., Främling, K. (eds.) Explainable and Transparent AI and Multi-agent Systems - Third International Workshop, EXTRAAMAS 2021. LNCS, vol. 12688, pp. 39–54. Springer, Cham (2021). https://doi.org/10.1007/978-3-030-82017-6_3
17. Guidotti, R., Monreale, A., Ruggieri, S., Turini, F., Giannotti, F., Pedreschi, D.: A survey of methods for explaining black box models. ACM Comput. Surv. (CSUR) **51**(5), 1–42 (2018). https://doi.org/10.1145/3236009
18. Hailesilassie, T.: Rule extraction algorithm for deep neural networks: a review. Int. J. Comput. Sci. Inf. Secur. **14**(7), 376–381 (2016). https://www.academia.edu/28181177/Rule_Extraction_Algorithm_for_Deep_Neural_Networks_A_Review
19. Hao, T., Li, X., He, Y., Wang, F.L., Qu, Y.: Recent progress in leveraging deep learning methods for question answering. Neural Comput. Appl. **34**(4), 2765–2783 (2022). https://doi.org/10.1007/s00521-021-06748-3
20. Hoover, J., et al.: Moral foundations Twitter corpus: a collection of 35k tweets annotated for moral sentiment. Soc. Psychol. Pers. Sci. **11**(8), 1057–1071 (2020). https://doi.org/10.1177/1948550619887662

21. Hopp, F.R., Fisher, J.T., Cornell, D., Huskey, R., Weber, R.: The extended moral foundations dictionary (eMFD): development and applications of a crowd-sourced approach to extracting moral intuitions from text. Behav. Res. Methods **53**, 232–246 (2021). https://doi.org/10.3758/s13428-020-01433-0

22. Ibrahim, M., Louie, M., Modarres, C., Paisley, J.: Global explanations of neural networks: mapping the landscape of predictions. In: Proceedings of the 2019 AAAI/ACM Conference on AI, Ethics, and Society, pp. 279–287 (2019). https://doi.org/10.1145/3306618.3314230

23. Jaume, G., et al.: Quantifying explainers of graph neural networks in computational pathology. In: IEEE Conference on Computer Vision and Pattern Recognition, CVPR 2021, Virtual, 19–25 June 2021, pp. 8106–8116. Computer Vision Foundation/IEEE (2021). https://doi.org/10.1109/CVPR46437.2021.00801

24. Kiesel, J., Alshomary, M., Handke, N., Cai, X., Wachsmuth, H., Stein, B.: Identifying the human values behind arguments. In: Proceedings of the 60th Annual Meeting of the Association for Computational Linguistics (vol. 1: Long Papers), pp. 4459–4471 (2022). https://doi.org/10.18653/v1/2022.acl-long.306

25. Kindermans, P.J., et al.: The (un)reliability of saliency methods. In: Samek, W., Montavon, G., Vedaldi, A., Hansen, L., Müller, K.R. (eds.) Explainable AI: Interpreting, Explaining and Visualizing Deep Learning, pp. 267–280. Springer, Cham (2019). https://doi.org/10.1007/978-3-030-28954-6_14

26. Kokalj, E., Škrlj, B., Lavrač, N., Pollak, S., Robnik-Šikonja, M.: BERT meets Shapley: extending SHAP explanations to transformer-based classifiers. In: Proceedings of the EACL Hackashop on News Media Content Analysis and Automated Report Generation, pp. 16–21 (2021)

27. Liscio, E., et al.: What does a text classifier learn about morality? An explainable method for cross-domain comparison of moral rhetoric. In: Proceedings of the 61st Annual Meeting of the Association for Computational Linguistics, pp. 1–12, Toronto (2023, to appear)

28. Liscio, E., Dondera, A., Geadau, A., Jonker, C., Murukannaiah, P.: Cross-domain classification of moral values. In: Findings of the Association for Computational Linguistics: NAACL 2022, pp. 2727–2745. Association for Computational Linguistics, Seattle, United States, July 2022. https://doi.org/10.18653/v1/2022.findings-naacl.209

29. Liu, G., et al.: Medical-VLBERT: medical visual language BERT for COVID-19 CT report generation with alternate learning. IEEE Trans. Neural Netw. Learn. Syst. **32**(9), 3786–3797 (2021). https://doi.org/10.1109/TNNLS.2021.3099165

30. Loh, W.Y.: Fifty years of classification and regression trees. Int. Stat. Rev. **82**(3), 329–348 (2014). https://doi.org/10.1111/insr.12016

31. Lundberg, S.M., Lee, S.I.: A unified approach to interpreting model predictions. In: Advances in Neural Information Processing Systems, vol. 30 (2017). https://proceedings.neurips.cc/paper_files/paper/2017/file/8a20a8621978632d76c43dfd28b67767-Paper.pdf

32. Luo, S., Ivison, H., Han, C., Poon, J.: Local interpretations for explainable natural language processing: a survey. arXiv preprint arXiv:2103.11072 (2021)

33. Madsen, A., Reddy, S., Chandar, S.: Post-hoc interpretability for neural NLP: a survey. ACM Comput. Surv. **55**(8), 1–42 (2022). https://doi.org/10.1145/3546577

34. Otter, D.W., Medina, J.R., Kalita, J.K.: A survey of the usages of deep learning for natural language processing. IEEE Trans. Neural Netw. Learn. Syst. **32**(2), 604–624 (2021). https://doi.org/10.1109/TNNLS.2020.2979670

35. Raffel, C., et al.: Exploring the limits of transfer learning with a unified text-to-text transformer. J. Mach. Learn. Res. **21**(1), 5485–5551 (2020). https://dl.acm.org/doi/abs/10.5555/3455716.3455856
36. Ramachandran, D., Parvathi, R.: Analysis of Twitter specific preprocessing technique for tweets. Procedia Comput. Sci. **165**, 245–251 (2019). https://doi.org/10.1016/j.procs.2020.01.083
37. Ribeiro, M.T., Singh, S., Guestrin, C.: "Why should I trust you?": Explaining the predictions of any classifier. In: Proceedings of the 22nd ACM SIGKDD International Conference on Knowledge Discovery and Data Mining, pp. 1135–1144 (2016). https://doi.org/10.18653/v1/N16-3020
38. Samek, W., Montavon, G., Lapuschkin, S., Anders, C.J., Müller, K.R.: Explaining deep neural networks and beyond: a review of methods and applications. Proc. IEEE **109**(3), 247–278 (2021). https://doi.org/10.1109/JPROC.2021.3060483
39. Sheikhalishahi, S., et al.: Natural language processing of clinical notes on chronic diseases: systematic review. JMIR Med. Inform. **7**(2), e12239 (2019). https://doi.org/10.2196/12239
40. Stahlberg, F.: Neural machine translation: a review. J. Artif. Intell. Res. **69**, 343–418 (2020). https://doi.org/10.1613/jair.1.12007
41. Sundararajan, M., Taly, A., Yan, Q.: Axiomatic attribution for deep networks. In: Precup, D., Teh, Y.W. (eds.) International Conference on Machine Learning. Proceedings of Machine Learning Research, vol. 70, pp. 3319–3328. PMLR, August 2017. https://proceedings.mlr.press/v70/sundararajan17a.html
42. Tay, Y., Bahri, D., Metzler, D., Juan, D.C., Zhao, Z., Zheng, C.: Synthesizer: rethinking self-attention for transformer models. In: Meila, M., Zhang, T. (eds.) Proceedings of the 38th International Conference on Machine Learning. Proceedings of Machine Learning Research, vol. 139, pp. 10183–10192. PMLR, July 2021. https://proceedings.mlr.press/v139/tay21a.html
43. Warstadt, A., Singh, A., Bowman, S.R.: Neural network acceptability judgments. Trans. Assoc. Comput. Linguist. **7**, 625–641 (2019). https://doi.org/10.1162/tacl_a_00290
44. Wu, Z., Nguyen, T.S., Ong, D.C.: Structured self-attention weights encode semantics in sentiment analysis. In: Proceedings of the Third Blackbox NLP Workshop on Analyzing and Interpreting Neural Networks for NLP, pp. 255–264. Association for Computational Linguistics (2020). https://doi.org/10.18653/v1/2020.blackboxnlp-1.24
45. Zhang, L., Wang, S., Liu, B.: Deep learning for sentiment analysis: a survey. Wiley Interdisc. Rev. Data Mining Knowl. Discov. **8**(4) (2018). https://doi.org/10.1002/widm.1253
46. Zhang, Z., Sabuncu, M.: Generalized cross entropy loss for training deep neural networks with noisy labels. In: Bengio, S., Wallach, H., Larochelle, H., Grauman, K., Cesa-Bianchi, N., Garnett, R. (eds.) Advances in Neural Information Processing Systems, vol. 31. Curran Associates, Inc. (2018). https://proceedings.neurips.cc/paper_files/paper/2018/file/f2925f97bc13ad2852a7a551802feea0-Paper.pdf
47. Zini, J.E., Awad, M.: On the explainability of natural language processing deep models. ACM Comput. Surv. **55**(5), 1–31 (2022). https://doi.org/10.1145/3529755

Bottom-Up and Top-Down Workflows for Hypercube- And Clustering-Based Knowledge Extractors

Federico Sabbatini[1]([⊠])[iD] and Roberta Calegari[2][iD]

[1] Department of Pure and Applied Sciences (DiSPeA), University of Urbino "Carlo Bo", Urbino, Italy
f.sabbatini1@campus.uniurb.it
[2] Department of Computer Science and Engineering (DISI), University of Bologna, Bologna, Italy
roberta.calegari@unibo.it

Abstract. Machine learning opaque models, currently exploited to carry out a wide variety of supervised and unsupervised learning tasks, are able to achieve impressive predictive performances. However, they act as black boxes (BBs) from the human standpoint, so they cannot be entirely trusted in critical applications unless there exists a method to extract symbolic and human-readable knowledge out of them.

In this paper we analyse a recurrent design adopted by symbolic knowledge extractors for BB predictors—that is, the creation of rules associated with hypercubic input space regions. We argue that this kind of partitioning may lead to suboptimum solutions when the data set at hand is sparse, high-dimensional, or does not satisfy symmetric constraints. We then propose two different knowledge-extraction workflows involving clustering approaches, highlighting the possibility to outperform existing knowledge-extraction techniques in terms of predictive performance on data sets of any kind.

Keywords: explainable AI · symbolic knowledge extraction · clustering

1 Introduction

Machine learning (ML) models in general – and (deep) artificial neural networks in particular – are nowadays exploited to draw predictions in almost every application area [26]. However, when facing *critical* domains – e.g., involving human health, wealth, or freedom – ML models behaving as opaque predictors are not an acceptable choice. The *opaque* nature of these models makes them unintelligible for humans and this is the reason why they are called *black boxes* (BBs). Nonetheless, explainability can be obtained from BBs via several strategies [17]. For instance, one can rely uniquely on *interpretable* models [27], or build explanations by applying reverse engineering to the BB behaviour [21]. The former

D. Calvaresi et al. (Eds.): EXTRAAMAS 2023, LNAI 14127, pp. 116–129, 2023.
https://doi.org/10.1007/978-3-031-40878-6_7

option is not always practicable, since interpretable models as linear regressors and shallow decision trees are not always prediction-effective as more complex models—for instance, random forests or deep artificial neural networks. On the other hand, the latter approach allows users to combine the impressive predictive capabilities of opaque predictors with the human readability proper of symbolic models.

The majority of present literature offers a wide variety of procedures explicitly designed to extract symbolic knowledge from opaque ML classifiers [2,13,14,23,24,40,43, for instance]. A smaller set of procedures is dedicated to BB regressors [20,28,32,35,38,39,41, to cite some], whereas few exceptions are able to consider both categories [6,7,9,11,22,36]. A large amount of available techniques depend on the existence of software libraries supporting ML in general (e.g., Python's Scikit-Learn[1] [25]) and symbolic knowledge extraction in particular (for instance, PSyKE[2] [10,29,31,33]).

Unfortunately, any method offers peculiar advantages, but at the same time, it is subject to drawbacks and limitations. In the following we focus on the issues deriving from the extraction of a particular kind of rules from BB classifiers and regressors. More in detail, we observe that opaque predictors in general – and regressors in particular – are often explained via a human-interpretable partitioning of the input space. A typical design choice is to identify hypercubic regions of the input feature space enclosing similar instances and then to associate symbolic knowledge to each region, for instance in the form of *first-order logic* rules [20,28,32]. We agree that rules associated with hypercubic regions are the best choice from the human-readability perspective since they enable the description of an input space region in terms of constraints on single dimensions (e.g., $0.3 < X < 0.6, 0.5 < Y < 0.75$ for a hypercube in a 2-dimensional space having features X and Y). However, this solution may lead to the creation of suboptimum clusters in several scenarios, for instance, if the partitioning into hypercubes is performed following some sort of top-down symmetric procedure on asymmetric data sets, or if the partitioning is bottom-up and performed on sparse data sets. In this paper we suggest exploiting clustering techniques on the data set used to train the BB before extracting knowledge from it, in order to preemptively find and distinguish relevant input regions with the corresponding boundaries. This theoretically allows extractors to: *(i)* automatically tune the number of output rules w.r.t. the number of relevant regions found; *(ii)* give priority to more relevant regions, for instance those containing more input instances or having the largest volume; *(iii)* avoid unsupervised partitioning of the input feature space, otherwise leading to suboptimum solutions in terms of readability and/or fidelity. Given that our proposed workflows are based on the training data sample distribution and require no knowledge of the adopted opaque predictor, it may be possible to build pedagogical knowledge-extraction algorithms adhering to them.

---

[1] https://scikit-learn.org/stable/index.html.
[2] https://github.com/psykei/psyke-python.

Accordingly, in Sect. 2 an overview of symbolic knowledge extraction in general and some techniques in particular is provided. In Sect. 3 the main drawbacks of existing knowledge-extraction algorithms based on hypercubes are highlighted and discussed. We then propose novel clustering-based workflows for knowledge extraction in Sect. 4 and our work is finally summarised in Sect. 5.

2 Related Works

A predictive model can be defined as *interpretable* if human users are able to easily understand its behaviour and outputs [12]. Since the majority of modern ML predictors store the knowledge acquired during their training phase in a *sub-symbolic* way, they behave and appear to the human perspective as unintelligible black boxes. The explainable artificial intelligence community has proposed a variety of methods to enrich BB predictions with corresponding interpretations/explanations without renouncing their superior predictive performance. Usually, the proposed methods consist of creating an interpretable, mimicking model by inspecting the underlying BB in terms of internal behaviour and/or input/output relationships. For instance, REFANN analyses the architecture of neural network regressors with one hidden layer to obtain information about the internal parameters and thus build human-readable *if-then* rules having a linear combination of the input features as postconditions [39]. This kind of technique is called *decompositional*. On the other hand, when the internal structure of the BB is not considered to build explanations, algorithms are classified as *pedagogical*.

Symbolic knowledge-extraction techniques are currently applied in several critical areas, such as healthcare, credit-risk evaluation, intrusion detection systems, and many others [3–5,8,16,18,19,34,37,42].

2.1 ITER

The ITER algorithm [20] is a pedagogical technique to extract symbolic knowledge from BB regressors. ITER induces a hypercubic partitioning of the input feature space following a bottom-up strategy, starting with points in the multidimensional space and expanding them until the final hypercubic output regions. According to the ITER design, all the produced hypercubes are non-overlapping and they do not exceed the input feature space.

After the hypercube expansion ITER associates an *if-then* rule to each cube, selecting as a postcondition the mean output value of all the instances contained in the cube. This behaviour may be relaxed to support classification tasks or regression tasks having outputs described through linear laws by adopting the generalisation proposed in [30].

The algorithm's main advantage is to be capable of constructing hypercubes having different sides' lengths. However, especially when dealing with high-dimensional data sets, it may present several criticalities related to the hypercubes' expansion.

2.2   GridEx and GridREx

The GridEx algorithm [32] is a different pedagogical technique to extract symbolic knowledge from BB regressors. It is applicable under the same conditions of ITER and outputs the same kind of knowledge, but they differ in the strategy adopted during the input space partitioning. GridEx is not a bottom-up algorithm; conversely, it follows a top-down strategy, starting from the whole input space and recursively partitioning it into smaller hypercubic regions, according to a user-defined threshold acting as a trade-off criterion between readability (in terms of the number of extracted rules) and fidelity of the output model (intended as its ability to mimic the underlying BB).

Also for GridEx it is possible to adopt a generalisation enabling its application for classification tasks [30]. On the other hand, the GridREx algorithm [28] is the extension of GridEx providing linear combinations of the input features as outputs associated with the identified hypercubic relevant regions.

Amongst the common advantages of GridEx and GridREx there is the ability to automatically refine the output regions according to the provided threshold, as well as to perform a merging step after each split, when possible. In particular, the merging step consists of the pairwise unification of adjacent hypercubes to reduce the number of output rules (to enhance human readability) and it is based on the similarity between the samples included in each cube (to avoid a predictive performance worsening). It is useful since the splitting phase may create excessive amounts of disjoint but adjacent, similar regions.

3   Limitations of Existing Knowledge-Extraction Methods

In order to point out the main limitations affecting existing symbolic knowledge-extraction techniques based on hypercubic partitioning a clear insight into how the partitioning is performed has fundamental importance. For this reason the behaviour of ITER, GridEx and GridREx is further detailed in the following with the aim of spotting their major drawbacks and providing possible solutions. Examples of these knowledge-extraction algorithms applied to real-world data sets are also reported to strengthen our argumentation. ITER's examples are considered on the Istanbul Stock Exchange data set[3] [1], describing a regression task with 7 continuous input features plus another input feature representing a timestamp. Examples for GridEx are based on the Wine Quality data set[4] [15], composed of 13 continuous input features.

3.1   ITER

The ITER procedure is exemplified for the Istanbul Stock Exchange data set in Fig. 1, where only the 2 most relevant input features are reported, i.e., the stock market return index of UK (FTSE) and the MSCI European index (EU). The

---

[3] https://archive.ics.uci.edu/ml/datasets/ISTANBUL+STOCK+EXCHANGE.
[4] https://archive.ics.uci.edu/ml/datasets/wine.

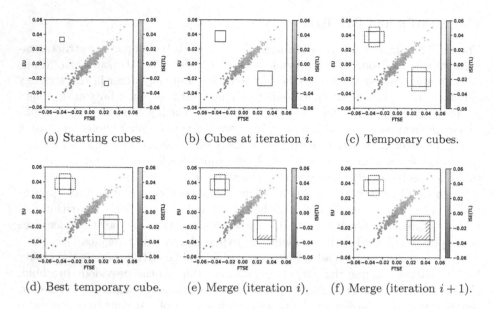

(a) Starting cubes.   (b) Cubes at iteration i.   (c) Temporary cubes.

(d) Best temporary cube.   (e) Merge (iteration i).   (f) Merge (iteration $i + 1$).

Fig. 1. Example of ITER hypercube expansion.

figure is a mere sketch depicting a possible undesired execution scenario of the algorithm, starting from random points belonging to negligible input regions.

During the execution of ITER a certain number of infinitesimally small hyper-cubes (i.e., multidimensional points) are created within the input feature space (cf. Fig. 1a) and they are iteratively expanded until some stopping criteria are met – i.e., the whole space is covered or it is not possible to expand the existing cubes nor to create new ones –, or after a specified amount of iterations.

The expansion follows this strategy, for a generic iteration i (cf. Fig. 1b):

1. build adjacent temporary cubes around the existing ones (2 temporary cubes per input dimension per existing cube; cf. Fig. 1c, temporary cubes are represented with dashed perimeter);
2. select the best temporary cube—i.e., the most similar to the corresponding adjacent existing cube (cf. Fig. 1d, the best temporary cube is highlighted with red perimeter);
3. merge the best temporary cube with the existing one (cf. Fig. 1e, the merged region is highlighted with red hatches);
4. repeat from step 1 for every successive iteration (cf. Fig. 1f).

The temporary cube selected to be merged may become instead a new independent cube if the similarity w.r.t. its adjacent cube is below a user-defined threshold. The number of starting cubes as well as the size of the temporary cubes and the maximum number of allowed iterations are other hyper-parameters to be provided by users. Conversely, the position of the initial cubes is randomly chosen, so it represents a source of non-determinism.

It is important to notice that at each algorithm iteration, only one hypercube amongst all the available temporary cubes is merged. The others are discarded and the majority of them are created again without modifications during the successive iteration. This may lead to an enormous waste of computational time and resources due to the repetition of (the same) useless calculations, other than the possibility to exceed the maximum allowed iterations without having convergence. The absence of convergence results in a non-exhaustive partitioning of the input feature space, which in turn implies the inability to predict output values of data samples belonging to regions that are not covered by a hypercube (i.e., there are no human-interpretable predictive rules associated with uncovered regions). Conversely, the coverage of the whole input feature space enables drawing predictions for any input instance.

Given the existence of the non-exhaustivity issue, it is of paramount importance the ability to focus on relevant input space subregions first, actually missing in the ITER design, since cubes are initialised randomly and expanded regardless of the subregion relevance. A naive notion of relevance for an input space subregion may be the amount of contained data set samples. If a training data set is representative of a certain task, it is admissible to envisage that if an input space subregion has no training instances, then probably there will not be data samples belonging to that same region to be predicted in the future, so the region is negligible, or at least not compelling. We highlight that it is not sufficient to build the starting cubes around existing training samples randomly chosen, since these may still be outliers. Conversely, the notion of region relevance should be intertwined with that of instance density and therefore the most relevant regions should be those containing more training instances. Since ITER is actually unaware of the sample density, we believe that this procedure could be improved by making the hypercube initialisation and expansion density-driven, or at least density-aware, without neglecting the pivotal similarity criterion of the original design.

3.2 GridEx and GridREx

The GridEx algorithm applied to the Wine Quality data set is visually shown in Fig. 2 to highlight its weaknesses. Only 2 out of 13 input features are reported in the figure to avoid a chaotic representation, i.e., the most relevant for the classification task.

GridEx considers the data point distribution during its execution, intended as the location of the data points inside the input feature space, but it neglects the output value of the training instances during the assignment of a priority to input feature space subregions. Indeed, it identifies 3 kinds of subregions:

negligible regions i.e., those without training instances belonging to them. These regions may be neglected without a noticeable impact on the overall extractor performance since the probability of existing instances enclosed by them is very low;

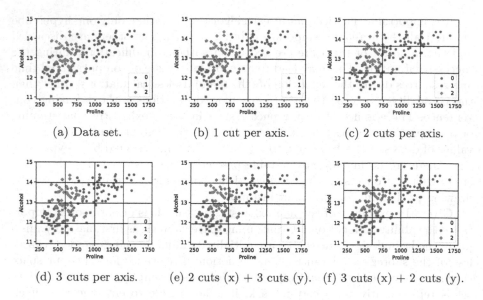

(a) Data set.   (b) 1 cut per axis.   (c) 2 cuts per axis.

(d) 3 cuts per axis.   (e) 2 cuts (x) + 3 cuts (y).   (f) 3 cuts (x) + 2 cuts (y).

Fig. 2. Example of GridEx hypercubic partitioning.

permament regions i.e., those containing training samples and for which the estimated predictive error is below the user-defined threshold. These regions have a satisfying predictive performance from the user standpoint and require no further partitioning;

eligible regions i.e., those containing training data but having an associated predictive error beyond the user-defined threshold. These regions require to be refined with further splitting since they hinder the overall predictive performance of the model.

By following this strategy the complete coverage of the input feature space is not granted. Nonetheless, very high coverage rates are achieved when predicting new unknown instances.

For the examples reported in Fig. 2 only actual GridEx instances performing a single iteration are considered. This means that the eligible regions do have not the possibility to be further refined. By observing Figs. 2b to 2f it is possible to notice that the cuts performed by GridEx are perpendicular to the axes and create for each dimension a set of partitions having the same size, even if the number of cuts may differ for each dimension (cf. Figs. 2e and 2f). To apply a different amount of cuts for each input dimension it is necessary to adopt the adaptive version of GridEx, selecting a number of slices that are proportional to the dimensions' relevance.

The symmetric strategy is the main disadvantage of GridEx, in its base and adaptive versions, and the same occurs with GridREx, since it follows the same hypercubic partitioning strategy of GridEx. This design choice may lead to a drop in the predictive performance if the identified hypercubes include portions of separated input regions. Given the symmetric nature of the algorithm, there is no certainty that the predictive error measured for the hypercubic regions can be reduced by augmenting the number of partitions, since a more fine-grained partitioning may lead in any case to a poor approximation on asymmetric data sets. We believe that GridEx and GridREx may be improved by performing slices that are not blindly symmetric, but somehow aware of the training data points' outputs. More in detail, cuts should avoid splitting into different regions training samples that are similar, as well as avoid creating regions containing too different training instances. To achieve this goal clustering may be performed before the cuts in order to identify different clusters of data and therefore avoid cuts in the proximity of the clusters' centroids. Conversely, cuts in correspondence of the intersection of two clusters should be encouraged.

4   Clustering-Based Approaches

Density and similarity estimates inside the training set input space may be obtained, for instance, via the application of clustering techniques to the available data. With reference to Fig. 1, three clusters may be identified: one enclosing the data points in the bottom-left part of the plot, one for the middle samples and the other for the top-right instances. On the other hand, for Fig. 2 the clusters may be defined according to the output expected classes and therefore also in this case three distinct clusters are identified.

An ideal extraction technique should be able to identify relevant clusters of data in a *fast* and *flawless* way, especially if clusters are linearly separable. An ideal procedure should also be able to approximate these clusters according to a human-interpretable format, for instance with hypercubic regions, enclosing the training data points without overlaps between approximated regions. We stress that this is only a desideratum, since real-world data sets seldom contain linearly separable classes of instances, but the design of a clustering-based knowledge extractor should take this ideal goal into account.

Bottom-up strategies like the one adopted by ITER generally result in time-consuming executions in the n-dimensional domain, with large n. This may result in a very slow convergence or even incomplete input space coverage. It is possible to fasten the ITER convergence by acting on the algorithm parameters, but at the expense of a coarser partitioning. This latter inconvenience is the same encountered by using GridEx, since it induces an equally spaced grid not always able to capture the properties of the data distribution inside the input feature space. If a grid cell only contains instances belonging to a single cluster, the predictive error will be small. Otherwise, it will be less or more large depending on the amount of contamination of each grid cell.

An optimum, fast bottom-up hypercubic-based extraction technique can be obtained by performing the following steps:

1. apply a clustering technique to the data set and therefore identify the different relevant regions;

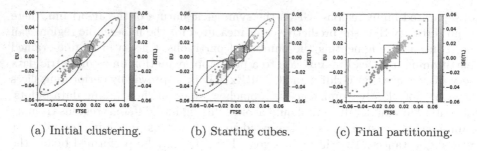

(a) Initial clustering.   (b) Starting cubes.   (c) Final partitioning.

Fig. 3. Example of an ideal bottom-up clustering- and hypercube-based symbolic knowledge extraction.

2. construct hypercubes to include the whole found regions, or a part of them, since this shape can be straightforwardly represented in a symbolic and human-readable format;
3. refine the hypercubes to enhance coverage and predictive performance of the explainable model, for instance by expanding the approximated regions;
4. remove the overlaps amongst the hypercubic regions, or impose an order to avoid ambiguity in evaluating the membership of instances to regions;
5. describe each hypercube in terms of the input features and then associate a corresponding human-interpretable output value to obtain the final explainable model.

The described workflow for an ideal bottom-up clustering-based extractor performing hypercubic partitioning of the input feature space is depicted in Fig. 3.

Conversely, an effective and fast top-down extraction technique based on hypercubic partitioning can be performed by substituting the middle steps of the workflow described above:

2. cut the input feature space in order to separate different clusters while avoiding spreading instances of a single cluster over multiple regions;
3. create hypercubic regions approximating the identified optimum cuts, avoiding overlapping cubes;
4. refine the hypercubes by recursively repeating the previous steps for each cube, to enhance the predictive performance of the explainable model.

The corresponding workflow for an ideal top-down knowledge-extraction procedure based on clustering is depicted in Fig. 4. Both Fig. 3 and Fig. 4 are conceptual sketches highlighting the fundamental steps of the aforementioned workflows, with the goal of guiding the future development of procedures adhering to the presented concepts.

4.1   Open Issues

Extraction techniques may surely benefit from cluster-aware partitioning methods. Accuracy in the selection of different clusters and in the construction

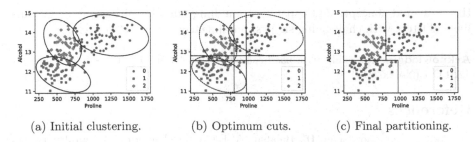

(a) Initial clustering.   (b) Optimum cuts.   (c) Final partitioning.

Fig. 4. Example of an ideal top-down clustering- and hypercube-based symbolic knowledge extraction.

of clustering-based hypercubes may enable the achievement of the following desiderata: extracting the minimum number of different predictive rules (one per cluster) having the lowest possible predictive error. However, a number of challenges arise from the aforementioned workflows. For instance:

i. How to select the correct number of clusters to identify, if unknown?
ii. How to handle outliers in the construction of hypercubes for the found regions and in deciding where to cut the input space?
iii. How to build hypercubes around clusters associated with overlapping hypercubic regions?
iv. How to discern amongst different clusters approximated by the same hypercubic region?

We believe that powerful strategies to describe non-trivial clusters may exploit *difference cubes* – e.g., regions of the input feature space having a non-cubic shape and described by the subtraction of cubic areas – and hierarchical clusters. We stress that the importance of adopting hypercubes to describe input regions depends on the possibility to define a hypercube in terms of single variables belonging to specific intervals, in a highly human-comprehensible form. This is not true when dealing with other representations—e.g., oblique rules, *M-of-N* rules.

5 Conclusions

In this paper we propose two clustering-based workflows to enhance the symbolic knowledge extraction from BB predictors in terms of computational complexity, fidelity, and predictive performance. The former is based on a bottom-up hypercubic approximation of the input feature space, resulting in a density-driven fast partitioning. Conversely, the latter is a top-down cutting of the input feature space providing hypercubic partitioning as well. Our methods can be exploited to build interpretable regions associated with human-readable logic rules based on an upstream clustering technique. In our future works we plan to implement and include in the PSyKE framework different knowledge extractors adhering to the presented concepts and capable of handling complex situations—e.g., outliers, clusters with challenging shapes, non-linearly separable clusters.

Acknowledgments. This work has been supported by the EU ICT-48 2020 project TAILOR (No. 952215).

References

1. Akbilgic, O., Bozdogan, H., Balaban, M.E.: A novel hybrid rbf neural networks model as a forecaster. Stat. Comput. **24**, 365–375 (2014)
2. Andrews, R., Geva, S.: RULEX & CEBP networks as the basis for a rule refinement system. In: Hallam, J. (ed.) Hybrid Problems, Hybrid Solutions, pp. 1–12. IOS Press (1995)
3. Azcarraga, A., Liu, M.D., Setiono, R.: Keyword extraction using backpropagation neural networks and rule extraction. In: The 2012 International Joint Conference on Neural Networks (IJCNN 2012), pp. 1–7. IEEE (2012). https://doi.org/10.1109/IJCNN.2012.6252618
4. Baesens, B., Setiono, R., De Lille, V., Viaene, S., Vanthienen, J.: Building credit-risk evaluation expert systems using neural network rule extraction and decision tables. In: Storey, V.C., Sarkar, S., DeGross, J.I. (eds.) ICIS 2001 Proceedings, pp. 159–168. Association for Information Systems (2001). http://aisel.aisnet.org/icis2001/20
5. Baesens, B., Setiono, R., Mues, C., Vanthienen, J.: Using neural network rule extraction and decision tables for credit-risk evaluation. Manage. Sci. **49**(3), 312–329 (2003). https://doi.org/10.1287/mnsc.49.3.312.12739
6. Barakat, N., Diederich, J.: Eclectic rule-extraction from support vector machines. Int. J. Comput. Inform. Eng. **2**(5), 1672–1675 (2008). https://doi.org/10.5281/zenodo.1055511
7. Benítez, J.M., Castro, J.L., Requena, I.: Are artificial neural networks black boxes? IEEE Trans. Neural Netw. **8**(5), 1156–1164 (1997). https://doi.org/10.1109/72.623216
8. Bologna, G., Pellegrini, C.: Three medical examples in neural network rule extraction. Phys. Medica **13**, 183–187 (1997). https://archive-ouverte.unige.ch/unige:121360
9. Breiman, L., Friedman, J., Stone, C.J., Olshen, R.A.: Classification and Regression Trees. CRC Press (1984)
10. Calegari, R., Sabbatini, F.: The PSyKE technology for trustworthy artificial intelligence **13796**, 3–16 (2023). https://doi.org/10.1007/978-3-031-27181-6_1, XXI International Conference of the Italian Association for Artificial Intelligence, AIxIA 2022, Udine, Italy, 28 November - 2 December, Proceedings (2022)
11. Castillo, L.A., González Muñoz, A., Pérez, R.: Including a simplicity criterion in the selection of the best rule in a genetic fuzzy learning algorithm. Fuzzy Sets Syst. **120**(2), 309–321 (2001). https://doi.org/10.1016/S0165-0114(99)00095-0
12. Ciatto, G., Calvaresi, D., Schumacher, M.I., Omicini, A.: An abstract framework for agent-based explanations in AI. In: El Fallah Seghrouchni, A., Sukthankar, G., An, B., Yorke-Smith, N. (eds.) 19th International Conference on Autonomous Agents and MultiAgent Systems, pp. 1816–1818. IFAAMAS (May 2020)
13. Craven, M.W., Shavlik, J.W.: Using sampling and queries to extract rules from trained neural networks. In: Machine Learning Proceedings 1994, pp. 37–45. Elsevier (1994). https://doi.org/10.1016/B978-1-55860-335-6.50013-1

14. Craven, M.W., Shavlik, J.W.: Extracting tree-structured representations of trained networks. In: Touretzky, D.S., Mozer, M.C., Hasselmo, M.E. (eds.) Advances in Neural Information Processing Systems 8, Proceedings of the 1995 Conference, pp. 24–30. The MIT Press (Jun 1996)
15. Forina, M., Leardi, R., Armanino, C., Lanteri, S., Conti, P., Princi, P.: Parvus: An extendable package of programs for data exploration, classification and correlation. J. Chemom. **4**(2), 191–193 (1988)
16. Franco, L., Subirats, J.L., Molina, I., Alba, E., Jerez, J.M.: Early breast cancer prognosis prediction and rule extraction using a new constructive neural network algorithm. In: Sandoval, F., Prieto, A., Cabestany, J., Graña, M. (eds.) IWANN 2007. LNCS, vol. 4507, pp. 1004–1011. Springer, Heidelberg (2007). https://doi.org/10.1007/978-3-540-73007-1_121
17. Guidotti, R., Monreale, A., Ruggieri, S., Turini, F., Giannotti, F., Pedreschi, D.: A survey of methods for explaining black box models. ACM Comput. Surv. **51**(5), 1–42 (2018). https://doi.org/10.1145/3236009
18. Hayashi, Y., Setiono, R., Yoshida, K.: A comparison between two neural network rule extraction techniques for the diagnosis of hepatobiliary disorders. Artif. Intell. Med. **20**(3), 205–216 (2000). https://doi.org/10.1016/s0933-3657(00)00064-6
19. Hofmann, A., Schmitz, C., Sick, B.: Rule extraction from neural networks for intrusion detection in computer networks. In: 2003 IEEE International Conference on Systems, Man and Cybernetics, vol. 2, pp. 1259–1265. IEEE (2003). https://doi.org/10.1109/ICSMC.2003.1244584
20. Huysmans, J., Baesens, B., Vanthienen, J.: ITER: an algorithm for predictive regression rule extraction. In: Tjoa, A.M., Trujillo, J. (eds.) DaWaK 2006. LNCS, vol. 4081, pp. 270–279. Springer, Heidelberg (2006). https://doi.org/10.1007/11823728_26
21. Kenny, E.M., Ford, C., Quinn, M., Keane, M.T.: Explaining black-box classifiers using post-hoc explanations-by-example: the effect of explanations and error-rates in XAI user studies. Artif. Intell. **294**, 103459 (2021). https://doi.org/10.1016/j.artint.2021.103459
22. König, R., Johansson, U., Niklasson, L.: G-REX: A versatile framework for evolutionary data mining. In: 2008 IEEE International Conference on Data Mining Workshops (ICDM 2008 Workshops), pp. 971–974 (2008). https://doi.org/10.1109/ICDMW.2008.117
23. Markowska-Kaczmar, U., Trelak, W.: Extraction of fuzzy rules from trained neural network using evolutionary algorithm. In: ESANN 2003, 11th European Symposium on Artificial Neural Networks, Bruges, Belgium, 23–25 April 2003, Proceedings, pp. 149–154 (2003). https://www.elen.ucl.ac.be/Proceedings/esann/esannpdf/es2003-9.pdf
24. Núñez, H., Angulo, C., Català, A.: Rule extraction based on support and prototype vectors. In: Diederich, J. (ed.) Rule Extraction from Support Vector Machines. SCI, vol. 80, pp. 109–134. Springer (2008). https://doi.org/10.1007/978-3-540-75390-2_5
25. Pedregosa, F., et al.: Scikit-learn: Machine learning in Python. J. Mach. Learn. Res. (JMLR) **12**, 2825–2830 (2011), https://dl.acm.org/doi/10.5555/1953048.2078195
26. Rocha, A., Papa, J.P., Meira, L.A.A.: How far do we get using machine learning black-boxes?. Int. J. Patt. Recogn. Artifi. Intell. **26**(02), 1261001-(1–23) (2012). https://doi.org/10.1142/S0218001412610010
27. Rudin, C.: Stop explaining black box machine learning models for high stakes decisions and use interpretable models instead. Nat. Mach. Intell. **1**(5), 206–215 (2019). https://doi.org/10.1038/s42256-019-0048-x

28. Sabbatini, F., Calegari, R.: Symbolic knowledge extraction from opaque machine learning predictors: GridREx & PEDRO. In: Kern-Isberner, G., Lakemeyer, G., Meyer, T. (eds.) Proceedings of the 19th International Conference on Principles of Knowledge Representation and Reasoning, KR 2022, Haifa, Israel, 31 July - 5 August (2022). https://doi.org/10.24963/kr.2022/57

29. Sabbatini, F., Ciatto, G., Calegari, R., Omicini, A.: On the design of PSyKE: a platform for symbolic knowledge extraction. In: Calegari, R., Ciatto, G., Denti, E., Omicini, A., Sartor, G. (eds.) WOA 2021–22nd Workshop "From Objects to Agents". CEUR Workshop Proceedings, vol. 2963, pp. 29–48, Bologna, Italy, 1–3 Sep, Proceedings, Sun SITE Central Europe, RWTH Aachen University (Oct 2021)

30. Sabbatini, F., Ciatto, G., Calegari, R., Omicini, A.: Hypercube-based methods for symbolic knowledge extraction: Towards a unified model. In: Ferrando, A., Mascardi, V. (eds.) WOA 2022–23rd Workshop "From Objects to Agents", CEUR Workshop Proceedings, vol. 3261, pp. 48–60. Sun SITE Central Europe, RWTH Aachen University (Nov 2022). http://ceur-ws.org/Vol-3261/paper4.pdf

31. Sabbatini, F., Ciatto, G., Calegari, R., Omicini, A.: Symbolic knowledge extraction from opaque ML predictors in PSyKE: Platform design & experiments. Intelligenza Artificiale 16(1), 27–48 (2022). https://doi.org/10.3233/IA-210120

32. Sabbatini, F., Ciatto, G., Omicini, A.: GridEx: an algorithm for knowledge extraction from black-box regressors. In: Calvaresi, D., Najjar, A., Winikoff, M., Främling, K. (eds.) EXTRAAMAS 2021. LNCS (LNAI), vol. 12688, pp. 18–38. Springer, Cham (2021). https://doi.org/10.1007/978-3-030-82017-6_2

33. Sabbatini, F., Ciatto, G., Omicini, A.: Semantic Web-based interoperability for intelligent agents with PSyKE. In: Calvaresi, D., Najjar, A., Winikoff, M., Främling, K. (eds.) Explainable and Transparent AI and Multi-Agent Systems. LNCS, vol. 13283, chap. 8, pp. 124–142. Springer (2022). https://doi.org/10.1007/978-3-031-15565-9_8

34. Sabbatini, F., Grimani, C.: Symbolic knowledge extraction from opaque predictors applied to cosmic-ray data gathered with LISA Pathfinder. Aeronau. Aerospace Open Access J. 6(3), 90–95 (2022). https://doi.org/10.15406/aaoaj.2022.06.00145

35. Saito, K., Nakano, R.: Extracting regression rules from neural networks. Neural Netw. 15(10), 1279–1288 (2002). https://doi.org/10.1016/S0893-6080(02)00089-8

36. Schmitz, G.P.J., Aldrich, C., Gouws, F.S.: ANN-DT: an algorithm for extraction of decision trees from artificial neural networks. IEEE Trans. Neural Netw. 10(6), 1392–1401 (1999). https://doi.org/10.1109/72.809084

37. Setiono, R., Baesens, B., Mues, C.: Rule extraction from minimal neural networks for credit card screening. Int. J. Neural Syst. 21(04), 265–276 (2011). https://doi.org/10.1142/S0129065711002821

38. Setiono, R., Leow, W.K.: FERNN: an algorithm for fast extraction of rules from neural networks. Appl. Intell. 12(1–2), 15–25 (2000). https://doi.org/10.1023/A:1008307919726

39. Setiono, R., Leow, W.K., Zurada, J.M.: Extraction of rules from artificial neural networks for nonlinear regression. IEEE Trans. Neural Netw. 13(3), 564–577 (2002). https://doi.org/10.1109/TNN.2002.1000125

40. Setiono, R., Liu, H.: NeuroLinear: from neural networks to oblique decision rules. Neurocomputing 17(1), 1–24 (1997). https://doi.org/10.1016/S0925-2312(97)00038-6

41. Setiono, R., Thong, J.Y.L.: An approach to generate rules from neural networks for regression problems. Eur. J. Oper. Res. 155(1), 239–250 (2004). https://doi.org/10.1016/S0377-2217(02)00792-0

42. Steiner, M.T.A., Steiner Neto, P.J., Soma, N.Y., Shimizu, T., Nievola, J.C.: Using neural network rule extraction for credit-risk evaluation. Int. J. Comput. Sci. Netw. Sec. **6**(5A), 6–16 (2006). http://paper.ijcsns.org/07_book/200605/200605A02.pdf
43. Thrun, S.B.: Extracting rules from artifical neural networks with distributed representations. In: Tesauro, G., Touretzky, D.S., Leen, T.K. (eds.) Advances in Neural Information Processing Systems 7, [NIPS Conference, Denver, Colorado, USA, 1994]. pp. 505–512. MIT Press (1994). http://papers.nips.cc/paper/924-extracting-rules-from-artificial-neural-networks-with-distributed-representations

Imperative Action Masking for Safe Exploration in Reinforcement Learning

Sumanta Dey[(⊠)], Sharat Bhat, Pallab Dasgupta, and Soumyajit Dey

Indian Institute of Technology Kharagpur, Kharagpur, India
{sumanta.dey,sharatbhat}@iitkgp.ac.in, {pallab,soumya}@cse.iitkgp.ac.in

Abstract. Reinforcement Learning (RL) needs sufficient exploration to learn an optimal policy. However, exploratory actions could lead the learning agent to safety hazards, not necessarily in the next state but in the future. Therefore, it is essential to evaluate each action beforehand to ensure safety. The exploratory actions and the actions proposed by the RL agent could also be unsafe during training and in the deployment phase. In this work, we have proposed the Imperative Action Masking Framework, a Graph-Plan-based method considering a finite and small look ahead to assess the safety of actions from the current state. This information is used to construct action masks on the run, filtering out the unsafe actions proposed by the RL agent (including the exploitative ones). The Graph-Plan-based method makes our framework interpretable, while the finite and small look ahead makes the proposed method scalable for larger environments. However, considering the finite and small look ahead comes with a cost of overlooking safety beyond the look ahead. We have done a comparative study against the probabilistic safety shield in Pacman and Warehouse environments approach. Our framework has produced better results in terms of both *safety* and reward.

Keywords: Graph-Plan Algorithm · Explainable/Interpretable Machine Learning · Reinforcement Learning · Exploration considering Safety

1 Introduction

Reinforcement Learning (RL) is a reward-based learning approach where the learning agent learns a state-to-action mapping policy based on the reward [19] accumulated during exploration in the underlying environment. RL, therefore, is suitable for many real-world scenarios where the underlying environment is not known a priori. On that account, the exploration is also necessary to capture the state-to-action-to-reward mapping to learn the highest rewarding policy or optimal policy. Therefore, while training, the RL agent takes a few actions that are not optimal according to the current policy, which is called exploratory actions. However, when RL applies in safety-critical environments, where some

© The Author(s), under exclusive license to Springer Nature Switzerland AG 2023
D. Calvaresi et al. (Eds.): EXTRAAMAS 2023, LNAI 14127, pp. 130–142, 2023.
https://doi.org/10.1007/978-3-031-40878-6_8

states are unsafe or hazardous, the goal is to learn an optimal policy while avoiding those unsafe states. Therefore actions that potentially could lead to an unsafe state, not necessarily in the very next state could be in the near future, should be avoided. Also, not only the exploratory actions but the action proposed by the RL agent could be unsafe. Therefore, all actions, whether exploratory or optimal, should be checked before taking for safety-critical environments.

For example, consider the Pacman Game layout as shown in Fig. 1. Based on the current position of the Pacman (yellow colored) and the ghost (blue colored), if the Pacman takes the right action from there, then in the next state, it will be in the green block, which might be safe for now as the ghost still not able to catch the Pacman. However, the Pacman will be trapped within the green-to-red block region, and eventually, the ghost will catch the Pacman. Therefore the right action from the current position is not safe.

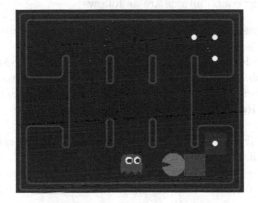

Fig. 1. Pacman Game Layout.

In general, several approaches to reinforcement learning through safe explo-ration can be broadly classified into Reward Shaping and Action Shaping. Reward Shaping techniques can be found in [2,6,10,18], where safety constraints are included in the reward function as a regularization term. On the other hand, works like [3,7,11,12,14,17] use action-shaping techniques to restrict the RL agent from taking unsafe actions. Apart from these methods, the possibilities of safe exploration using safe baselines/backup policies whenever a safety vio-lation is detected have been demonstrated in [4,13]. On the other hand, in [1], the author proposes a teacher advice-based technique where the RL agent seeks expert or teacher advice whenever it detects any unknown/unsafe situation.

Among the action shaping techniques, [14] and [3] are better choices when safety is the major concern, as these provide a formal safety guarantee even though they suffer from scalability issues. A scalable implementation of [3] can be found in [16]. In [16], the authors use a probabilistic model checker to deter-mine the safety probabilities of the actions, which are later used to filter out the actions having unsafe probabilities beyond a threshold. Using a probabilistic model checker helps to improve scalability but with the cost of safety and interpretability.

In this context, we propose the *Imperative Action Masking framework*, a scalable and interpretable action-shaping technique. Action Mask filters out unsafe actions from the current state. Our framework uses the classic Graph Plan algorithm [9] to determine the action mask considering the state space up to a small finite look ahead. The Graph Plan algorithm is used to improve interpretability, as the outcomes of this algorithm are easy to interpret by a human. In contrast, the small finite look ahead helps to increase the scalability, but with the cost of omitting safety hazards that could occur beyond the look ahead. Therefore, our framework is suitable for ensuring safety for environments with safety violations in the near future; however, it can also be used to reduce the chances of safety violations for environments with very far safety consequences.

In summary, our contributions are as follows:

- We develop a classic Graph Plan-based method to determine the membership probabilities of a state in each Action Specific Robust Set.
- These probabilities are later used to compute the Action-Masks. The Action Masks are applied on top of the RL Agent's decision to ensure safety by filtering out potential unsafe actions. Thus our method can be applied for safe exploration in Reinforcement Learning in various discrete state and action environments.
- The use of the Graph Plan-based method makes the Imperative Action Masking framework interpretable.
- We demonstrate the effectiveness of our method in Pacman and Warehouse Environments. We also compare the results of our method over the Probabilistic Shield presented in [16].

The paper is organized as follows. Section 2 outlines the overall problem statement. Section 3 presents the proposed Imperative Action Masking framework. Experimental environment configuration and results are presented in Sect. 4 and Sect. 5, respectively. Concluding remarks are given in Sect. 6.

2   Problem Formalization

Given a Constrained MDP (CMDP) as defined in [15], $< \mathcal{S}, \mathcal{A}, \mathcal{P}, \gamma, \mathcal{R}, \mu, \mathcal{C} >$, where $\mathcal{S}$ and $\mathcal{A}$ represent the set of states and the set of actions, respectively. $\mathcal{P}$ is the transition probability function, γ is the discount factor, and $\mathcal{R}$ is the reward function. μ is the start state distribution, and $\mathcal{C} = \{(c_i : \mathcal{S} \to \{0,1\}) | i \in \mathbb{Z}\}$ is the set of given binary safety constraints; $c_i : s$ can take either 0 (safe) or 1 (unsafe). $\mathcal{C} : s$ is considered unsafe if any of the $c_i : s$ is 1.

We use X to denote the set of safe states where, $\{\forall i, c_i : X \to 0\}$ and X_u to denote the set of unsafe states where, $\{\exists i, c_i : X \to 1\}$. There can be states ($s_k$ in Fig. 2) in the MDP which does not violate the safety constraints; however, all the possible combinations of legitimate actions will eventually lead to unsafe states. We consider this set of states as pseudo-unsafe states X_p and should be avoided just like the unsafe states. We have defined another two sets of safe states, namely Action Specific Robust Safe Sets ($X_I^{a_i}$) and Robust Safe Set (X_I) that is a subset of $\{X \setminus X_p\}$. Figure 2 depicts a schematic relation among these sets.

Definition 1 (Robust Safe Set). Given an MDP having a set of safe states X and action set $\mathcal{A}$, the Robust Safe Set is defined as:

$$X_I = \{s \mid \forall s \in X_I, \exists a \in \mathcal{A}, \forall s' \in \mathcal{P}(s, a), \ s' \in X_I\}$$

Definition 2 (Action Specific Robust Safe Set). Given an MDP having a set of safe states X and action set $\mathcal{A}$, the Action Specific Robust Safe Set is defined as:

$$X_I^{a_i} = \{s \mid \forall s \in X_I^{a_i}, \ s' \in \mathcal{P}(s, a_i), \ s' \in X_I\}$$

Hence, the Robust Safe Set (X_I) is $\bigcup_{\forall a \in \mathcal{A}} X_I^a$.

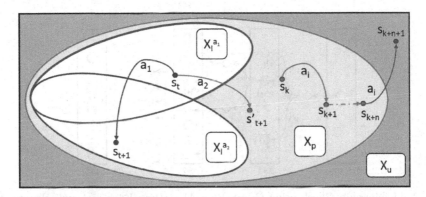

Fig. 2. The yellow bordered region represents the set of safe states(X). The blue and green encircled regions represent the Action Specific Robust Safe Sets for action a_1 and a_2, respectively. The white, yellow, and gray colored regions represent the Robust Safe Set (X_I), Pseudo Unsafe Set X_p, and set of unsafe states X_u, respectively. (Color figure online)

We consider using the action masking method to keep the RL agent within X. The mask value for unsafe actions' will be zero and omitted from the masked action list. Hence, the action is chosen from the masked distribution $\mathcal{M} : \pi$ will always be safe, where M denotes the masking function. Our Imperative Action Masking problem can be formulated as follows:

Problem 1 *(Imperative Action Masking Problem)*
Given a CMDP, a state $s_t \in X$ and, current policy π, the Online Imperative Action Masking Problem is to find an action mask $\mathcal{M}_{s_t}$ for the state s_t such that the next state s_{t+1} for any action a_t sampled from the masked distribution $\mathcal{M}_{s_t} : \pi(s_t)$ belongs to X, i.e., $\mathcal{P}(s_t, a_t) \to s_{t+1} \in X$.

3   Imperative Action Masking Framework

To construct Action Mask $(\mathcal{M}_{s_t})$ for a state (s_t) requires information about the state's membership in different Action Specific Robust Safe Sets. The masking value for an action is decided as follows.

$$\mathcal{M}_{s_t}[a_t] = \begin{cases} 1, \ if \ s_t \in X_I^{a_t}; \\ 0, \ Otherwise. \end{cases}$$

However, it is not computationally feasible for MDPs with huge state space to calculate the Robust Safe Set for the entire state space, covering all future time steps. It is also hard to accurately determine future states beyond a particular horizon if the MDP is not fully known. Therefore, we approximate the member-ship values of a state s_t in an Action Specific Robust Safe Set $X_I^{a_t}$ by checking all the possible states up to a small look ahead l. If the next state s_{t+1} for the action a_t belongs to X and there exist a trace $\tau = s_{t+2}, ..., s_{t+l}$ from s_{t+1}, such that $i = 2..l \ s_{t_i} \in X$ then we consider that the state $s_t \in X_I^{a_t}$.

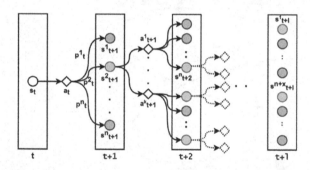

Fig. 3. Membership Check using Graph Plan Algorithm. Each rectangle denotes the state space of Graph-Plan for a time instance. Whereas Circles denote the states and Diamonds represent the action. Red circles denote that the state is unsafe or $\notin X$ and Green circles denote that the state is safe. As shown in the figure, only the safe states are expanded for the next time instances of the Graph-Plan Algorithm. (Color figure online)

We consider the well-established Graph Plan algorithm [9] to efficiently check for such traces τ. The actions of the MDP are considered the actions for the Graph Plan algorithm, and the effect of the actions are the next states as shown in Fig. 3. Two preconditions ($\neg explored$, $s \in X$) for all the actions are also considered to improve efficiency. To determine the membership of a state s_t in the Action Specific Robust Safe Set of action a_t in the first stage of the Graph Plan Algorithm, we apply action a_t only. All the actions that satisfy the preconditions are applied in the later stages. For non-deterministic MDP, the membership function $\mathcal{F}$ also becomes probabilistic $\mathcal{F}(X_I^a, s) \rightarrow [0, 1]$. In this case, the membership probability of only the maximum reachability probability

---

Algorithm 1: Determining Membership using GraphPlan

Input: Action List $\mathcal{A}$, Safe States X, Current State s_t, Given Action a_t, Look Ahead l

1 **Function** Membership(A, X, s_t, a_t, l):
2   if $s_t \notin X$ then
3 | return 0
4     $\mathcal{F}_0(s_t) = 1$
5     $GPS = \{s_t\}$
6     $GPA = \{a_t\}$
7   for $i = 1$ to l do
8   if $\forall s \in GPS : s \notin X$ then
9 | return 0
10        $GPS \leftarrow ExpandGraph(GPS, GPA)$
11   for $s' \in GPS$ do
12   if $s' \notin X$ then
13 | $\mathcal{F}_i(s') = 0$
14   else
15 | $\mathcal{F}_i(s') = max\left[\forall s, a \in Parent(s'), \ \mathcal{F}_i(s) \times \mathcal{P}(s, a, s')\right]$
16   end
17        $GPA \leftarrow \mathcal{A}$
18   end
19   return $max[\mathcal{F}_l]$
20 **Comments:**
21 GPS: Current State Set of the Graph-Plan Algorithms
22 GPA: Current Action set for the Graph-Plan Algorithm
23 $ExpandGraph$: Returns the set of next states from the states in GPS by taking actions in GAP while satisfying the preconditions
24 $Parent(s')$: Return all the state, action combination used in GPS_{t-1} to reach s' in GPS_t.

---

to all safe states presented in the $(t + l)$ time step of the Graph Plan algorithm is considered the membership probability. The membership probability $\mathcal{F}(X_I^{a_t}, s_t)$ is determined as follows:

$$\mathcal{F}(X_I^{a_t}, s_t) = max\left[\forall s_{t+l} \in X, max\left[\forall \tau, \ \tau[-1] = s_{t+l}, \prod_{(s,a,s')\in\tau} \mathcal{P}(s, a, s')\right]\right]$$

Here, $\tau[-1]$ returns the last state of the trajectory τ. In the end, it returns maximum $\mathcal{F}$ values among the states of Graph-Plan l-th time step. This membership value is used to decide the action mask for the state s_t.

$$\mathcal{M}_{s_t}[a_t] = \begin{cases} 1, \ if \ \mathcal{F}(X_I^{a_t}, s_t) \geq \mathcal{H}; \\ 0, \ Otherwise. \end{cases}$$

Here, $\mathcal{H}$ is the safety threshold range between [0,1]. The $\mathcal{H}$ value is a design choice. However, with the higher value of $\mathcal{H}$ exploration process becomes more conservative. Algorithm 1 outlines the overall process flow to find the membership values. Line 2 of the algorithm checks if the state s_t is in X or else it returns the membership value as zero (Line 3). If the state s_t is in X, the membership value is initialized with one (Line 4). The Graph Plan State list (GPS) and Graph Plan Action list (GPA) are also initialized with the current state and action. Then the graph is expanded (Line 10) up to depth l (look ahead value) (Line 7–18). However, after each level expansion, the new states of that level are checked if those states are in X, or else the membership for those states is assigned to zero (Line 13). If the states are in X, then their temporary membership values up to layer i are updated with the maximum value of their parent states $(i-1$ layer) value times corresponding transition probability, as shown in Line 15. Finally, in Line 19, the maximum temporary membership value among the last graph plan layer or $l-th$ layer states are returned as the membership value of state s_t in the corresponding Robust Set of Action a_t.

Graph Plan [9] is a simple and intuitive graph-based planning algorithm. The use of the Graph Plan-based algorithm for determining whether the action is safe or unsafe in our framework also makes the process intuitive and interpretable. For example, consider the Pacman example shown in Fig. 1. Here based on the current situation, our Imperative Action Masking Framework is able to identify the Left action as Safe even though the Left action may lead the agent to a trap or dead-end (colored in red) from which it cannot run away from an optimal ghost, but it also has a safe state (colored in green) from which it can run away from the ghost. Determining all these action labels (safe/unsafe) is quite intuitive and interpretable.

4   Environment Setup

In order to demonstrate the working of our method, we have used two different environments: (1) Pacman and (2) Warehouse. We have also presented a performance comparison analysis of our method with the Probabilistic Safety Shield for both environments. Table 1 describes the values of the hyperparameters used in the experiments, and we consider the default reward settings for all the environments. All the hyperparameter values used here are chosen empirically while keeping the environment settings or parameters like reward settings as default. Our implementation is uploaded as a public GitHub repository[1].

4.1   Pacman

In this environment, [5], the Pacman (Learning Agent) navigates through the grid as shown in Fig. 4 aims to collect all the pellets/food in the maze in the least possible steps without getting caught by the chasing ghosts (Adversarial Agents). The Pacman receives a positive reward of +10 for collecting a pellet and a negative reward of -1 at each time step. A large positive reward of +500

---

[1] https://github.com/sumantasunny/ImperativeActionMasking.git.

Table 1. Hyper-parameter values used in the experiments.

Hyper-Parameters	Values
Discount Factor γ	0.8
Learning Rate α	0.2
Initial/Min ϵ	0.90/0.05
ϵ-Decay	0.90
Max Steps/Episode	10000
Safety Threshold $\mathcal{H}$	1

is received on collecting all the dots, and on getting caught by the ghosts, a large negative reward of -500 is received. Then the game restarts. We have used directional ghosts (Probability of following the Pacman = 0.9), thus enabling a more severe safety-critical environment. We have chosen two types of layouts - Grid Trap (Fig. 4b) is a layout with dead-ends (traps), while Original Classic (Fig. 4a) and Medium Classic (Fig. 5a) layouts are without dead-end (no traps) for better comparison.

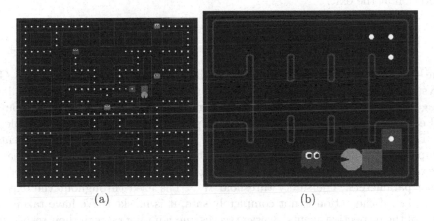

(a)   (b)

Fig. 4. (a) Original Classic Pacman Layout. (b) Grid Trap Pacman Layout. The yellow bot is the Pacman, controlled by the RL Agent. (Color figure online)

4.2   Warehouse

This environment, as shown in Fig. 5b, consists of a warehouse floor plan containing packages on the shelves and an exit. The yellow fork-lifter (Learning Agent) has to collect all the packages from the respective shelves and deliver them at the exit by navigating through the narrow corridors. It has to make sure that it does not collide with other fork-lifters, similar to the setup in [8].

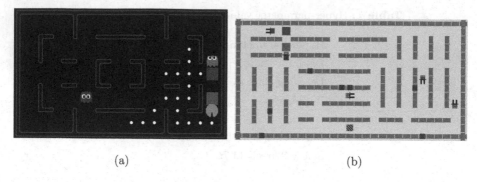

(a)   (b)

Fig. 5. (a)Medium Classic Pacman Layout. (b) Warehouse Grid Layout. The RL Agent controls the yellow robot. (Color figure online)

The yellow fork-lifter receives a positive reward of +20 on respectively loading and delivering each package and a negative reward of -1 at each time step. On delivering all the packages, a large positive reward of +500 is received, and on collision with other fork lifters, it gets a large negative reward of -500. All the fork lifters have only one exit in the entire floor plan, creating a safety-critical condition at the exit.

5   Results

The experimental results were produced in a Ubuntu 20.04 OS with 16 GB physical memory and Intel Core-i7 8th Gen processor. As a programming language, we use Python 3.7. As both environments are discrete, we use Tabular Q-Learning-based RL Agents. The used hyper-parameters are given in Table 1.

The RL algorithm has been trained for 300 episodes in each environmental setup, and each episode runs till termination. The discount factor $\gamma = 0.8$, learning rate $\alpha = 0.2$, and $\epsilon = 0.05$ in the ϵ-greedy exploration policy are used as hyper-parameters. The safety threshold $\mathcal{H} = 1$ is used throughout the experiment, i.e., if the action is not completely safe, it is masked. We have taken the plots of the respective agents' average scores and winning rates in their respective environmental setups during training over different look ahead values, which can be seen in Fig. 6, 8. The comparison of the training scores over several episodes in each environmental setup using various action-masking methods can be seen in Fig. 7, 9.

5.1   Pacman

We can observe from Fig. 6a and Fig. 6b that for all grids, the training scores and winning rates are very low when the look ahead is either very small or very large. This is because, for a smaller look ahead, the action masker fails to mask those actions that could be potentially hazardous in the future, if not immediately.

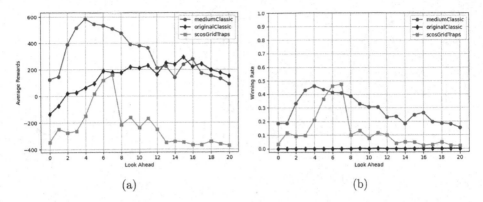

(a)   (b)

Fig. 6. Training Results on different look ahead for different Pacman grids. (a) Pacman's average training score over different look ahead. (b) Pacman's average winning rate over different look ahead.

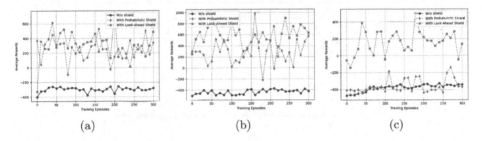

(a)   (b)   (c)

Fig. 7. Training Results on different Pacman grids. (a) Training Scores on original classic Pacman. (b) Training Scores on medium classic Pacman. (c) Training Scores on Pacman grid containing traps.

On the other hand, for a larger look ahead, the action masker becomes over-conservative and thus may end up masking even those actions which practically may not be unsafe. As a result, we see an optimal look ahead, leading to the optimal performance of the Pacman in each of the grids.

Figure 7 compares the scores of RL approaches using our method of GraphPlan-based look ahead shield to the unshielded RL approaches and RL approaches using probabilistic shield constructed via model checker as in [16]. Figure 7a and Fig. 7b correspond to the grid without traps (dead ends), while Fig. 7c corresponds to the grid with traps (dead ends). While outperforming the unshielded RL method in the first two scenarios, our method performed better in the third scenario, whereas the model checker-based probabilistic shield method could not. Where our method is able to accumulate an average of +200 reward, the model checker-based probabilistic shield performs similarly to the unshielded RL method with an average of -400 reward.

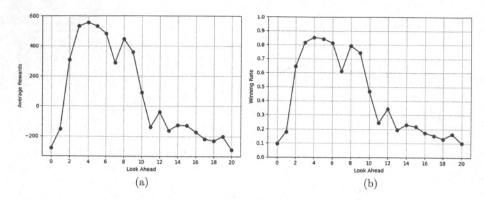

Fig. 8. Training Results on different look ahead for the warehouse floor plan (a) Fork-lifter's average training score over different look ahead. (b) Fork-lifter's average winning rate over different look aheads.

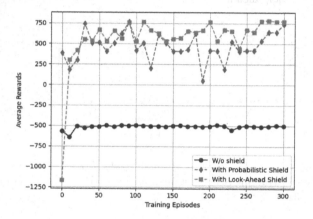

Fig. 9. Training Scores on warehouse environment.

5.2   Warehouse

Just like in the Pacman environment, as we can see from the overall trend of the graphs in Figures Fig. 8a and Fig. 8b, the reward and winning rate increases with the look ahead. However, after reaching a peak, the reward and winning rate start decreasing with the further increase of look ahead length. A large look ahead leads to overprotective scenarios and may end up masking not-so-unsafe action, as shown in Fig. 8a and Fig. 8b.

The comparison of the scores of the RL approach using Graph Plan-based look ahead shield of look ahead (l) 4 to the unshielded RL approach and RL approach using probabilistic shield constructed via model checker as in [16] has been shown in Fig. 9 where it is clearly visible that both the shield approach performs much better than the unshielded RL method. Also, our RL with the look ahead-based shielding approach performs slightly better in terms of rewards than the RL with the probabilistic shield.

6   Conclusion

We have presented an interpretable safe exploration method in safety-critical environments with discrete state action in RL. We use the classic Graph Plan algorithm to calculate the membership probabilities in each Action Specific Robust Safe Set for all states up to a small finite look ahead. The Graph Plan algorithm helps to improve the interpretability where the small finite look ahead helps to increase the scalability. Using membership probabilities to construct Action Masks in our proposed framework helps ensure safety with a certain probability. We have also presented various experiments that empirically validate that our method incurs fewer safety incidents while achieving higher rewards than the Probabilistic Safety Shield technique.

References

1. Abbeel, P., Coates, A., Ng, A.: Autonomous helicopter aerobatics through apprenticeship learning. Int. J. Robot. Res. **29**, 1608–1639 (2010)
2. Achiam, J., Held, D., Tamar, A., Abbeel, P.: Constrained policy optimization. In: International Conference on Machine Learning (2017)
3. Alshiekh, M., Bloem, R., Ehlers, R., Könighofer, B., Niekum, S., Topcu, U.: Safe reinforcement learning via shielding. In: AAAI Conference on Artificial Intelligence (2017)
4. Amodei, D., Olah, C., Steinhardt, J., Christiano, P.F., Schulman, J., Mané, D.: Concrete problems in AI safety. arXiv abs/1606.06565 (2016)
5. Berkeley, U.: UC Berkeley CS188 Intro to AI reinforcement learning. http://ai.berkeley.edu/reinforcement.html Accessed 14 Jun 2023
6. Berkenkamp, F., Moriconi, R., Schoellig, A.P., Krause, A.: Safe learning of regions of attraction for uncertain, nonlinear systems with gaussian processes. In: 2016 IEEE 55th Conference on Decision and Control (CDC), pp. 4661–4666 (2016)
7. Bharadhwaj, H., Kumar, A., Rhinehart, N., Levine, S., Shkurti, F., Garg, A.: Conservative safety critics for exploration. arXiv abs/2010.14497 (2020)
8. Bit-Monnot, A., Leofante, F., Pulina, L., Ábrahám, E., Tacchella, A.: Smartplan: a task planner for smart factories. arXiv abs/1806.07135 (2018)
9. Blum, A., Furst, M.L.: Fast planning through planning graph analysis. In: International Joint Conference on Artificial Intelligence (1995)
10. Chow, Y., Nachum, O., Duéñez-Guzmán, E.A., Ghavamzadeh, M.: A lyapunov-based approach to safe reinforcement learning. In: Neural Information Processing Systems (2018)
11. Dey, S., Dasgupta, P., Dey, S.: Safe reinforcement learning through phasic safety oriented policy optimization. In: SafeAI@AAAI Conference on Artificial Intelligence (2023)
12. Dey, S., Mujumdar, A., Dasgupta, P., Dey, S.: Adaptive safety shields for reinforcement learning-based cell shaping. IEEE Trans. Netw. Serv. Manage. **19**, 5034–5043 (2022)

13. Feghhi, S., Aumayr, E., Vannella, F., Hakim, E.A., Iakovidis, G.: Safe reinforcement learning for antenna tilt optimisation using shielding and multiple baselines. arXiv abs/2012.01296 (2020)
14. Fulton, N., Platzer, A.: Safe reinforcement learning via formal methods: Toward safe control through proof and learning. In: AAAI Conference on Artificial Intelligence (2018)
15. García, J., Fernández, F.: A comprehensive survey on safe reinforcement learning. J. Mach. Learn. Res. **16**, 1437–1480 (2015)
16. Jansen, N., Könighofer, B., Junges, S., Serban, A.C., Bloem, R.: Safe reinforcement learning using probabilistic shields (invited paper). In: International Conference on Concurrency Theory (2020)
17. Nikou, A., Mujumdar, A., Orlic, M., Feljan, A.V.: Symbolic reinforcement learning for safe ran control. In: Adaptive Agents and Multi-Agent Systems (2021)
18. Perkins, T.J., Barto, A.G.: Lyapunov design for safe reinforcement learning. J. Mach. Learn. Res. 3(null), 803–832 (mar 2003)
19. Sutton, R.S., Barto, A.G.: Reinforcement learning: An introduction. MIT press (2018)

Reinforcement Learning in Cyclic Environmental Changes for Agents in Non-Communicative Environments: A Theoretical Approach

Fumito Uwano[1]([✉])[iD] and Keiki Takadama[2][iD]

[1] Okayama University, 3-1-1 Tsushima-naka, Kita-ku, Okayama, Japan
uwano@okayama-u.ac.jp
[2] The University of Electro-Communications, 1-5-1 Chofugaoka, Chofu-shi, Tokyo, Japan
keiki@inf.uec.ac.jp

Abstract. Multi-agent Reinforcement Learning is required to adapt to the dynamic of the environment by transferring the learning outcomes in the case of the non-communicative and dynamic environment. Profit minimizing reinforcement learning with the oblivion of memory (PMRL-OM) enables agents to learn a co-operative policy using learning dynamics instead of communication information. It enables the agents to adapt to the dynamics of the other agents' behaviors without any design of the relationship or communication rules between agents. It helps easily to add robots to the system with keeping co-operation in a multi-robot system. However, it is available for long-term dynamic changes, but not for the short-them changes because it used the outcome with enough trials. This paper picked up cyclic environmental changes as short-term changes and aimed to improve the performance in cyclic environmental changes and analyze theoretically the rationality of this approach. Specifically, we extend PMRL-OM based on an analysis of the PMRL-OM approach. Our experiments evaluated the performance of the proposed method for a navigation task in a maze-type environment undergoing cyclic environmental change, with the results showing that the proposed method gave an enhanced performance. Our method also enabled the adaptation to cyclic change to occur sooner than for the existing PMRL-OM method. In addition, the theoretical analysis not only investigates the PMRL-OM rationality but also suggests optimal parameter values for the proposed method. The proposed method contributed to XAI by showing the precise profits of the agents and the approach with rationality.

Keywords: Reinforcement Learning · Multi-agent System · Reward Design · Non-communication · Dynamic Environment

This research was supported by JSPS KAKENHI Grant Number JP21K17807.

D. Calvaresi et al. (Eds.): EXTRAAMAS 2023, LNAI 14127, pp. 143–159, 2023.
https://doi.org/10.1007/978-3-031-40878-6_9

1   Introduction

Multi-agent reinforcement learning (MARL) is an important technique for controlling agents when learning optimal policies for gaming AI, path planning, robot control, and data mining [1,8]. MARL simulates the above problems with agents, environments, and rewards, and enable the agents to communicate with each other, thereby achieving co-operative control [4]. There are two directions of research, focusing on the scale of the environment called "small world" and "large world". This paper focuses on co-operative learning in a "small world" with its scarce information being used as effectively as possible. In a small world, the agents have to learn an appropriate policy for decision-making both optimally and strategically to achieve co-operation, considering action selection at each step [7,9]. On the other hand, in a "large world", the agents are also required to search the whole space and assign an acquired profit for the action in each episode because the large world has its own problems, including sparse rewards and the curse of dimensionality [6,14]. For example, transportation has two types of domain: area-to-area in the large world and base-to-clientele in the small world [3]. A large-scale MARL uses scarce information to make agents' learning affordable, while a small-scale MARL requires sufficient information to learn a strategic policy. A small-scale MARL generally allows the agent-communication, which might decrease the flexibility in the case of the multi-robot system where the number of robots is. For example, any number of robots can use ad-hoc communication for avoiding collision [2], but it is difficult for many agents to learn strategic policy. Therefore, developing learning techniques with no information in a small world is a great challenge.

Uwano et al. proposed a new MARL method for two agents in a maze-type environment with non-communication and evaluated its rationality via theoretical analysis [12,13]. It is considered valuable for that two agents learn an optimal policy using only learning dynamics in the absence of contextual information, even for simple target problems. It enables the agents to adapt to the dynamics of the other agents' behaviors without any design of the relationship or communication rules between agents. It helps easily to add robots to the system with keeping co-operation in a multi-robot system. Furthermore, considering instability in the real world, they proposed profit-minimizing reinforcement learning with oblivion of memory (PMRL-OM) as a new MARL method for adapting the dynamics in response to environmental change in a non-communicative environment [11]. This approach should enable agents to learn a co-operative policy that can adapt its dynamics autonomously, after having learned for sufficient episodes. However, PMRL-OM has the prerequisite that the agent adaptively relearns sufficiently often after experiencing environmental changes. PMRL-OM has yet to be analysed theoretically. Therefore, this paper analyses theoretically the rationality of the mechanisms in PMRL-OM used for co-operation when adapting to environmental change. Moreover, the analysis suggests a way of removing the PMRL-OM prerequisite. In our experiments, we compare our extended PMRL-OM method with an existing method for navigation tasks in a maze-type environments with cyclic environmental change, where the features of the environment are changed randomly after a certain period. The main contribution of this paper is a proposed MARL method that can adapt to dynamic changes in the non-communicative environment with insufficient learning episodes.

This paper is organized as follows. Section 2 describes the problem addressed in this work. The PMRL-OM method is introduced in Sect. 3, and its analysis and extension are described in Sect. 4. Section 5 then describes the details of our experiments and discusses the results. We conclude this work in Sect. 6.

2 Preliminaries

2.1 Problem Domain

We characterize the problem domain as a decentralized, partially observable, Markov decision process (Dec-POMDP) [10], where agents cannot access fully observed data, including other agents' locations and goal locations. Note that we designed a new Dec-POMDP along with the setting in [11]. Equations (1) to (5) are symbolic expressions of the problem, defined in terms of the tuple $\langle S, A, \tau, O, \Omega, R \rangle$. Let S, A, Ω, and R be a set of states, an action, an observed state, and a reward function, respectively. The agent observes Ω_i decided by the observation function O_i, where i is an identification index for an agent, and selects the action a_i. The agent transitions to its next state via Eq. (3), which is based on A, the Cartesian product of the actions of all agents. During the learning process, if it has reached the goal, the agent acquires the reward expressed by Eq. (5), where $-i$ indicates an agent other than any agent i. The training has two termination conditions. The first is when all agents have reached their goals and the second is when a given number of steps have been executed. This paper refers to this number of training steps as "one episode".

$$S = \{s_1, s_2, ..., s_N\} \tag{1}$$
$$A = a_1 \times a_2 \times ... \times a_I \tag{2}$$
$$\tau : S \times A \times S \rightarrow 1 \tag{3}$$
$$O_i : S \times A \times S \times \Omega_i \rightarrow 1 \tag{4}$$
$$R_i = \{r | s_i \in S_{goal} \wedge s_i \neq s_{-i}\} \tag{5}$$

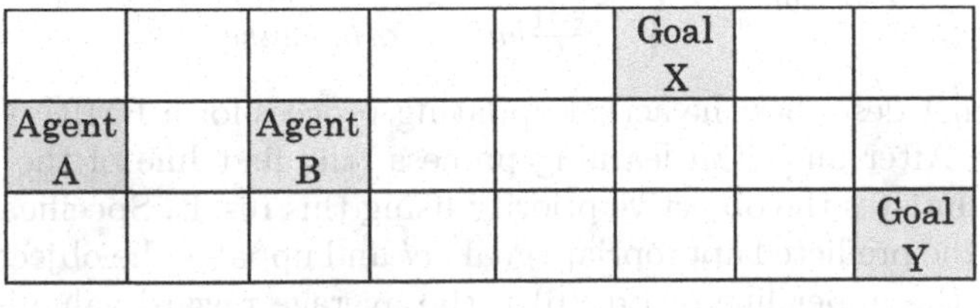

				Goal X		
Agent A		Agent B				
						Goal Y

Fig. 1. Example of a maze-type environment for two agents.

2.2   Navigation Task in Dynamic Grid World Maze

This paper considers the navigation task in a grid-world maze-type environment, as exemplified in Fig. 1, where two agents "A" and "B" depart from the states labelled "Agent A" and "Agent B", respectively, and aim to reach the goals "Goal X" and "Goal Y", respectively. Each agent can observe only its own location, and can move up, down, left, or right. An agent acquires the reward after reaching the goal faster than the other agent.

The environment, which is defined in terms of the start and goal locations, can change. An environmental change is assumed to be executed regularly, after a certain number of episodes, because this paper focuses on such cyclic changes. Note that the locations are changed randomly. Let S_{start} and S_{goal} be the sets having the states indicating the starts and goals for all agents, the system replaces the states in S_{start} or S_{goal} randomly and exclusively after a certain-length term and then repeats it.

3   Profit Minimizing Reinforcement Learning with Oblivion of Memory

A PMRL-OM enables agents to learn a co-operative policy using learning dynamics instead of communication information. In particular, PMRL-OM updates its learning parameters in response to environmental changes and learns its co-operative policy without inter-agent communications [11].

3.1   Objective Priority

A PMRL-OM updates its objective priority when deciding its optimal objective, such as the appropriate goal in a maze-type environment. Equation (6) expresses the calculation of the objective priority, where bid_g^i indicates the objective priority for agent i and goal g, with t_g^i indicating the minimum number of steps until agent i reaches goal g. The parameter ξ indicates the rate of change in previous updates of the objective priority. PMRL-OM updates the objective priority, which converges to the minimum number of steps t_g^i required to acquire the reward. Otherwise, the updates converge to zero. Using these priorities, PMRL-OM leads the agent toward the most distant of its first-to-reach goals:

$$bid_g^i \leftarrow \begin{cases} \frac{\xi-1}{\xi}bid_g^i + \frac{t_g^i}{\xi} & if\ R_i = r \\ \frac{\xi-1}{\xi}bid_g^i & otherwise \end{cases} . \qquad (6)$$

Algorithm 1 describes the actual updating process for a PMRL-OM's objective priority. After an initial learning process (the first line of the algorithm), PMRL-OM updates the objective priority using this result. Specifically, PMRL-OM decides the predicted appropriate goal sel and updates the objective priority by following the upper line of Eq. (6) if the average reward value is above the threshold TH_R and the reached goal g_o matches the selected goal sel. Next, the selection sel is changed by following the objective priority $\boldsymbol{bid}$, while sel is changed randomly with an arbitrary probability.

Algorithm 1. Objective Priority Estimation in PMRL-OM

1. Learning for one episode.
2. **if** $AveR_{g_o} > TH_R$ and g_o is *sel* **then**
3. $bid[sel] = \frac{\xi-1}{\xi} bid[sel] + \frac{t[sel]}{\xi}$
4. **else**
5. $bid[sel] = \frac{\xi-1}{\xi} bid[sel]$
6. **end if**
7. $sel = \arg\max \boldsymbol{bid}$
8. Setting random value to *sel* with arbitrary probability

3.2   Internal Reward and Learning

PMRL-OM estimates its Q-table based on an internal reward instead of an external reward. The internal reward is calculated for the agent reaching the goal selected by the objective priority. Figure 2 shows the internal reward setting for PMRL-OM, where agent B calculates that internal reward ir_{BY} is larger than the other ir_{BX} that yields the goal X for agent A. Because the agent learns its policy according to the goal associated with the largest expected reward per step, the internal reward can control learning together with the objective priority.

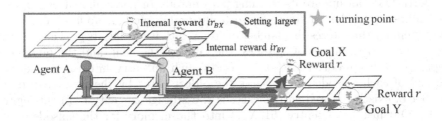

Fig. 2. Internal reward.

Equation (7) expresses the calculation of the internal reward and Eq. (8) expresses the update calculation for the Q-value using the internal reward as follows:

$$ir(g) = \begin{cases} \max\limits_{g_o \in G, g_o \neq g} r_{g_o} \gamma^{t_{g_o} - t_g} + \delta \ if \ g = sel \wedge R_i = r_g \\ r_g \qquad\qquad\qquad if \ g \neq sel \wedge R_i = r_g \\ 0 \qquad\qquad\qquad\quad otherwise \end{cases} \tag{7}$$

$$Q(s,a) \leftarrow (1-\alpha)Q(s,a) + \alpha \left[ir(g) + \gamma \max\limits_{a' \in A} Q(s',a') \right], \tag{8}$$

where $ir(g)$ indicates the internal reward for the goal g, r_{g_o} indicates the external reward for the arbitrary goal g_o, γ indicates the discount rate, and t_g indicates the minimum number of steps to reach the goal g. In Eq. (7), the agent does not acquire the reward before having reached the goal and acquires the internal reward if the reached goal is selected via the objective priority and if the agent acquires the external reward. The agent acquires the external reward if it acquires the reward by having reached the unselected goal.

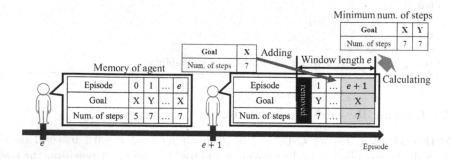

Fig. 3. Saving the minimum number of steps in PMRL-OM.

3.3 Updating of Minimum Number of Steps

A PMRL-OM can update its learning parameters. In particular, it can update the minimum number of steps. Figure 3 illustrates such an update. The agent has a memory that stores the number of steps until it reaches its goal as an entry consisting of the number of episodes, the label of the goal, and the number of steps, for each episode. PMRL-OM inserts such an entry into the memory at the end of each episode and replaces the oldest entry by the new entry if the number of existing stored entries exceeds the storage capacity e. In this figure, PMRL-OM inserts the entry $(51, X, 7)$ into the memory for the episode $e + 1$, and removes the oldest entry. PMRL-OM then creates a table for the minimum number of steps, as shown at the top right in the figure. Using the latest entries, PMRL-OM updates the minimum number of steps for each goal.

4   Improvement of Objective Value Update and Theoretical Analysis

4.1   Analysis of the Existing Objective Value Update

Here, we analyse the objective priority update. The update function is expanded as follows:

$$bid_g^i = \sum_{m=1}^{n_g} \left(\frac{\xi - 1}{\xi}\right)^{n_g - m} \frac{t_g^i}{\xi} \tag{9}$$

$$= \frac{t_g}{\xi} \cdot \frac{1 - \left(\frac{\xi-1}{\xi}\right)^{n_g}}{1 - \frac{\xi-1}{\xi}} = t_g \left\{1 - \left(\frac{\xi - 1}{\xi}\right)^{n_g}\right\}, \tag{10}$$

where m is an arbitrary variable, n_g is the estimated number of steps for the agent to reach the goal g, and ξ is a positive constant value. Because $0 < \frac{\xi-1}{\xi} < 1$ holds, if n_g is sufficiently large, the objective priority bid_g^i in Eq. (9) converges to the constant value t_g, following Eq. (10). The quantitative relationship of the objective priority values do not change. Therefore, it is revealed that PMRL-OM keeps the rationality of the objective priority. However, it is the case that PMRL-OM cannot update the objective priority close to the minimum number of steps for the early learning episodes after changing the environment. Naturally, changing the variable ξ to a smaller value would make the convergence faster, but it would be more sensitive to the environmental changes.

4.2   Improvement of Objective Value Updating

After sufficient learnings, the agent will have an optimal policy for reaching the goal in a minimum number of steps. Therefore, if the minimum number of steps t_g^i is in error, PMRL-OM should use the average number of steps instead of the minimum number of steps because it cannot calculate a true minimum number of steps. Furthermore, PMRL-OM may require more than e entries to update the minimum number of steps to true values, even though it retains only the number of steps for the previous e episodes. However, omitting the calculation of the minimum number of steps makes PMRL-OM adapt to the environmental changes both smoothly and rapidly. Given that Eq. (9) approximates the average value of the minimum number of steps for traversing episodes, PMRL-OM can use this average instead of the minimum number of steps. The learning makes the shortest estimate of the number of steps and the equation resists any errors in the estimate by averaging the number of steps. We can therefore improve the objective priority update equation as follows:

$$mbid_g^i \leftarrow \begin{cases} \frac{\xi-1}{\xi}mbid_g^i + \frac{l_g^i}{\xi} & if \ AveR_i > TH_R \\ \frac{\xi-1}{\xi}mbid_g^i + \frac{\max_{g_o} mbid_{g_o}^i}{\xi} & else \ if \ nothing \ is \ l_g^i \ , \\ \frac{\xi-1}{\xi}mbid_g^i & otherwise \end{cases} \quad (11)$$

where $mbid$ is the variable indicating the improved objective priority, and l_g^i is the number of steps for the agent i to reach the goal g in the latest episode. Equation (11) is derived from Eq. (6) by replacing the minimum number of steps t_g^i by l_g^i. At the beginning of the training, l_g^i might not exist, i.e., the agent may have reached an unselected goal in the first episode. In this situation, the proposed method proceeds via a different approach.

Algorithm 2 shows the objective-priority estimation algorithm for the proposed method. The coloured text indicates an alternative non-PMRL-OM process. Specifically, the proposed method stores the goal label and the number of steps for one tuple into the array l (Line 2), and removes the oldest entry after the length of the array exceeds the threshold TH_L (Lines 3–5). The proposed

---

Algorithm 2. Objective priority estimation in the proposed method.

1. Learning for one episode.
2. $l.add\,(goal, t)$
3. **if** $l.length > TH_L$ **then**
4.   Removing $l[0]$
5. **end if**
6. **if** l_{sel} is empty **then**
7. $bid[sel] = \frac{\xi-1}{\xi}bid[sel] + \frac{\max bid}{\xi}$
8. **else if** $AveR_{g_o} > TH_R$ and g_o is sel **then**
9. $bid[sel] = \frac{\xi-1}{\xi}bid[sel] + \frac{l_{sel}[l_{sel}.length-1]}{\xi}$
10. **else**
11. $bid[sel] = \frac{\xi-1}{\xi}bid[sel]$
12. **end if**
13. $sel = \arg\max bid$
14. Setting random value to sel by arbitrary probability

---

method then updates the objective priority if l has no entry for the selected goal sel (Lines 6 and 7). The proposed method updates via the maximum number of objective priorities used to promote the agent toward the goal because no entry means the agent has not reached the goal recently. In addition, the proposed method updates the objective priority using the latest entry of l for the selected goal sel (Line 9).

4.3 Theoretical Analsis of the Proposed Method

Here, we discuss the effectiveness of l_g^i via theoretical analysis. Let $d_g^i(n)$ be the function calculating the difference between the number of steps and the minimum for the number n when reaching the goal. We can transform Eq. (9) to Eq. (12) for the modified objective priority $mbid_g^i$ as follows. Because $l_g^i = t_g^i + d_g^i(m)$ holds, the latest number of steps uses both the minimum number of steps and the error.

$$mbid_g^i = \sum_{m=1}^{n_g} \left(\frac{\xi-1}{\xi}\right)^{n_g-m} \frac{t_g^i + d_g^i(m)}{\xi} \tag{12}$$

Expanding Eq.(12), we obtain the following Eq. (15).

$$mbid_g^i = bid_g^i + \sum_{m=1}^{n_g} \left(\frac{\xi-1}{\xi}\right)^{n_g-m} \frac{d_g^i(m)}{\xi} \tag{13}$$

$$\approx t_g^i + \frac{(\xi-1)^{n_g-1} \cdot d_g^i(1)}{\xi^{n_g}} + ... + \frac{\xi^{n_g-1} \cdot d_g^i(n_g)}{\xi^{n_g}} \tag{14}$$

$$\approx t_g^i + \frac{\sum_{m=1}^{n_g} d_g^i(m)}{\xi} \tag{15}$$

Before arriving at Eq. (15), we first transform to Eq. (13), adding the accumulated error to the objective priority of PMRL-OM. Because $bid_g^i \approx t_g^i$ holds (see Sect. 4.1), we can expand Eq. (13) to Eq. (14). If n_g is not sufficiently large (i.e., at the beginning of the training) and if ξ is sufficiently large, because $\xi \approx \xi - 1$ holds, the objective priority can be calculated by Eq. (15), which calculates the average value of all errors in the number of steps for the previous episodes. On the other hand, if n_g remains large after several episodes, $\xi \approx \xi - 1$ will no longer hold, and Eq. (15) then calculates via the weighted averaged value. Setting an upper limit to the weight as *weight*, we can calculate values for the parameters ξ and n_g as follows:

$$\left(\frac{\xi - 1}{\xi}\right)^{n_g - 1} \leq weight, \tag{16}$$

$$n_g \leq \frac{\ln weight}{\ln(\xi - 1) - \ln \xi} + 1. \tag{17}$$

If we set $\xi = 500$ for *weight* $= 0.001$, the weight is non-zero and $n_g \leq 3451$ holds. The accumulated error then approximates to zero at episode 3451. After that episode, the weight decreases and the objective priority of $\xi = 500$ is then the minimum number of steps plus the average error over the latest 3500 episodes. Continuing the training, the objective priority is approximated by the minimum number of steps because the error becomes smaller. Therefore, the proposed method has robustness toward instability in the number of steps.

4.4   Algorithm

Algorithm 3 shows the algorithm for the proposed method. The variables l and **bid** are the arrays storing the number of steps and the objective priority, respectively. The start state is s_0, and the goal states are contained in the array s_{end}. First, the agent observes the start state s_0 and selects its appropriate action via the Q-values. After executing the action a, the agent receives the reward and transits to the next state s'. The agent learns its policy using the internal reward ir. If the next state is the goal state or if the number of steps is equivalent to $MaxStep$, the training is terminated. After the termination, the proposed method updates l and **bid** and then selects the appropriate goal sel via the objective priority **bid**. With a certain probability (this paper used the value 0.15, following [12]), the selected goal sel is changed randomly.

Algorithm 3. The proposed method.

1. $Q(s, a)$ is initialized $\forall s \in S, \forall a \in A$
2. The sets of steps l, goal values $\boldsymbol{bid}$, and goals s_{end} are initialized.
3. Setting the initial state s_0.
4. **for** $iteration = 1$ to $MaxIteration$ **do**
5. $s = s_0$
6.   **for** $step = 1$ to $MaxStep$ **do**
7. $a = ActionSelect(Q, s)$
8.   Executing the action a, acquiring the reward r, and transiting to the next state s'.
9.   Calculating the internal reward ir to reach the goal sel
10. $Q(s, a) = (1 - \alpha)Q(s, a) + \alpha \left[ir + \gamma \max_{a' \in A} Q(s', a') \right]$
11. $step = step + 1$
12.   **if** $s' \in s_{end}$ **then**
13.   break
14.   **end if**
15.   **end for**
16.   **if** Having reached the goal g **then**
17. $l.add(g, step)$
18.   **if** $l.length > TH_L$ **then**
19.   Removing $l[0]$
20.   **end if**
21.   **end if**
22.   **if** l_{sel} is empty **then**
23. $\boldsymbol{bid}[sel] = \frac{\xi - 1}{\xi} \boldsymbol{bid}[sel] + \frac{\max \boldsymbol{bid}}{\xi}$
24.   **else if** $AveR_{g_o} > TH_R$ and g_o is sel **then**
25. $\boldsymbol{bid}[sel] = \frac{\xi - 1}{\xi} \boldsymbol{bid}[sel] + \frac{l_{sel}[l_{sel}.length - 1]}{\xi}$
26.   **else**
27. $\boldsymbol{bid}[sel] = \frac{\xi - 1}{\xi} \boldsymbol{bid}[sel]$
28.   **end if**
29. $sel = \arg \max \boldsymbol{bid}$
30.   Assign random value to sel with arbitrary probability
31. **end for**

5 Experiment: Learning in Cyclic Change

5.1 Experimental Setup

To investigate its adaptability, we compared the proposed method with PMRL-OM and profit sharing (PS) [5] in navigation tasks in 8×3 mazes where the cyclic changes of environment involved random changes to start locations and goal locations. We investigated the performance for cyclic lengths of 1000, 1500, 2000, 2500, 3000, 3500, 4000, 4500, 5000, 5500, 6000, 6500, 7000, and 7500. For a fair comparison, we modify the reward function for PS by replacing r by $\frac{r}{N}$ where N is the number of steps for the current episode.

5.2   Evaluation Criteria

We evaluated the number of steps for all agents until goals were reached and rewards acquired for 30 trials with the various seeds. The methods ran training modes and evaluation modes one by one, where the agents operated using their policy without learning in the evaluation mode. If an agent could not reach the goal, the number of steps was recorded as the maximum number. We used the configuration setting in [11]. The methods ran 50000 episodes for training, where the agent can operate for a maximum of 100 steps in each episode. The initialized Q-value was zero, the external reward value was 10, and the learning rate and discount rate were 0.1 and 0.9, respectively. The constant value δ was 10, the parameter ξ was 300, and the memory length e was 300. The threshold TH_L was 300, and TH_R was 3. The last four parameters, ξ, e, TH_L, and TH_R, are set for the excellent performance in this experiment that we checked beforehand.

5.3   Results

Figure 4 shows the results for the number of steps required for all agents to reach their goals and Fig. 5 shows the results for rewards acquired. The vertical axis indicates the quantity evaluated, and the horizontal axis indicates the cycle length. The results for the number of steps and acquired rewards were averaged across all episodes for all 30 trials, indicating that the results are better if the method learns its appropriate policy efficiently, i.e., the agents are able to acquire their reward rapidly. These results demonstrate that the proposed method gave the best performance in all cases. The difference in the results between the proposed method and PMRL-OM (for cycle lengths of 1000 and 7500) was smaller

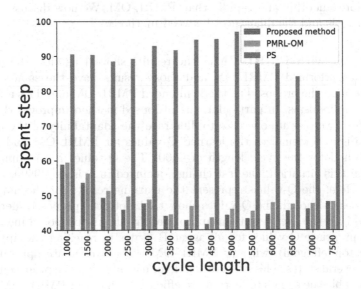

Fig. 4. Results for steps spent to reach the goal. The lower bar is better.

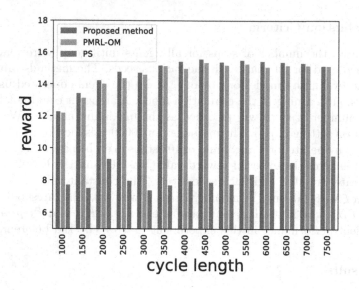

Fig. 5. Results for acquired rewards. The higher bar is better.

than for other cycle lengths. This is because the proposed method finds it hard to adapt to the environmental change at a cycle length of 1000 and PMRL-OM works well with a cycle length of 7500.

5.4   Discussion

These results demonstrate that the proposed method adapts to environmental change more smoothly and rapidly than PMRL-OM. We now discuss our theoretical analysis and the differences between methods.

Comparison with PMRL-OM. The results showed us that the proposed method outperformed PMRL-OM and those results have the same property. That is because the proposed method enhanced PMRL-OM and does not change the basic mechanisms. In particular, the proposed method improves the adaptability. Here, we pick up one case to illustrate the adaptability of the proposed method. Figure 6 visualizes the learned Q-values for PMRL-OM and the proposed method for the cycle length of 4500. The Q-values are summarized in Table 1. In this situation, the last change occurred in episode 49500, as shown in Fig. 7. Here, the Q-values represent the learning results for the most recent 500 episodes. Although the Q-values show that both the proposed method and PMRL-OM led the agents to their appropriate goals, the proposed method estimated appropriate Q-values for all agent B states in reaching the appropriate goal (the top-left goal), whereas PMRL-OM did not estimate appropriate Q-values in several states. This observation demonstrates that the proposed method tends to enable the agents to learn more efficiently than can PMRL-OM. In particular, PMRL-OM estimated a Q-value of around 3.5 for each state in the right

(a) Agent A in PMRL-OM.

(b) Agent B in PMRL-OM.

(c) Agent A in the proposed method.

(d) Agent B in the proposed method.

Fig. 6. Final Q-values visualization for environmental changes every 4500 episodes. The green, orange, and blue squares represent a normal state, goal state, and start state, respectively. The arrow indicates the action, and its thickness indicates the Q-value (the Q-values are shown in Table 1 (Color figure online)).

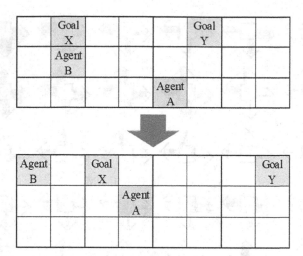

Fig. 7. The final environmental change. The locations of the agents and the goals are changed from the top to the bottom at episode 49500.

Table 1. Q-values in Fig. 6. The first column shows the state as a pair with the row number and column number. The symbol "–" indicates non-value because no action exists.

States (coordinates)	PMRL-OM								Proposed method							
	Agent A				Agent B				Agent A				Agent B			
	Actions				Actions				Actions				Actions			
	up	down	left	right	up	down	left	right	up	down	left	right	up	down	Left	right
(1,1)	–	2.810	–	2.834	–	7.690	–	9.000	–	3.380	–	3.392	–	8.151	–	8.479
(2,1)	–	2.803	2.857	3.304	–	8.872	6.749	10.000	–	3.443	3.442	4.026	–	8.717	8.315	9.722
(3,1)	–	9.904	11.659	8.814	–	10.640	10.779	10.842	–	7.347	3.485	7.894	–	7.226	6.867	5.164
(4,1)	–	6.425	10.189	8.119	–	3.935	10.040	3.862	–	4.300	6.348	13.098	–	9.211	8.811	9.158
(5,1)	–	7.009	6.147	14.803	–	3.492	5.993	3.472	–	8.021	7.538	15.672	–	7.828	9.161	7.752
(6,1)	–	11.864	7.744	16.961	–	3.552	3.805	3.562	–	12.393	9.274	17.609	–	7.693	10.462	5.695
(7,1)	–	15.278	15.155	18.745	–	3.598	3.579	3.289	–	14.874	14.677	19.825	–	3.555	7.946	3.557
(8,1)	–	6.976	8.717	–	–	7.694	3.651	–	–	7.445	8.989	–	–	3.533	3.974	–
(1,2)	2.793	2.789	–	2.808	5.572	2.877	–	8.711	3.349	3.348	–	3.414	8.315	8.625	–	8.634
(2,2)	2.822	2.812	2.806	3.351	6.546	3.291	6.206	9.396	3.393	3.310	3.325	5.430	8.703	8.974	8.942	9.094
(3,2)	4.107	2.841	2.781	9.961	10.093	5.894	5.489	4.595	5.420	3.484	3.537	10.256	10.543	9.439	9.373	9.193
(4,2)	8.031	8.227	6.985	12.690	6.987	3.133	5.782	3.310	9.817	7.862	7.260	12.731	9.168	9.231	9.277	9.269
(5,2)	11.643	9.517	10.874	13.915	3.486	3.374	3.422	3.416	12.103	9.761	10.768	14.213	9.026	9.032	8.985	8.843
(6,2)	14.916	10.452	12.057	15.240	3.525	3.538	3.481	3.511	14.633	10.276	11.516	15.835	6.335	4.149	10.216	4.799
(7,2)	16.768	11.660	13.516	13.719	3.610	3.571	3.619	3.581	17.667	10.103	12.630	14.115	4.327	3.498	6.333	3.471
(8,2)	18.072	4.002	8.421	–	5.887	3.645	3.654	–	18.746	4.194	8.717	–	5.164	3.489	3.478	–
(1,3)	2.775	–	–	2.820	4.840	–	–	2.964	3.347	–	–	3.328	8.812	–	–	8.785
(2,3)	2.812	–	2.807	2.825	3.819	–	2.835	6.070	3.500	–	3.326	3.375	8.947	–	8.937	9.065
(3,3)	2.855	–	2.804	5.368	8.924	–	3.501	3.011	5.761	–	3.343	3.663	9.310	–	9.268	9.214
(4,3)	10.277	–	3.207	8.071	3.862	–	3.046	3.177	5.776	–	3.689	10.100	9.402	–	9.313	9.273
(5,3)	12.426	–	3.087	6.072	3.327	–	3.303	3.340	12.831	–	4.666	8.255	9.309	–	9.390	5.007
(6,3)	13.607	–	4.725	6.327	3.457	–	3.457	3.512	13.754	–	4.565	4.772	7.389	–	6.144	3.350
(7,3)	14.827	–	6.556	4.098	3.558	–	3.580	3.578	13.719	–	6.458	4.59	4.321	–	3.712	3.414
(8,3)	6.443	–	7.884	–	3.609	–	3.605	–	8.112	–	4.161	–	3.451	–	3.459	–

half of the maze for agent B, with the difference between the Q-values being less than 0.1. On the other hand, the proposed method estimated a Q-value for the state that was double that estimated by PMRL-OM, even when it was located five states away from the goal state. It is clear that the proposed method can promote learning both efficiently and rapidly.

Validation of the Theoretical Analysis. From the theoretical analysis described in Sect. 4.3, setting the upper limit of the weight as $weight = 0.001$ gives $n_g \leq 2070$ using Eq. (17). This implies that the objective priority should be influenced by the number of steps for the most recent 2070 episodes, but should probably not be influenced by the number of steps before episode 2070. Furthermore, Figs. 4 and 5 show that the reward results increase and the step results decrease, eventually converging. The difference, together with the cycle length, decreases at episode 2500. That is because the current setting causes the proposed method to approximate the minimum number of steps from the actual number of steps in the most recent 2070 episodes. This confirms that the theoretical analysis can be validated from an empirical viewpoint.

6 Conclusion

PMRL-OM enables agents to learn a co-operative policy using learning dynamics instead of communication information. It enables the agents to adapt to the dynamics of the other agents' behaviors without any design of the relationship or communication rules between agents. It helps easily to add robots to the system with keeping co-operation in a multi-robot system. However, it is available for long-term dynamic changes, but not for the short-them changes because it used the outcome with enough trials. This paper picked up cyclic environmental changes as short-term changes and proposes a new MARL method for adapting to cyclic changes in the environment without communication for real-world problems. Specifically, we analysed the objective priority update mechanism in PMRL-OM and proposed an improved PMRL-OM method based on this analysis. The experiment compared the new method with PMRL-OM and PS in navigation tasks in the dynamic environment with cyclic environmental changes. The term-length was set from 1000 to 7500 by 500. Our experimental results validated empirically the theoretical analysis and showed that the proposed method worked well for navigation tasks in maze-type environments with varying cycle intervals.

From our results, neither the proposed method nor PMRL-OM operated optimally because the step results did not reflect the minimum number of steps and the reward result was not 20. However, they can learn their policies while avoiding instability. In future work, we should seek improved performance by investigating their optimality. Furthermore, we would propose a new MARL method and show that rationality in the dynamic environment where the number of agents changes for future robotics research.

References

1. Bargiacchi, E., Verstraeten, T., Roijersk, D.M., Nowé, A., van Hasselt, H.: Learning to coordinate with coordination graphs in repeated single-stage multi-agent decision problems. In: The 35th International Conference on Machine Learning, vol. 80, 482–490 (2018)
2. Chen, L., et al.: Multiagent path finding using deep reinforcement learning coupled with hot supervision contrastive loss. IEEE Trans. Industr. Electron. **70**(7), 7032–7040 (2023). https://doi.org/10.1109/TIE.2022.3206745
3. Ding, S., Aoyama, H., Lin, D.: Combining multiagent reinforcement learning and search method for drone delivery on a non-grid graph. In: Advances in Practical Applications of Agents, Multi-Agent Systems, and Complex Systems Simulation. The PAAMS Collection: 20th International Conference, PAAMS 2022, L'Aquila, Italy, July 13–15, 2022, Proceedings, pp. 112–126. Springer-Verlag, Berlin, Heidelberg (2022)
4. Du, Y., et al.: Learning correlated communication topology in multi-agent reinforcement learning. In: Proceedings of the 20th International Conference on Autonomous Agents and MultiAgent Systems, pp. 456–464. AAMAS '21, International Foundation for Autonomous Agents and Multiagent Systems, Richland, SC (2021)
5. Grefenstette, J.J.: Credit assignment in rule discovery systems based on genetic algorithms. Mach. Learn. **3**(2), 225–245 (1988). https://doi.org/10.1023/A:1022614421909
6. Li, J., Shi, H., Hwang, K.S.: Using fuzzy logic to learn abstract policies in large-scale multiagent reinforcement learning. IEEE Trans. Fuzzy Syst. **30**(12), 5211–5224 (2022). https://doi.org/10.1109/TFUZZ.2022.3170646
7. Raileanu, R., Denton, E., Szlam, A., Fergus, R.: Modeling others using oneself in multi-agent reinforcement learning. In: Dy, J., Krause, A. (eds.) Proceedings of the 35th International Conference on Machine Learning. Proceedings of Machine Learning Research, vol. 80, pp. 4257–4266. PMLR (10–15 Jul 2018). https://proceedings.mlr.press/v80/raileanu18a.html
8. Rashid, T., Samvelyan, M., Schroeder, C., Farquhar, G., Foerster, J., Whiteson, S.: QMIX: Monotonic value function factorisation for deep multi-agent reinforcement learning. In: The 35th International Conference on Machine Learning, vol. 80, pp. 4295–4304 (2018). http://proceedings.mlr.press/v80/rashid18a.html
9. Rashid, T., Farquhar, G., Peng, B., Whiteson, S.: Weighted qmix: Expanding monotonic value function factorisation for deep multi-agent reinforcement learning. In: Proceedings of the 34th International Conference on Neural Information Processing Systems. NIPS'20, Curran Associates Inc., Red Hook, NY, USA (2020)
10. Sigaud, O., Buffet, O.: Markov Decision Processes in Artificial Intelligence. Wiley-IEEE Press (2010)
11. Uwano, F., Takadama, K.: Utilizing observed information for no-communication multi-agent reinforcement learning toward cooperation in dynamic environment. SICE J. Contr. Measure. Syst. Integr. **12**(5), 199–208 (2019). https://doi.org/10.9746/jcmsi.12.199
12. Uwano, F., Tatebe, N., Tajima, Y., Nakata, M., Kovacs, T., Takadama, K.: Multi-agent cooperation based on reinforcement learning with internal reward in maze problem. SICE J. Contr., Measure. Syst. Integr. **11**(4), 321–330 (2018). https://doi.org/10.9746/jcmsi.11.321

13. Uwano, F., Takadama, K.: Directionality reinforcement learning to operate multi-agent system without communication (2021). 10.48550/ARXIV.2110.05773, arXiv:2110.05773
14. Zhou, Z., Xu, H.: Decentralized adaptive optimal tracking control for massive autonomous vehicle systems with heterogeneous dynamics: A stackelberg game. IEEE Trans. Neural Netw. Learn. Syst. **32**(12), 5654–5663 (2021). https://doi.org/10.1109/TNNLS.2021.3100417

Inherently Interpretable Deep Reinforcement Learning Through Online Mimicking

Andreas Kontogiannis[1,2]([✉]) and George A. Vouros[1]

[1] University of Piraeus, Piraeus, Greece
andr.kontog@gmail.com
[2] National Technical University of Athens, Athens, Greece

Abstract. Although deep reinforcement learning (DRL) methods have been successfully applied in challenging tasks, their application in real-world operational settings - where transparency and accountability play important roles in automation - is challenged by methods' limited ability to provide explanations. Among the paradigms for explainability in DRL is the interpretable box design paradigm, where interpretable models substitute inner closed constituent models of the DRL method, thus making the DRL method "inherently" interpretable. In this paper we propose a generic paradigm where interpretable DRL models are trained following an online mimicking paradigm. We exemplify this paradigm through XDQN, an explainable variation of DQN that uses an interpretable model trained online with the deep Q-values model. XDQN is challenged in a complex, real-world operational multi-agent problem pertaining to the demand-capacity balancing problem of air traffic management (ATM), where human operators need to master complexity and understand the factors driving decision making. XDQN is shown to achieve high performance, similar to that of its non-interpretable DQN counterpart, while its abilities to provide global models' interpretations and interpretations of local decisions are demonstrated.

Keywords: Deep Reinforcement Learning · Mimic Learning · Explainability

1 Introduction

Deep Reinforcement Learning (DRL) has mastered decision making policies in various difficult control tasks [11], games [19] and real-time applications [31]. Despite the remarkable performance of DRL models, the knowledge of mastering these tasks remains implicit in deep neural networks (DNNs). Thus, its application in real-world operational settings is challenged by methods' limited ability to provide explanations at global (policy) and local (individual decisions) levels. This lack of interpretability makes it difficult for human operators to understand DRL solutions, which can be important for solving safety-critical

D. Calvaresi et al. (Eds.): EXTRAAMAS 2023, LNAI 14127, pp. 160–179, 2023.
https://doi.org/10.1007/978-3-031-40878-6_10

real-world tasks. Additionally, DRL models are unable to provide information about the evolution of models during the training process, which is useful to gain insights about the models' accumulated knowledge through time.

To address the aforementioned challenges, one may follow different paradigms for the provision of explanations: the interpretable box design paradigm is one of them where interpretable models substitute or are integrated to inner closed-box components of DRL [29]. Mimic learning has been proposed so as to infer interpretable models that mimic the behavior of well-trained DNNs [1,4]. In the DRL case, to improve interpretability, mimicking models can replace closed-box DRL models [16,29]. To do so, mimicking models must achieve performance that is comparable to closed-box models, also optimizing their *fidelity* [29], so that interpretable and closed models take the same decisions for the same reasons, in any specific circumstance. To extract knowledge from close-box models, recent works (e.g. [2,16]) have applied mimic learning using decision trees: In this case, criteria used for splitting tree nodes provide a tractable way to explain the predictions made by the controller.

Typically, mimic learning approaches require well-trained networks (which we refer to as *mature* networks), whose behaviour are mimicking towards interpretability. In doing so, interpretability of the model during the training process is totally ignored. In real-world settings this could be quite impractical, since the training overhead required to train the mimic models can often be a sample-inefficient and time-consuming process, especially for large state-action spaces and multi-agent settings. More importantly, the training process can be "unsafe", given that mimicking models' decisions may diverge from that of the original policy models: though fidelity can be measured at the end of the training process, we need to ensure high fidelity *during* training. In conclusion, while the mimic learner can provide explanations about the decisions of an inferred DRL model, it neither allows examining the knowledge accumulated throughout the training process, nor ensures fidelity during the training process.

To deal with these challenges, in this paper we propose a generic interpretable DRL paradigm, where interpretable models are trained in interplay with the original DRL models without requiring these models to be mature: the original model trains the mimicking model and the mimicking model drives the subsequent updates of the original model offering target value estimators during the DRL training process. This is what we call the *online* (i.e. during training and in interplay with other models) training approach. This approach assures fidelity of decisions of the mimicking model, w.r.t. those of the original model. We conjecture that such a paradigm is effective in many DRL methods and can be applied to value based, policy based or actor-critic methods, depending on the model that is mimicked, over discrete or continuous state-action spaces.

To exemplify the proposed paradigm and provide evidence for its applicability, this paper proposes the *eXplainable Deep Q-Network* (*XDQN*) method, an explainable variation of the well-known DQN [19] method, with the goal to provide inherent explainability of DQN via mimic learning in an online manner. By following an online mode of training the mimicking model, XDQN does not require the existence of a well-trained model to train an interpretable one: The mimic learner is trained and updated while the DQN Q-value model is trained and updated, supporting the maintenance of multiple "snapshots" of the model while it evolves through time, offering interpretability on intermediate models, and insights about the patterns and behaviors that DQN learns during training.

To evaluate the effectiveness of XDQN, this is challenged in complex, real-world multi-agent tasks, where thousands of agents aim to solve airspace congestion problems. As it is shown in other works, the use of independent DQN agents has reached unprecedented performance [13] and therefore such tasks provide a suitable real-world testbed for the proposed method. Agents in this setting are trained via parameter sharing following the centralized training decentralized execution (CTDE) schema.

We summarize the main contributions of this paper below:

- To our knowledge, this work is the first that provides *inherent* interpretability through online mimicking of DRL models, without requiring the existence of a well-trained DRL model: As far as we know, there is not any work that supports this straightforward paradigm for interpretability.
- XDQN exemplifies this paradigm and is proposed as an explainable variation of DQN, in which an interpretable mimic learner is trained online, in interplay with the Q-network of DQN, playing the role of the target Q-network.
- Experimentally, it is shown that XDQN can perform similarly to DQN, demonstrating good play performance and fidelity to DQN decisions in complex real-world multi-agent problems.
- The ability of our method to provide global (policy) and local (in specific circumstances) explanations regarding agents' decisions, also while models are being trained, is demonstrated in a real-world complex setting.

The structure of this article is as follows: Sect. 2 provides background definitions on DRL, deep Q-networks, mimicking approaches, and clarifies the paradigm introduced. Section 3 presents related work. Section 4 exemplifies the proposed paradigm with XDQN, and Sect. 5 provides details on the experimental setting and results, as well as concrete examples of local and global explainability. Section 6 concludes the article.

2 Background

We consider a sequential decision-making setup, in which an agent interacts with an environment E over discrete time steps. At a given timestep, the agent perceives features regarding a state $s_t \in S$, where S is the set of all states in agent's environment (state space). The agent then chooses an action a_t from its repertoire of actions A (i.e., the actions that it can perform in any state), and gets the reward r_t generated by the environment.

The agent's behavior is determined by a policy π, which maps states to a probability distribution over the actions, that is $\pi \colon S \to P(A)$. Apart from an agent's policy, the environment E may also be stochastic. We model it as a Markov Decision Process (MDP) with a state space S, action space A, an initial state distribution $p(s_1)$, transition dynamics $p(s_{t+1}|s_t, a_t)$ and a reward $r(s_t, a_t, s_{t+1})$, for brevity denoted as r_t.

The agent aims to learn a stochastic policy π^* which drives it to act so as to maximize the expected discounted cumulative reward $G_t = \sum_{\tau=t}^{\infty} \gamma^{\tau-t} r_\tau$, where $\gamma \in (0, 1)$ is a discount factor.

2.1 Deep Reinforcement Learning with Interpretable Models.

To deal with a high dimensional state space, a policy can be approximated by exploiting a DNN with weight parameters θ. DRL methods learn and exploit different models (e.g. objectives model, value models, models of the environment), which support updating the policy model to fit to the sampled experience generated while interacting with the environment.

In this work we introduce a paradigm for providing DRL methods with inherent interpretability, by replacing closed-box models with interpretable ones. This follows the interpretable box design paradigm specified in [29].

But, how to train interpretable models? Interpretable models can be trained either during or after training the DRL models. Being trained during the training of the DRL models, the interpretable model evolves as the DRL model evolves, and can be used to explain how the training process affects agent's responses. However, this may result to instability and inefficiency of the training process since, interpretable models may aim to reach a moving target and may suffer from high variance. Such pitfalls can be mitigated by means of the "interplay" of interpretable and original models: The interpretable model is trained by specific instances of the DRL model, and its decisions affect subsequent updates of the original model. As an important implication of the online training, the fidelity of the mimicking models with respect to the original DRL models is empirically assured.

Such an online training scheme is followed, for instance, by DRL architectures exploiting a target network (succinctly presented below). A target network provides a stable objective in the learning procedure, and allows a greater coverage of the training data. Target networks, in addition to the benefits they provide in the learning procedure, they can support interpretability of the policy models, given that these can be replaced by interpretable models that are trained with the deep networks in an online manner.

The introduced paradigm for inherently interpretable DRL through online mimicking can be applied to different closed-box DRL models. Here, we exemplify and test this idea to Deep Q-networks, training interpretable Q-value models online with closed-box Q-value models through mimicking. Q-value closed-box and/or interpretable models can be used to extract a DRL policy.

2.2 Deep Q-Networks

Considering that an agent acts under a stochastic policy π, the Q-function (state-action value) of a pair (s, a) is defined as follows

$$Q^{\pi}(s, a) = \mathbb{E}\left[G_t \mid s_t = s, a_t = a, \pi\right] \tag{1}$$

which can also be computed recursively with bootstrapping:

$$Q^{\pi}(s, a) = \mathbb{E}\left[r_t + \gamma \mathbb{E}_{a \sim \pi(s_{t+1})}[Q^{\pi}(s_{t+1}, a)] \mid s_t = s, a_t = a, \pi\right] \tag{2}$$

The Q-function measures the value of choosing a particular action when the agent is in this state. We define the optimal policy π^* under which the agent receives the optimal $Q^*(s, a) = max_{\pi}Q^{\pi}(s, a)$. For a given state s, under the optimal policy π^*, the agent selects action $a = argmax_{a' \in A}Q^*(s, a')$. Therefore, it follows that the optimal Q-function satisfies the Bellman equation:

$$Q^*(s, a) = \mathbb{E}\left[r_t + \gamma \max_a Q^*(s_{t+1}, a) \mid s_t = s, a_t = a, \pi\right]. \tag{3}$$

In Deep Q-Networks (DQN), to estimate the parameters θ of the Q-values model, at iteration i the expected mean squared loss between the estimated Q-value of a state-action pair and its temporal difference target, produced by a fixed and separate *target* Q-network $Q(s, a; \theta^-)$ with weight parameters θ^-, is minimized. Formally:

$$L_i(\theta_i) = \mathbb{E}\left[Y_i^{DQN} - Q(s, a; \theta)\right], \tag{4}$$

with

$$Y_i^{DQN} = r_t + \gamma \max_{a \in A} Q(s_{t+1}, a; \theta^-) \tag{5}$$

In order to train DQN and estimate θ, we could use the standard Q-learning update algorithm. Nevertheless, the Q-learning estimator performs very poorly in practice. To stabilize the training procedure of DQN, Mnih et al. [19] freezed the parameters, θ^-, of the target Q-network for a fixed number of training iterations while updating the closed Q-network with gradient descent steps with respect to θ.

The direct application of the online mimicking approach in DQN uses an interpretable target DQN model to mimic the online Q-network, and thus, the decisions of the original policy model.

In addition to the target network, during the learning process, DQN uses an experience replay buffer [19], which is an accumulative dataset, D_t, of state transition samples - in the form of (s, a, r, s') - from past episodes. In a training step, instead of only using the current state transition, the Q-Network is trained by sampling mini-batches of past transitions from D uniformly, at random. Therefore, the loss can be written as follows:

$$L_i(\theta_i) = \mathbb{E}_{(s,a,r,s') \sim U(D)}\left[(Y_i^{DQN} - Q(s, a; \theta))^2\right]. \tag{6}$$

As it is well known, the main advantage of using an experience replay buffer is that uniform sampling reduces the correlation among the samples used for training the Q-network. The replay buffer improves data efficiency through reusing the experience samples in multiple training steps. Instead of sampling mini-batches of past transitions uniformly from the experience replay buffer, a further improvement over DQN results from using a prioritized experience replay buffer [24]. This aims at increasing the probability of sampling those past transitions from the experience replay that are expected to be more useful in terms of absolute temporal difference error.

3 Related Work

Explainability in Deep Reinforcement Learning (DRL) is an emergent area whose necessity is related to the fact that DRL agents solve sequential tasks, acting in the real-world, in operational settings where safety, criticality of decisions and the necessity for transparency (i.e. explainability with respect to real-world pragmatic constraints [29]) is the norm. However, DRL methods use closed-boxes whose functionality is intertwined and are not interpretable. This may hinder DRL methods explainability. In this paper we address this problem by proposing an interpretable DRL method comprising two models which are trained jointly in an online manner: An interpretable mimicking model and a deep model. The later offers training samples to the mimicking one and the former interpretable model offers target action values for the other to improve its predictions. At the end of the training process, the mimicking model has the capacity to provide high-fidelity interpretations to the decisions of the deep model and thus, it can replace the deep model. This proposal is according to the interpretable box design paradigm, which follows the conjecture (stated for instance in [22]) that there is high probability that decisions of closed-boxes can be approximated by well designed interpretable models.

There are many proposals for interpreting Deep Neural Networks (DNNs) through mimicking approaches. These approaches differ in several dimensions: (a) the targeted representation (e.g., decision trees in DecText [5], logistic model trees (LMTs) in reference [8], or Gradient Boosting Trees in reference [6]), (b) to the different splitting rules used towards learning a comprehensive representation, (c) to the actual method used for building the interpretable model (e.g., [8] uses the LogiBoost method, reference [5] proposes the DecText method, while the approach proposed in reference [6] proposes a pipeline with an external classifier, (d) on the way of generating samples to expand the training dataset. These methods can be used towards interpreting constituent individual DRL models employing DNNs. Distillation could be another option [23], but it typically requires mature DRL models: Online distillation of models, as far as we know, has not been explored. The interested reader is encouraged to read a thorough review on these methods provided in [3,12,20,22].

Recent work on mimic learning [6,16] has shown that rule-based models, like decision trees, or shallow feed-forward neural networks can mimic a not linear function inferred by a DNN with millions of parameters. The goals here is to train a mimic model with efficiency, resulting into a high performance model, which takes decisions in high-fidelity with respect to the decisions of the original model.

For DRL, authors in [16] introduce Linear Model U-trees (LMUTs) to approximate predictions for DRL agents. An LMUT is learned by an algorithm that is well-suited for an active play setting. The use of LMUTs is compared against using CART, M5 with regression tree, Fast Incremental Model Tree (FIMT) and with Adaptive Filters (FIMT-AF). The use of decision trees as interpretable policy models trained through mimicking has been also investigated in [18], in conjunction to using a causal model representing agent's objectives and opportunity chains. However, the decision tree in this work is used to infer the effects of actions approximating the causal model of the environment. Similarly to what we do here, the decision tree policy model is trained concurrently with the RL policy model, assuming a model-free RL algorithm and exploiting state-action samples using an experience replay buffer. In [7] authors illustrate how Soft Decision Trees (SDT) [10] can be used in spatial settings as interpretable policy models. SDT are hybrid classification models of binary trees of predetermined depth, and neural networks. However their inherent interpretability is questioned given their structure. Other approaches train interpretable models other than trees, such as the Abstracted Policy Graphs (APGs) proposed in [28], assuming a well-trained policy model. APGs concisely summarize policies, so that individual decisions can be explained in the context of expected future transitions.

In contrast to the above mentioned approaches, the proposed paradigm, exemplified by means of the proposed XDQN algorithm, can be applied to any setting with arbitrary state features, where the interpretable model is trained jointly to the deep model through online mimicking.

It is worth noting that, instead of the Gradient Boosting Regressors mimic learner for XDQN, we also tested naturally interpretable Linear Trees (such as LMUTs [16]); i.e. decision trees with linear models in their leaves). However, such approaches demonstrated quite low play performance with very low fidelity in the real-world complex experimental setting considered.

Regarding mimicking the Q-function of a DRL model, two known settings are the experience training and the active play settings.

In the experience training setting [6,16], all the state-action pairs $\langle s, a \rangle$ of a DRL training process are collected in a time horizon T. Then, to obtain the corresponding Q-values, these pairs are provided as input into a DRL model. The final set of samples $\{(\langle s_1, a_1 \rangle, Q_1), ...(\langle s_T, a_T \rangle, Q_T)\}$ is used as the experience training dataset. The main problem with the experience training is that suboptimal samples are collected through training, making it more difficult for a learner to mimic the behavior of the DRL model.

Active play [16] uses a mature DRL model to generate samples to construct the training dataset of an active mimic learner. The training data is collected in an online manner through queries, in which the active learner selects the actions, given the states, and the mature DRL model provides the estimated Q-values. These Q-values are then used to update the active learner's parameters on minibatches of the collected dataset. While the pitfall of suboptimal samples is addressed here, active play cannot eliminate the need for generating new trajectories to train the mimic models, which can be computationally prohibitive for real-world controllers.

Rather than following an experience or an active play training scheme, in this paper we use online training, that collects samples generated from a DRL model in a horizon of training timesteps, without requiring these samples to be generated from a mature DRL model. Samples are gathered from intermediate DRL models, during the DRL training process.

As for the provision of explanations, we opted for features' contributions to the Q-values, in a rather aggregated way, using the residue of each Gradient Boosting Regressor node, as done in [9]. This approach, as shown in [9], reports advantages over using well known feature importance calculation methods, avoiding linearity assumptions made by LIME [21] and bias in areas where features have high variance. It also avoids taking all tree paths into account in case of outliers, as done, for instance, by SHAP [17].

4   Explainable DRL Through Online Mimicking: The Deep Q-Network (XDQN) Example

To demonstrate the inherent interpretability of deep Q-learning through online mimicking of the Q-network, we propose *eXplainable Deep Q-Network (XDQN)*[1], which is an explainable variation of DQN [19]. XDQN aims at inferring a mimic learner, trained in an online manner, substituting the target Q-network of DQN.

Formally, let θ be the parameters of the Q-network and $\tilde{\theta}$ be the parameters of the mimic learner. In XDQN, the mimic learner estimates the state-action value function and selects the best action for the next state playing the role of the XDQN target:

$$Y_i^{XDQN} = r_t + \gamma \max_{a \in A} Q\left(s_{t+1}, a; \tilde{\theta}\right) \tag{7}$$

Similar to DQN, $\tilde{\theta}$ are updated every T_u number of timesteps. The full training procedure of XDQN is presented in Algorithm 1.

In contrast to DQN in which the parameters θ of the Q-network are simply copied to the target Q-network, here we perform mimic learning on $Q(s, a, \theta)$ (steps 17–20). To update $\tilde{\theta}$ we train the mimic learner on minibatches of the experience replay buffer B by minimizing the Mean Squared Error (MSE) loss function using $Q(s, a, \theta)$ to estimate the soft labels (Q-values) of the state-action pairs in the minibatches. The problem for optimizing $\tilde{\theta}$ at each update can be written as:

$$\min_{\tilde{\theta}} \mathbb{E}_{(s,a) \sim B}\left[\left(Q(s, a; \tilde{\theta}) - Q(s, a; \theta)\right)^2\right] \tag{8}$$

---

[1] The implementation code will be made available in the final version of the manuscript.

Algorithm 1 eXplainable Deep Q-Network (XDQN)

1: Initialize replay buffer B with capacity N
2: Initialize θ and $\tilde{\theta}$
3: Initialize timestep count $c = 0$
4: **for** episode 1, M **do**
5:     Augment $c = c + 1$
6:     Initialize state s_1
7:     With probability ϵ select a random action a_t, otherwise $a_t = argmax_a Q(s_t, a; \theta)$
8:     Execute action a_t and observe next state s_{t+1} and reward r_t
9:     Store transition (s_t, a_t, s_{t+1}, r_t) in B
10:     Sample a minibatch of transitions (s_i, a_i, s_{i+1}, r_i) from B
11:     **if** s_{i+1} not terminal **then**
12:         Set $Y_i^{XDQN} = r_i + \gamma \max_{a \in A} Q\left(s_{i+1}, a; \tilde{\theta}\right)$
13:     **else**
14:         Set $Y_i^{XDQN} = r_i$
15:     **end if**
16:     Perform a gradient descent step on $\left(Y_i^{XDQN} - Q(s_i, a_i; \theta)\right)^2$ w.r.t. θ
17:     **if** $c \bmod T_u = 0$ **then**
18:         Initialize $\tilde{\theta}$
19:         Sample a minibatch of transitions (s_i, a_i, s_{i+1}, r_i) from B that were stored at most $c - K$ timesteps before
20:         Perform mimic learning update on $\left(Q(s, a; \tilde{\theta}) - Q(s, a; \theta)\right)^2$ w.r.t $\tilde{\theta}$
21:     **end if**
22: **end for**

---

where, B is the prioritized experience replay buffer [24], as described in Sect. 2. Similarly to active play, when updating $\tilde{\theta}$, to ensure that the samples in minibatches provide up-to-date target values with respect to θ, we use records from the replay buffer that were stored during the K latest training steps.

It is worth noting that the hyperparameter K plays a similar role as the discounted factor γ plays for future rewards, but from the mimic learner's perspective. Building upon the experience training and active play schemes, the online training scheme leverages the benefits of both of them, aiming to minimize the required sample complexity for training the mimic model without simulating new trajectories of a mature DRL model. In particular, hyperparameter K manages the trade-off between experience training and active play in XDQN. If K is large, the mimic model learns from samples that may have been collected through more suboptimal instances of θ; deploying however data-augmented versions of Q-value. On the other hand, if K is small, it learns from the most recent instances of θ; making use of up-to-date Q-values. Nevertheless, opting for very small values of K could lead to less stable mimic training, due to the smaller number of minibatches that can be produced for updating $\tilde{\theta}$, while using large K can result in a very slow training process.

From all the above, we note that θ (Q-network) and $\tilde{\theta}$ (mimic learner) are highly dependent. To update θ, Q-network uses the mimic learner model with $\tilde{\theta}$

to compute the target soft labels (target Q-values), while to update $\tilde{\theta}$ the mimic learner uses the original Q-network with parameters θ to compute the respective target soft labels (Q-values). It is conjectured that through this dependency the interpretable target model converges to Q-values that are close to the values of the Q-network, and thus to nearly the same policy, which is an inherent feature of the DQN algorithm.

Since XDQN produces different instances of $\tilde{\theta}$ throughout training, it can eventually output multiple interpretable mimic learner models (up to the number of $\tilde{\theta}$ updates), with each one of them corresponding to a different training timestep. Since all these mimic learner instances are interpretable models, XDQN can also provide information on how the Q-network evolves towards solving the target task.

Finally, after Q-network (θ) and mimic learner ($\tilde{\theta}$) have been trained, we can discard the closed-box Q-network and use the mimic learner model as the controller. Therefore, in testing, given a state, the interpretable mimic learner selects the action with the highest Q-value, being able to also provide explainability.

5   Experimental Setup

This section demonstrates the effectiveness of XDQN through experiments in real-world settings pertaining to the demand-capacity balancing (DCB) problem of air traffic management (ATM) domain. XDQN uses a Gradient Boosting Regressor (GBR) [30] mimic learner, whose boosting ability supports effective learning by exploiting instances generated by the deep Q-network. We opted for GBR, as it usually results into robust and accurate models compared to other decision tree - based models.

Although the boosting structure of GBR makes it very difficult to provide explainability, following the work in [9] we are able to measure the contribution of state features to the predicted Q-values. In so doing, the mimic learner is expected to give local and global explanations on its decisions.

Overall, we are interested in demonstrating the proposed paradigm and show the importance of online training of mimic interpretable models. In so doing, the performance of XDQN is compared to that of DQN in complex real-world DCB problems.

5.1   Real-World Demand-Capacity Problem Setting

The current ATM system is based on time-based operations resulting in DCB [15] problems. To solve the DCB issues at the pre-tactical stage of operations, the ATM system opts for methods that generate delays and costs for the entire system. In ATM, the airspace consists of a set of 3D sectors where each one has a specific capacity. This is the number of flights that cross the sector during a specific period (e.g. of 20 min). The challenge of dealing with the DCB problem is to reduce the number of congestion cases (DCB issues, or *hotspots*), where the demand of airspace use exceeds its capacity, with small delays to an - as much as possible - low number of flights.

Recent work has transformed the DCB challenge to a multi-agent RL problem by formulating the setting as a multi-agent MDP [15]. We follow the work and the experimental setup of [13–15, 25, 26] and encourage the reader to see the problem formulation [15] in details. In this setting, we consider a society of agents, where each agent is a flight (related to a specific aircraft) that needs to coordinate its decisions regarding minutes of delay to be added in its existing delay, so as to resolve hotspots that occur, together with the other society agents. Agents' local states comprise 81 state variables related to: (a) the existing minutes delay, (b) the number of hotspots in which the agent is involved in, (c) the sectors that it crosses, (d) the minutes that the agent is within each sector it crosses, (e) the periods in which the agent joins in hotspots in sectors, and (f) the minute of the day that the agent takes off. The tuple containing all agents' local states is the joint global state. Q-learning [27] agents have been shown to achieve remarkable performance on this task [13]. In our experiments, all agents share parameters and replay buffer, but act independently.

A DCB scenario comprises multiple flights crossing various airspace sectors in a time horizon of 24 h. This time horizon is segregated into simulation time steps. At each simulation time step (equal to 10 min of real time), given only the local state, each agent selects an action which is related to its preference to add ground delay regulating its flight, in order to resolve hotspots in which it participates. The set of local actions for each agent contains |maxDelay + 1| actions, at each simulation time step. We use maxDelay = 10. The joint (global) action is a tuple of local actions selected by the agents. Similarly, we consider local rewards and joint (global) rewards. The local reward is related to the cost per minute within a hotspot, the total duration of the flight (agent) in hotspots, as well as to the delay that a flight has accumulated up to the simulation timestep [13].

5.2   Evaluation Metrics and Methods

For the evaluation of the proposed method, first, we make use of two known evaluation metrics: (a) *play performance* [16] of the deep Q-network, and (b) *fidelity* [29] of the mimic learner. Play performance measures how well the deep Q-network performs with the mimic learner estimating its temporal difference targets, while fidelity measures how well the mimic learner matches the predictions of the deep Q-network.

As far as play performance is concerned, in comparison with results reported in [13], we aim at minimizing the number of *hotspots*, the *average delay per flight* and the number of *delayed flights*. As for fidelity, we use two metric scores: (a) the *mean absolute error (MAE) of predicted Q-values* and (b) the *mimicking accuracy* score. Given a minibatch of states, we calculate the MAE of this minibatch for any action as the mean absolute difference between the Q-values estimated by the mimic learner and the Q-values estimated by the deep Q-network for that action. More formally, for a minibatch of states D_s, the MAE_i of action a_i is denoted as:

$$MAE_i = \frac{1}{|D_s|} \sum_{s \in D_s} |Q(s, a_i; \tilde{\theta}) - Q(s, a_i; \theta)| \tag{9}$$

It is worth noting that minimizing the MAE of the mimic learner is very important for training XDQN. Indeed, training a mimic model to provide the target Q-values, large MAEs can lead the deep Q-network to overestimate bad states and understimate the good ones, and thus, find very diverging policies that completely fail to solve the task.

Given a minibatch of samples, mimicking accuracy measures the percentage of the predictions of the two models that agree with each other, considering that models select the action with the highest estimated Q-value.

Second, we illustrate XDQN's *local* and *global* interpretability. We focus on providing aggregated interpretations, focusing on the contribution of state features to local decisions and to the overall policy: This, as suggested by ATM operators, is beneficial towards understanding decisions, helping them to increase their confidence to the solutions proposed, and mastering the inherent complexity in such a multi-agent setting, as solutions may be due to complex phenomena that are hard to be traced [13]. Specifically, in this work, local explainability measures state features' importance on a specific instance (i.e. a single state-action pair), demonstrating which features contribute to the selection of a particular action over the other available ones. Global explainability aggregates features' importance on particular action selections over many different instances, to explain the overall policy of the mimic learner.

Finally, the evolution of the DRL model throughout the training process is demonstrated through GBR interpretability.

5.3   Experimental Scenarios and Settings

Experiments were conducted on three in total scenarios. Each of these scenarios corresponds to a date in 2019 with heavy traffic in the Spanish airspace. In particular, the date scenarios, on which we assess our models, are 20190705, 20190708 and 20190714. However, to bootstrap the training process we utilize a deep Q-network pre-trained in various scenarios, also including 20190705 and 20190708, as it is done in [15]. In the training process, the deep Q-network is further trained according to the method we propose. The experimental scenarios were selected based on the number of hotspots and the average delay per flight generated in the ATM system within the duration of the day, which shows the difficulty of the scenario. Table 1 presents information on the three experimental scenarios. In particular, the flights column indicates the total number of flights (represented by agents) during the specific day. The initial hotspots column indicates the number of hotspots appearing in the initial state of the scenario. The flights in hotspots column indicates the number of flights in at least one of the initial hotspots. Note that all three scenarios display populations of agents (flights) of similar size, within busy summer days. For each scenario we ran five separate experiments and average results.

Table 1. The three experimental scenarios.

Scenario	Flights	Initial Hotspots	Flights in Hotspots
20190705	6676	100	2074
20190708	6581	79	1567
20190714	6773	92	2004

The implementation of XDQN utilizes a deep multilayer perceptron as the deep Q-network. The maximum depth of the Gradient Boosting Regressor is set equal to 45 and the number of minimum samples for a split equal to 20. We use the mean squared error as the splitting criterion. To train a single decision tree for all different actions, a non binary splitting rule of the root is used, based on the action size of the task, so that the state-action pairs sharing the same action match the same subtree of the splitting root. XDQN uses an ϵ-greedy policy, which at the start of exploration has ϵ equal to 0.9 decaying by 0.01 every 15 episodes until reaching the minimum of 0.04 during exploitation. The total number of episodes are set to 1600 and the update target frequency is set to 9 episodes. The memory capacity of the experience replay for the online training of the mimic learner, i.e. the hyperparameter K, is set equal to the 1/20 of the product of three other hyperparameters, namely the total number of timesteps per episode (set to 1440), the update target frequency (set to 9) and the number of agents (set to 7000). Thus, K is set to 4536000 steps.

5.4   Evaluation of Play Performance

Table 2. Comparison of performance of XDQN and DQN on the three experimental ATM scenarios (*FH*: Final Hotspots, *AD*: Average Delay, *DF*: Delayed Flights).

Scenario	DQN			XDQN		
	FH	AD	DF	FH	AD	DF
20190705	38.4	13.04	1556.5	39.0	13.19	1618.54
20190708	4.6	11.4	1387.2	6.0	11.73	1331.58
20190714	4.8	10.72	1645.2	7.0	13.46	1849.49

Table 2 demonstrates the play performance of DQN and XDQN on the three experimental scenarios. The final hotspots column indicates the number of unresolved hotspots in the final state: It must be noted that these hotspots may have emerged due to delays assigned to flights and may be different than the hotspots at the beginning of each scenario. The average delay per flight column shows the total minutes of delay imposed to all flights (when the delay is more that 4 min, according to operational practice), divided by the number of flights in the specific scenario. The delayed flights column indicates the number of flights affected by more than 4 min of delay.

We observe that XDQN performs similarly to DQN in all three evaluated metric scores, reducing considerably the number of hotspots in the scenarios, and assigning delays to the same proportion of flights. In particular, while DQN slightly outperforms XDQN in terms of the final hotspots and average delay in all three scenarios, XDQN decreases the number of the delayed flights in one scenario, while it demonstrates competitive performance on the others. This demonstrates the ability of XDQN to provide qualitative solutions while offering transparency in decision making, in contrast to DQN, which offers slightly better solutions that, however, are difficult (or even - due to their complexity - impossible) to be understood by humans [13].

5.5   Evaluation of Fidelity

As discussed in Subsect. 4.2, for the fidelity evaluation we measure the mean absolute error (MAE) between models' predicted Q-values and the mimicking accuracy score of the interpretable model. Given the DCB experimental scenarios, we train three different mimic models; namely X0705, X0708 and X0714. Table 3 reports the average MAE for each decided action over all mimic learning updates. We observe that all errors are very small, given that in testing, the absolute Q-values hovered around 200. This is very important for stabilizing the training process of XDQN, since mimic-model Q-value predictions should be ideally equal to the ones generated by the deep Q-network.

Table 3. Average Mean Absolute Errors (MAE) of the mimic model over three updates.

Action (Delay Option)	XDQN mimic model update		
	X0705	X0708	X0714
0	0.279	0.237	0.291
1	1.766	1.971	1.942
2	0.910	0.928	1.002
3	0.575	0.661	0.640
4	0.639	0.748	0.725
5	1.893	2.096	2.121
6	1.590	1.766	1.715
7	1.610	1.816	1.733
8	0.449	0.514	0.497
9	0.740	0.849	0.823
10	1.292	1.525	1.461

To further assess the fidelity of XDQN mimic learner, Table 4 illustrates the average mimicking accuracy scores over all mimic learning updates and the corresponding accuracy scores of the final mimic model. Since a Gradient Boosting Regressor mimic learner is a boosting algorithm, it produces sequential decision trees that can successfully separate the state space and approximate well the predictions of the deep Q-network function. We observe that the mimic learner and the deep Q-network agree with each other to a very good extent; namely from approximately 81% to 92%, including the final mimic model. Therefore, we expect the mimic learner to be able to accumulate the knowledge from the deep Q-network with high fidelity.

Table 4. The accuracy scores of the mimic models.

Scenario	mimicking accuracy (%) (average over training steps)	mimicking accuracy (%) (final model)
20190705	88.45	87.53
20190708	81.89	82.39
20190714	90.88	92.63

5.6   Local and Global Explainability

In the DCB setting, it is important for the human operator to understand how the system reaches decisions on regulations (i.e. assignment of delays to flights): as already pointed out, this should be done at a level of abstraction that would allow operators to understand the rationale behind decisions towards increasing their confidence to the solutions proposed, mastering the inherent complexity of the setting, and further tune solutions when necessary, without producing side-effects that will increase congestion and traffic. Therefore, operators are mainly interested in receiving succinct explanations about which state features contribute to the selection of delay actions (i.e. actions larger than 1 min) over the no-delay action (i.e. action equal to 0 min).

First, we demonstrate the ability of the mimic learner to provide local explainability. As already said, local explainability involves showing which state features contribute to the selection of a particular action over the other available ones in a specific state. To this aim, for any pair of actions - a_1 and a_2 - we calculate the differences of feature contributions in selecting a_1 and a_2 in a single state. To highlight the most significant differences, we focus only on those features whose absolute differences are above a threshold. Empirically, we set this threshold equal to 0.5. Figure 1 illustrates local explainability on a given state in which action "2" was selected: It provides the differences of feature contributions to the estimation of Q-values when selecting action "0" against selecting action "2" (denoted by "0–2"). We observe that the features that contributed more to the selection of the delay action "2" were those with index 32 (i.e. The sector in which the last hotspot occurs), 2 (i.e. the sector in which the first hotpot occurs)

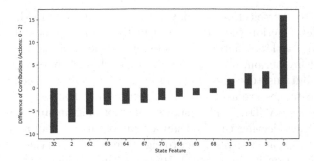

Fig. 1. Illustration of significant differences of feature contributions in selecting action "0" and action "2" in a single state, in which action "2" was selected. Positive differences mean that the respective state features have a greater contribution to Q-value when action "0" is selected, rather than when action "2" is selected. Negative differences have the opposite meaning.

Table 5. The most significant state features in terms of average contribution difference (ACD) in selecting the no-delay action versus a delay action. A positive ACD means that the corresponding state feature on average contributes more to the selection of the no-delay action "0". On the contrary, a negative ACD means that the corresponding state feature on average contributes to the selection of a delay action "1–10".

Feature Index	Feature Meaning	ACD
0	Delay the flight has accumulated up to this point	Positive
1	Total number of hotspots the flight participates in	Positive
2	The sector in which the first hotspot the flight participates occurs	Negative
3	The sector in which the second hotspot the flight participates occurs	Positive
32	The sector in which the last hotspot the flight participates occurs	Negative
62	The minutes the flight remains in the last sector it crosses	Negative
63	The minute of day the flight takes off given the delay (CTOT)	Negative
64	The minutes the flight remains in the first sector it crosses	Negative
68	The minutes the flight remains in the fifth sector it crosses	Negative

and 62 (i.e. the minutes that the flight spends crossing the last sector). Similarly to the explanations provided in [29], the arguments in favor of receiving additional minutes of delay concern the hotspots in which the flight participates, as well as the duration of the time span in which the flight crosses congested sectors (and mainly the first sector), as well as the delay that the flight has accumulated up to this point (if this is low). On the contrary, the arguments against receiving delay concern the delay that the flight has accumulated up to this point (if this is somehow high), and the small duration of the time span the flight spends in congested sectors.

Finally, we demonstrate XDQN's global explainability by aggregating the importance of features on particular action selections over many different state-action instances. In particular, we are interested in measuring the state feature

contributions to the selection of delay actions (i.e. actions in the range [1, 10])
over the no-delay action (i.e. action "0") in the overall policy. To this aim, all
pairs of actions, with one action always being the no-delay action and the other
one being a delay action, are considered. For each such pair, the differences of
feature contributions to estimating the actions' Q-values over many different
state-action instances are averaged: This results into features' Average Contribution Difference (ACD). Table 5 shows the most significant state features in
terms of ACD in selecting the no-delay action versus a delay action. The most
significant features, are (currently set to eight) features with the highest absolute
ACD, for each action in the range [1, 10]) over the no-delay action. We observe
that features with index 0, 3 contribute more to the selection of the no-delay
action. On the contrary, features with indexes 32, 2, 62 contribute more to the
selection of a delay action.

Last but not least, we demonstrate how global explainability evolves through
the training process, addressing the question of how a DRL model learns to solve
the target task. To this aim, features' contribution to selecting an action are averaged. This results into Average Features' Contribution (AFC). The (currently
set to eight) most significant features, i.e. those with the highest absolute AFC,
are those considered in explanations. Then the AFC to predicting the Q-value of
a selected action at different training episodes is provided for the most significant
features: Fig. 2 illustrates the evolution of global explainability for selecting the
no-delay action (left) and a delay action (right) through 5 representative training
episodes (360th, 720th, 1100th, 1400th and 1600th), in terms of eight features
with the highest AFC values in the final model (episode 1600). We observe that
for both evaluated actions most of the features show an increasing/decreasing
trend in their average contribution to Q-value over time, such as those with
indices 0, 1 and 63. It is worth noting that although the features 0 and 1 have

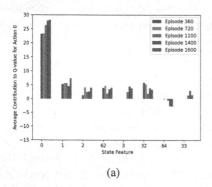 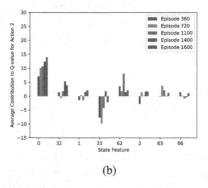

(a)   (b)

Fig. 2. Illustration of the evolution of features' contributions for selecting the no-delay
action ("0") and a delay one ("2") through 5 representative training episodes (360th,
720th, 1100th, 1400th and 1600th) in terms of average feature contribution (AFC) to
Q-value for the eight features with highest absolute AFC values in the final model
(episode 1600) in the selection of the aforementioned actions.

been highlighted as the most significant for the selection of the no-delay action, they have also significant but less contribution to selecting a delay action, as well.

6 Conclusion and Future Work

This work shows how interpretable models can be trained through online mimicking and substitute closed-box DRL models, providing inherent DRL interpretability, or be used for the provision of DRL methods interpretability. This generic paradigm follows the interpretable box design paradigm for explainable DRL and is exemplified by means of an explainable DQN (XDQN) method, where the target model has been substituted by an interpretable model that provides interpretations regarding the importance of state features in decisions. XDQN, utilizing a Gradient Boosting Regressor as the mimic learner, has been evaluated in challenging multi-agent demand-capacity balancing problems pertaining to air traffic management. Experimentally, we have shown that the XDQN method performs on a par with DQN in terms of play performance, whereas demonstrating high fidelity and global/local interpretability.

Overall, this work provides evidence for the capacity of the proposed paradigm to provide DRL methods with inherent interpretability, with high play performance and high-fidelity to the decisions of original DRL models, through online training of interpretable mimicking models.

Further work is necessary to explore how this paradigm fits into different types of DRL architectures, utilizing interpretable models that are trained to mimic different DRL closed-box models. In this line of research, we need to benchmark our methodology, utilizing state-of-the-art DRL in various experimental settings.

Regarding XDQN, further work is to design, evaluate and compare various explainable mimic models that can effectively substitute the target Q-Network.

Acknowledgements. This work has been supported by the TAPAS H2020-SESAR2019-2 Project (GA number 892358) Towards an Automated and exPlainable ATM System and it is partially supported by the University of Piraeus Research Center.

References

1. Ba, J., Caruana, R.: Do deep nets really need to be deep? In: Ghahramani, Z., Welling, M., Cortes, C., Lawrence, N., Weinberger, K. (eds.) Advances in Neural Information Processing Systems, vol. 27. Curran Associates, Inc. (2014). https://proceedings.neurips.cc/paper/2014/file/ea8fcd92d59581717e06eb187f10666d-Paper.pdf
2. Bastani, O., Pu, Y., Solar-Lezama, A.: Verifiable reinforcement learning via policy extraction. In: Proceedings of the 32nd International Conference on Neural Information Processing Systems, NIPS 2018, pp. 2499–2509. Curran Associates Inc., Red Hook, NY, USA (2018)

3. Belle, V., Papantonis, I.: Principles and practice of explainable machine learning. Front. Big Data **4**, 39 (2021)
4. Boz, O.: Extracting decision trees from trained neural networks. In: KDD 2002, pp. 456–461. Association for Computing Machinery, New York, NY, USA (2002). https://doi.org/10.1145/775047.775113
5. Boz, O.: Extracting decision trees from trained neural networks. In: Proceedings of the Eighth ACM SIGKDD International Conference on Knowledge Discovery and Data Mining, pp. 456–461 (2002)
6. Che, Z., Purushotham, S., Khemani, R., Liu, Y.: Interpretable deep models for ICU outcome prediction. In: AMIA Annual Symposium Proceedings 2016, pp. 371–380, February 2017
7. Coppens, Y., et al.: Distilling deep reinforcement learning policies in soft decision trees. In: Proceedings of the IJCAI 2019 Workshop on Explainable Artificial Intelligence, pp. 1–6 (2019)
8. Dancey, D., Bandar, Z.A., McLean, D.: Logistic model tree extraction from artificial neural networks. IEEE Trans. Syst. Man Cybern. Part B (Cybern.) **37**(4), 794–802 (2007)
9. Delgado-Panadero, Á., Hernández-Lorca, B., García-Ordás, M.T., Benítez-Andrades, J.A.: Implementing local-explainability in gradient boosting trees: feature contribution. Inf. Sci. **589**, 199–212 (2022). https://doi.org/10.1016/j.ins.2021.12.111, https://www.sciencedirect.com/science/article/pii/S0020025521013323
10. Frosst, N., Hinton, G.: Distilling a neural network into a soft decision tree. arXiv preprint arXiv:1711.09784 (2017)
11. Gu, S., Holly, E., Lillicrap, T., Levine, S.: Deep reinforcement learning for robotic manipulation with asynchronous off-policy updates. In: 2017 IEEE International Conference on Robotics and Automation (ICRA), pp. 3389–3396 (2017). https://doi.org/10.1109/ICRA.2017.7989385
12. Guidotti, R., Monreale, A., Ruggieri, S., Turini, F., Giannotti, F., Pedreschi, D.: A survey of methods for explaining black box models. ACM Comput. Surv. (CSUR) **51**(5), 1–42 (2018)
13. Kravaris, T., et al.: Explaining deep reinforcement learning decisions in complex multiagent settings: towards enabling automation in air traffic flow management. Appl. Intell. (Dordrecht, Netherlands) 53, 4063–4098 (2022)
14. Kravaris, T., et al.: Resolving congestions in the air traffic management domain via multiagent reinforcement learning methods. arXiv:abs/1912.06860 (2019)
15. Kravaris, T., Vouros, G.A., Spatharis, C., Blekas, K., Chalkiadakis, G., Garcia, J.M.C.: Learning policies for resolving demand-capacity imbalances during pretactical air traffic management. In: Berndt, J.O., Petta, P., Unland, R. (eds.) MATES 2017. LNCS (LNAI), vol. 10413, pp. 238–255. Springer, Cham (2017). https://doi.org/10.1007/978-3-319-64798-2_15
16. Liu, G., Schulte, O., Zhu, W., Li, Q.: Toward interpretable deep reinforcement learning with linear model U-Trees. In: Berlingerio, M., Bonchi, F., Gärtner, T., Hurley, N., Ifrim, G. (eds.) ECML PKDD 2018, Part II. LNCS (LNAI), vol. 11052, pp. 414–429. Springer, Cham (2019). https://doi.org/10.1007/978-3-030-10928-8_25
17. Lundberg, S.M., Lee, S.I.: A unified approach to interpreting model predictions. In: Advances in Neural Information Processing Systems, vol. 30 (2017)
18. Madumal, P., Miller, T., Sonenberg, L., Vetere, F.: Explainable reinforcement learning through a causal lens. In: Proceedings of the AAAI Conference on Artificial Intelligence, vol. 34, pp. 2493–2500 (2020)

19. Mnih, V., et al.: Human-level control through deep reinforcement learning. Nature **518**(7540), 529–533 (2015). https://doi.org/10.1038/nature14236

20. Murdoch, W.J., Singh, C., Kumbier, K., Abbasi-Asl, R., Yu, B.: Definitions, methods, and applications in interpretable machine learning. Proc. Nat. Acad. Sci. **116**(44), 22071–22080 (2019)

21. Ribeiro, M.T., Singh, S., Guestrin, C.: "Why should i trust you?": Explaining the predictions of any classifier. In: Proceedings of the 22nd ACM SIGKDD International Conference on Knowledge Discovery and Data Mining, KDD 2016, pp. 1135–1144. Association for Computing Machinery, New York, NY, USA (2016). https://doi.org/10.1145/2939672.2939778

22. Rudin, C., Chen, C., Chen, Z., Huang, H., Semenova, L., Zhong, C.: Interpretable machine learning: fundamental principles and 10 grand challenges (2021). https://doi.org/10.48550/ARXIV.2103.11251, arXiv:2103.11251

23. Rusu, A.A., et al.: Policy distillation (2015). https://doi.org/10.48550/ARXIV.1511.06295, arXiv:1511.06295

24. Schaul, T., Quan, J., Antonoglou, I., Silver, D.: Prioritized experience replay (2015). https://doi.org/10.48550/ARXIV.1511.05952, arXiv:1511.05952

25. Spatharis, C., Bastas, A., Kravaris, T., Blekas, K., Vouros, G., Cordero Garcia, J.: Hierarchical multiagent reinforcement learning schemes for air traffic management. Neural Comput. Appl. **35**, 147–159 (2021). https://doi.org/10.1007/s00521-021-05748-7

26. Spatharis, C., et al.: Multiagent reinforcement learning methods to resolve demand capacity balance problems. In: Proceedings of the 10th Hellenic Conference on Artificial Intelligence, SETN 2018. Association for Computing Machinery, New York, NY, USA (2018). https://doi.org/10.1145/3200947.3201010

27. Tan, M.: Multi-agent reinforcement learning: Independent versus cooperative agents. In: ICML (1993)

28. Topin, N., Veloso, M.: Generation of policy-level explanations for reinforcement learning. In: Proceedings of the AAAI Conference on Artificial Intelligence, vol. 33, pp. 2514–2521 (2019)

29. Vouros, G.A.: Explainable deep reinforcement learning: state of the art and challenges. ACM Comput. Surv. **55**, 1–39 (2022). https://doi.org/10.1145/3527448,just Accepted

30. Zemel, R.S., Pitassi, T.: A gradient-based boosting algorithm for regression problems. In: Proceedings of the 13th International Conference on Neural Information Processing Systems, NIPS 2000, pp. 675–681. MIT Press, Cambridge, MA, USA (2000)

31. Zhao, X., et al.: DEAR: deep reinforcement learning for online advertising impression in recommender systems. In: Proceedings of the AAAI Conference on Artificial Intelligence, vol. 35(1), pp. 750–758, May 2021. https://ojs.aaai.org/index.php/AAAI/article/view/16156

Counterfactual, Contrastive, and Hierarchical Explanations with Contextual Importance and Utility

Kary Främling[1,2]([⊠]) (iD)

[1] Computing Science, Umeå University, 901 87 Umeå, Sweden
kary.framling@cs.umu.se
[2] Department of Industrial Engineering and Management, Aalto University,
Maarintie 8, 00076 Aalto, Finland
kary.framling@aalto.fi

Abstract. Contextual Importance and Utility (CIU) is a model-agnostic method for post-hoc explanation of prediction outcomes. In this paper we describe and show new functionality in the R implementation of CIU for tabular data. Much of that functionality is specific to CIU and goes beyond the current state of the art.

Keywords: Contextual Importance and Utility · Explainable AI · Open source · Counterfactual · Contrastive

1 Introduction

Contextual Importance and Utility (CIU) was presented by Kary Främling in 1992 [1] for explaining recommendations or outcomes of decision support systems (DSS) in a model-agnostic way. CIU was presented formally in [2,3] and more recent developments have been presented *e.g.* in [5]. This paper presents new functionality of CIU that is implemented in the R package for tabular data, available at https://github.com/KaryFramling/ciu. An earlier version of the package was presented at the Explainable Agency in Artificial Intelligence Workshop of the AAAI conference in 2021 [4].

CIU has a different mathematical foundation than the state-of-the-art XAI methods SHAP and LIME. CIU is not limited to "feature influence" and therefore offers richer explanation possibilities than the state-of-the-art methods.

After this Introduction, Sect. 2 resumes the core theory of CIU. Section 3 shows the new functionality, followed by Conclusions that include a brief discussion about CIU versus comparable state-of-the-art XAI methods.

2 Contextual Importance and Utility

Contextual Importance (CI) expresses to what extent modifying the value of one or more **feature(s)** $x_{\{i\}}$ can affect the output value y_j (or rather the *output utility* $u_j(y_j)$). CI is expressed formally as:

The work is partially supported by the Wallenberg AI, Autonomous Systems and Software Program (WASP) funded by the Knut and Alice Wallenberg Foundation.

D. Calvaresi et al. (Eds.): EXTRAAMAS 2023, LNAI 14127, pp. 180–184, 2023.
https://doi.org/10.1007/978-3-031-40878-6_16

$$CI_j(c, \{i\}, \{I\}) = \frac{umax_j(c, \{i\}) - umin_j(c, \{i\})}{umax_j(c, \{I\}) - umin_j(c, \{I\})}, \tag{1}$$

where c is the studied context/instance, $\{i\} \subseteq \{I\}$ and $\{I\} \subseteq \{1, \ldots, n\}$ and n is the number of features. $umin_j$ and $umax_j$ are the minimal and maximal output utility values that can be achieved by varying the value(s) of feature(s) $x_{\{i\}}$ while keeping all other feature values at those of c.

In classification tasks we have $u_j(y_j) = y_j \in [0, 1]$ and for regression tasks where $u_j(y_j) = Ay_j + b$ (which applies to most regression tasks) we can write:

$$CI_j(c, \{i\}, \{I\}) = \frac{ymax_j(c, \{i\}) - ymin_j(c, \{i\})}{ymax_j(c, \{I\}) - ymin_j(c, \{I\})}, \tag{2}$$

Contextual Utility (CU) expresses to what extent the current **value(s)** of given feature(s) contribute to obtaining a high output utility u_j. CU is expressed formally as:

$$CU_j(c, \{i\}) = \frac{u_j(c) - umin_j(c, \{i\})}{umax_j(c, \{i\}) - umin_j(c, \{i\})} \tag{3}$$

When $u_j(y_j) = Ay_j + b$, this can again be written as:

$$CU_j(c, \{i\}) = \left| \frac{y_j(c) - yumin_j(c, \{i\})}{ymax_j(c, \{i\}) - ymin_j(c, \{i\})} \right|, \tag{4}$$

where $yumin = ymin$ if A is positive and $yumin = ymax$ if A is negative.

Contextual influence expresses how much feature(s) influence the output value (utility) relative to a *reference value* or *baseline*, here denoted $neutral.CU \in [0, 1]$. Contextual influence is conceptually similar to Shapley value and other additive feature attribution methods. Formally, Contextual influence is:

$$\phi = CI \times (CU - neutral.CU) \tag{5}$$

where "$_j(c, \{i\}, \{I\})$" has been omitted for easier readability.

It is worth noting that CI and CU are values in the range $[0, 1]$ by definition, which makes it possible to assess whether a value is high or low. Contextual influence also has a maximal amplitude of one, where the range is $[-neutral.CU, 1 - neutral.CU]$. CIU calculations require identifying $ymin_j$ and $ymax_j$ values, which can be done in many ways. The approach used for the moment is described in [4] and is omitted here due to space constraints.

All CIU equations apply to each feature separately as well as to coalitions of features $\{i\}$ versus other coalitions of features $\{I\}$, where $\{i\} \subseteq \{I\}$ and $\{I\} \subseteq \{1, \ldots, n\}$. Such coalitions can be used to form *Intermediate Concepts*, which name a given set of inputs $\{i\}$ or $\{I\}$. Such Intermediate Concepts make it possible to define arbitrary explanation vocabularies with abstraction levels that can be adapted to the target user.

3   New Explanation Functionality with CIU

The source code for producing the results shown here is published at https:// github.com/KaryFramling/EXTRAAMAS2023. The R package source code is available at https://github.com/KaryFramling/ciu and on CRAN.

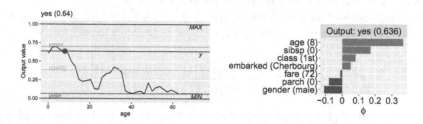

Fig. 1. Left: Generated illustration of CIU calculations. Right: Contextual influence barplot explanation for "Johnny D".

To begin, we use the Titanic data set, a Random Forest model and an instance "Johnny D", as in https://ema.drwhy.ai. "Johnny D" is an 8-year old boy that travels alone. The model predicts a survival probability of 63.6%. 63.6% is good compared to the average 32.5%, which is what we want to explain. Figure 1 illustrates how CI, CU and Contextual influence is calculated for the feature "age" and a Contextual influence plot for "Johnny D" with $neutral.CU = 0.325$.

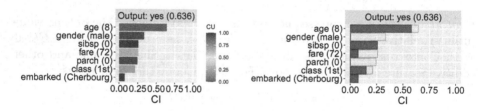

Fig. 2. Left: "Old" visualisation of CI and CU, with CU illustrated using colors. Right: New visualisation that answers more exactly the counterfactual "what-if" question.

Counterfactual Explanations answer the question "What if?". Figure 2 shows an older visualisation with CI as the bar length and CU illustrated with a color. The new visualisation illustrates CI as a transparent bar and CU as a solid bar. When $CU = 0$ (worst possible value), the solid bar has length zero. When $CU = 1$ (best possible value), the solid bar covers the transparent bar. This is called "counterfactual" because it indicates what feature(s) have the greatest potential to improve the result. In Fig. 2 we can see that being accompanied by at least one parent (feature "parch") could increase the probability of survival.

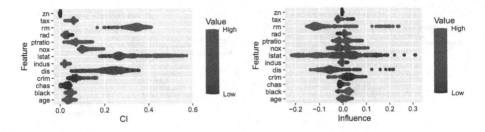

Fig. 3. Beeswarm visualisation of CI and Contextual influence for Boston data set.

Beeswarm Visualisation. Beeswarms give an overview of an entire data set by showing CI/CU/influence values of every feature and every instance. As in https://github.com/slundberg/shap, we use the Boston data set and a Gradient Boosting model. The dot color in Fig. 3 represents the feature value. The CI beeswarm in Fig. 3 reveals for example that the higher the value of "lstat" (% lower status of the population), the higher is the CI (contextual/instance-specific importance) of "lstat". The influence plot reveals that a high "lstat" value lowers the predicted home price and is nearly identical to the one produced for Shapley values. We use $neutral.CU = 0.390$, which corresponds to the average price so the reference value is the same as for the Shapley value.

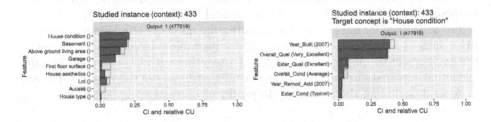

Fig. 4. Left: Top-level explanation for why Ames instance 433 is expensive. Right: Detailed explanation for Intermediate Concept "House condition".

Intermediate Concepts. Ames housing is a data set with 2930 houses described by 81 features. A gradient boosting model was trained to predict the sale price based on the 80 other features. With 80 features a "classical" bar plot explanation becomes unreadable. Furthermore, many features are strongly correlated, which causes misleading explanations because individual features have a small importance, whereas the joint importance can be significant. Intermediate Concepts solve these challenges, as illustrated in Fig. 4 that shows the top-level explanation and an explanation for one of the Intermediate Concepts for an expensive house. Here, the vocabulary has been constructed based on common-sense knowledge about houses but it could even be provided by the explainee.

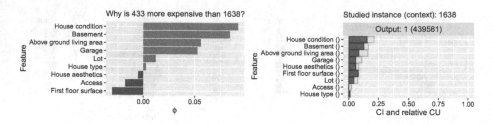

Fig. 5. Left: Contrastive "Why?" explanation for two expensive Ames houses. Right: Top-level counterfactual explanation for Ames instance 1638.

Contrastive Explanations. Contrastive explanations answer questions such as "Why alternative A rather than B" or "Why not alternative B rather than A". Any value in the range $[0, 1]$ can be used for *neutral.CU* in Eq. 5, including CU values of an instance to compare with. Figure 5 shows a contrastive explanation for why Ames instance #433 ($477919, see Fig. 4) is predicted to be more expensive than instance #1638 ($439581). Contrastive values are in the range $[-1, 1]$ by definition, so the differences between the compared instances in Fig. 5 are small.

4   Conclusion

CIU enables explanations that are not possible or available with current state-of-the-art methods. Notably, Shapley value and LIME are limited to "influence" values only. Even for influence values, Contextual influence offers multiple advantages such as a known maximal range and adjustable reference value. However, the emphasis of the paper is to show how CI together with CU can provide counterfactual explanations and give a deeper understanding of the model behaviour in general, including the possibility to produce contrastive explanations.

References

1. Främling, K.: Les réseaux de neurones comme outils d'aide à la décision floue. D.E.A. thesis, INSA de Lyon (1992)
2. Främling, K.: Explaining results of neural networks by contextual importance and utility. In: Andrews, R., Diederich, J. (eds.) Rules and networks: Proceedings of Rule Extraction from Trained Artificial Neural Networks Workshop, AISB'96 Conference. Brighton, UK (1–2 April 1996)
3. Främling, K.: Modélisation et apprentissage des préférences par réseaux de neurones pour l'aide à la décision multicritère. Phd thesis, INSA de Lyon (Mar 1996)
4. Främling, K.: Contextual Importance and Utility in R: the 'CIU' Package. In: Madumal, P., Tulli, S., Weber, R., Aha, D. (eds.) Proceedings of 1st Workshop on Explainable Agency in Artificial Intelligence Workshop, 35th AAAI Conference on Artificial Intelligence, pp. 110–114 (2021)
5. Främling, K.: Contextual importance and utility: a theoretical foundation. In: Long, G., Yu, X., Wang, S. (eds.) AI 2022. LNCS (LNAI), vol. 13151, pp. 117–128. Springer, Cham (2022). https://doi.org/10.1007/978-3-030-97546-3_10

Cross-Domain Applied XAI

Explanation Generation via Decompositional Rules Extraction for Head and Neck Cancer Classification

Victor Contreras[✉][ID], Andrea Bagante[✉], Niccolò Marini[ID],
Michael Schumacher[ID], Vincent Andrearczyk[ID], and Davide Calvaresi[ID]

University of Applied Sciences Western Switzerland (HES-SO), Delèmont,
Switzerland
{victor.contrerasordonez,davide.calvaresi}@hevs.ch

Abstract. Human papillomavirus (HPV) accounts for 60% of head and
neck (H&N) cancer cases. Assessing the tumor extension (tumor grading) and determining whether the tumor is caused by HPV infection
(HPV status) is essential to select the appropriate treatment. Therefore,
developing non-invasive, transparent (trustworthy), and reliable methods
is imperative to tailor the treatment to patients based on their status.
Some studies have tried to use radiomics features extracted from positron
emission tomography (PET) and computed tomography (CT) images to
predict HPV status. However, to the best of our knowledge, no research
has been conducted to explain (e.g., via rule sets) the internal decision
process executed on deep learning (DL) predictors applied to HPV status
prediction and tumor grading tasks. This study employs a decompositional rule extractor (namely DEXiRE) to extract explanations in the
form of rule sets from DL predictors applied to H&N cancer diagnosis.
The extracted rules can facilitate researchers' and clinicians' understanding of the model's decisions (making them more transparent) and can
serve as a base to produce semantic and more human-understandable
explanations.

Keywords: Local explainability · Global explainability · Feature
ranking · rule extraction · HPV status explanation · TNM explanation

1 Introduction

Despite recent advances in head-and-neck (H&N) cancer diagnosis and staging,
understanding the relationship between human papillomavirus (HPV) status and
such cancers is still challenging. An early diagnosis of HPV could dramatically
improve the patient's prognosis and enable targeted therapies for this group,
enhancing their life quality and treatment effectiveness [15]. Moreover, consolidating the diagnosis of cancer staging made by doctors for cancer growth and
spread could help better understand how to treat the specific patient, adapting
the therapy to the severity of the disease and HPV status.

© The Author(s), under exclusive license to Springer Nature Switzerland AG 2023
D. Calvaresi et al. (Eds.): EXTRAAMAS 2023, LNAI 14127, pp. 187–211, 2023.
https://doi.org/10.1007/978-3-031-40878-6_11

The term "head and neck cancer" describes a wide range of cancers that develop from the anatomical areas of the upper aerodigestive tract [35]. Considering the totality of H&N malignancies, this type of cancer is the 7th leading cancer by incidence [7]. The typical patients affected by H&N cancers are older adults who have used tobacco and alcohol extensively. Recently, with the ongoing progressive decrease in the use of these substances, the insurgences of such cancers in older adults is slowly declining [13]. However, occurrences of HPV-associated oropharyngeal cancer are increasing among younger individuals (e.g., in North America and Northern Europe) [17].

Diagnosis of HPV-positive oropharyngeal cancers in the United States grew from 16.3% to more than 71.7% in less than 20 years [8]. Fortunately, patients with HPV-positive oropharyngeal cancer have a more favorable prognosis than HPV-negative ones since the former are generally healthier, with fewer coexisting conditions, and typically have better responses to chemotherapy and radiotherapy. Therefore, promptly detecting HPV-related tumors is crucial to improve their prognosis and tailor the treatments [30]. The techniques used in clinical practice to test the presence of HPV have obtained promising results. Still, they are affected by drawbacks, including a high risk of contamination, time consumption, high costs, invasiveness, and possibly inaccurate results [5].

The TNM staging technique is used to address the anatomic tumor extent using the "tumor" (T), "lymph node" (N), and "metastasis" (M) attributes, where "T" denotes the size of the original tumor, "N" the extent of the affected regional lymph nodes, and "M" the absence or presence of distant metastasis [42]. Accurate tumor staging is essential for treatment selection, outcome prediction, research design, and cancer control activities [24]. TNM staging is determined employing diagnostic imaging, laboratory tests, physical exams, and biopsies [25]. Radiomics relies on extracting quantitative metrics (radiomic features) within medical images that capture tissue and lesion characteristics such as heterogeneity and shape [33].

In recent decades, the employment of radiomics has shown great benefits in personalized medicine [16]. Radiomics is adopted as a substitute for invasive and unreliable methods, and they are applied in many contexts, including H&N cancer, with a particular interest in tumor diagnostic, prognostic, treatment planning, and outcome prediction. While the application of radiomics to predict TNM staging has never been addressed in the literature, the possibility of using them to predict the presence or absence of HPV has been recently explored. For example, recent research has shown that it is possible to predict HPV status by using deep learning (DL) techniques that exploit radiomics features [6].

Nevertheless, to the best of our knowledge, no study has yet focused on explaining internal DL predictors' behavior through a rule extraction process, investigating and assessing the roles of the features and how they compose the rules leading the DL predictor decision. Therefore, there is a need for more investigations to fully understand the main tumor characteristics leading DL models to generate their prediction. To explain DL predictors' behavior, this study investigates the use of a tool for decompositional rules generation in deep neural networks (namely DEXiRE [9]) to generate rules from Positron Emission Tomography (PET) and Computed Tomography (CT) images in the context of H&N cancers.

In the context of ML models, transparency can be defined as the degree of understanding of the models' internal decision mechanisms, and overall behaviors can be simulated [18,45,52]. To increase transparency in ML models in general and in DL models in particular, we have employed decompositional rule extractors (a.k.a. DEXiRE) because this method can express neural activations in terms of logical rules that both human and artificial agents can understand, thus improving the understanding of the internal decision process executed by the model. The main contribution of this work is to apply the DEXiRE, explainable artificial intelligence (XAI) technique to explain through rule sets the internal decision process executed by DL predictors. Thus, clinicians and researchers can understand the predictors' behavior to improve them in terms of performance and transparency.

In particular, we extracted radiomics features from PET-CT scans and trained several machine learning (ML) and deep learning (DL) predictors in classification tasks. In turn, we leverage DEXiRE to extract rule sets from a DL model trained on radiomics features extracted offline from PET-CT images. DEXiRE determines the most informative neurons in each layer that lead to the final classification (henceforth, which features and in which combination have contributed to the final decision). Finally, we have assessed and discussed the rule sets.

The rest of the paper is organized as follows: Sect. 2 presents the state of the art of H&N cancer diagnosis using radiomics features and DL models. Section 3 describes the proposed methodology. Section 4 presents and analyses the results. Section 5 discusses the overall study. Finally, Sect. 6 concludes the paper.

2   State of the Art

The metabolic response captured on PET images enables tumors' localization and tissues' characterization. PET images are frequently employed as a first-line imaging tool for studying H&N cancer [27]. Moreover, PET is widely used in the early diagnosis of neck metastases. Indeed, PET highlights the metabolic response of the tumors since their early stages — which cannot be seen with other imaging techniques [1]. Thus, PET and CT scans are often used for several applications in the context of H&N cancer.

The most relevant study concerning tumor segmentation includes Myronenko et al. [39], which in the HECKTOR challenge (3rd edition), created an automatic pipeline for the segmentation of primary tumors and metastatic lymph nodes, obtaining the best result on the challenge with an average aggregate Dice Similarity Coefficient (DSC_{agg}) of 0.79. [2].

Rebaud et al. [46] predicted the risk of cancer recurrence's degree using radiomics features and clinical information, obtaining an encouraging concordance index score of 0.68. Among the classification contributions, it is worth mentioning Pooja Gupta et al. [21], who developed a DL model to classify CT scans as tumoral (or not), reaching 98.8% accuracy. Martin Halicek et al. [22] developed a convolutional neural network classifier to classify excised, squamous-cell carcinoma, thyroid cancer, and standard H&N tissue samples using Hyperspectral imaging (with an 80% accuracy). Konstantinos P. Exarchos et al. [12] used features extracted from CT and MRI scans in a classification scheme to predict potential diseases' reoccurrence, reaching 75.9% accuracy. Only recently, researchers have also focused on HPV status prediction. Ralph TH Leijenaar et al. [31] predict HPV status in oropharyngeal squamous cell carcinoma using radiomics extracted from computed tomography images (with an Area Under the Curve value of 0.78). Bagher-Ebadian et al. [6] construct a classifier for the prediction of HPV status using radiomics features extracted from contrast-enhanced CT images for patients with oropharyngeal cancers (The Generalized Linear Model shows an AUC of 0.878). Bolin Song et al. [51] develop and evaluate radiomics features within (intratumoral) and around the tumor (peritumoral) on CT scans to predict HPV status (obtaining an AUC of 0.70). Chong Hyun Suh et al. [53] investigated the ability of machine-learning classifiers on radiomics from pre-treatment multiparametric magnetic resonance imaging (MRI) to predict HPV status in patients with oropharyngeal squamous cell carcinoma (logistic regression, the random forest, XG boost classifier, mean AUC values of 0.77, 0.76, and 0.71, respectively). However, to the best of our knowledge, no studies have yet involved XAI techniques to unveil the underlying rules, mechanisms, and features leading the ML/DL predictors to their outcomes in such a context. Explaining why/how the models have been obtained is imperative, especially in the medical (diagnosis and decision support systems) domains. Having transparent (i.e., explainable models) models foster understandability, transparency, and trust.

Henceforth, contributions to the XAI field aim to explain the decision-making process carried out by AI algorithms to increase their transparency and trustworthiness [3,11]. XAI is fundamental in safe-critical domains like medicine, where clinicians and patients require a thorough understanding of decision processes carried out by automatic systems to trust them [38].

AI algorithms, including decision trees, linear models, and rule-based systems, are *explainable-by-design*, meaning that predictions can be expressed as rules, thresholds, or linear combinations of the input features making the decision process transparent and interpretable [36]. However, algorithms like DL models and support vector machines with non-linear kernels are characterized by non-linear relationships between the input and the output, which improves performance and generalization — making the explanation process more challenging [19]. Therefore, a post-hoc approach is necessary to explain the decision-making process in complex and non-linear algorithms non-explainable-by-design (a.k.a. black-boxes). The post-hoc explanation approach is a third-party method that uses the model structure and input-output relationship to explain AI models [37,50]. Post-hoc explanations can be classified into *local* and *global*. The former interprets one sample at a time — see methods based on sensibility analysis like Local Interpretable Model-agnostic Explanation (LIME) [34], local feature importance and utility CIU [14,28], and methods based on local surrogate models [40]. While local explanations have the drawback of being valid only for one example or a small set of examples in the input space, global explanation methods aim to explain the overall predictor's behavior covering as many samples as possible. Global explanations methods include global surrogate models [47,57], global feature importance and attribution [20,43], and rule extraction methods [4].

Rule extraction methods can follow three main approaches. First, *decompositional* methods look inside the predictors' structure to induce rules; algorithms like FERNN [49], ECLAIRE [56], and DEXiRE [9] are examples of this approach. Second, *pedagogical* approaches extract rules based on the relationship between input features and predictions (e.g., TREPAN [10]). Finally, *eclectic* methods combine decompositional and pedagogical methods to produce explanations (e.g., Recursive Rule-extraction (RX) [23]).

The following section describes the methodology used to produce post-hoc explanations of DL models through binarizing neurons and rule induction methods in the medical domain.

3   Methodology

This section presents the approach undertaken to explain the decision-making process carried out by a DL predictor trained on radiomics features to predict the HPV status and the TNM staging through rules. In particular, we executed experiments in two classification tasks:

T1: HPV diagnosis targets the binary variable *HPV_status*, which describes if a given patient has an HPV tumor or not (HPV_status = 1/0).

T2: Cancer staging consists of assigning to a tumor a grade value (an integer number between 1 and 4) based on the TNM tumor grading system, which is composed of three measures: Tumor primary size and extent, Nearby lymph nodes infiltrated, and Metastasis [26,48,54]. Tumor grading can be modeled as a multiclass classification task for machine learning.

Figure 1 schematizes the experimental pipeline employed to extract the underlying rules leading the predictors trained in T1 and T2 to their outcomes. The experimental pipeline starts with the feature extraction process from PET-CT images. Then, these features have been preprocessed, and exploratory data analysis (EDA) is required to understand the data and the task and choose the appropriate predictors. In turn, predictors are trained and fined tuned using 5-fold cross-validation. Next, the rule set extraction takes place using the training set and pre-trained DL predictor. Finally, the rule set is evaluated using the test set and compared with the baseline predictors' performance.

3.1   Experimental Pipeline

Below, a brief description of the data set used in this study.

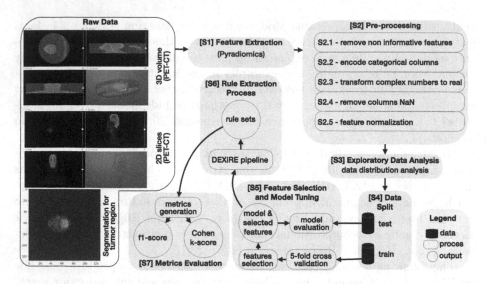

Fig. 1. Overall experimental pipeline.

Dataset Description. The dataset used in this study is the HEad and neCK TumOR (HECKTOR 2022) dataset. The dataset was introduced to compare and rank the top segmentation and outcome prediction algorithms using the same guidelines and a comparable, sufficiently large, high-quality set of medical data [2,41]. The patients populating the data set have histologically proven oropharyngeal H&N cancer and have undergone/planned radiotherapy and/or chemotherapy and surgery treatments. The data originates from FluoroDeoxyGlucose (FDG) and low-dose non-contrast-enhanced CT images (acquired with combined PET-CT scanners) of the H&N region. The primary tumor (GTVp) and lymph nodes (GTVn) segmentations were also provided with the images. The records from this dataset have been collected in eight different centers and contain 524 training examples and 359 for testing. For this study, for task T1 (HPV status prediction), we have removed all those samples with unknown values or without HPV status values (target variable), obtaining a value of 527 patients in total (train + test sets). For Task T2, TNM staging, we have chosen only patients from the TNM 7th edition, obtaining a value of 640 patients in total (train + test sets).

Experimental Pipeline Description. The steps composing the experimental pipeline shown in Fig. 1 are characterized as follows:

S1 Feature extraction: The features have been extracted from 3D PET-CT volumes using the Python library Pyradiomics, a tool to calculate radiomics features from 2D and 3D medical images [55]. In this study, the features for T1 and T2 have been extracted from a bounded box surrounding the interest area in the image and not from the whole image, reducing the computational cost of the feature extraction process and producing meaningful features that allow to classify and grade the tumors.

S2: Preprocessing: The extracted features have been preprocessed to make them suitable for training ML/DL predictors.

The feature preprocessing is structured as follows:

S2.1 Remove non-informative features: Non-informative features like duplicated, empty and constant ones have been removed. Additionally, features like the position of the bonding box have been removed because they are not informative for tasks T1 and T2. Features nb_lymphonodes and nb_lessions have also been removed, given that such features are directly related to the target tumor grade. Inferring the target variable from them is trivial, and it yields to ignore other radiomics features.

S2.2 Encode categorical features: Categorical features like *gender* have been encoded into numerical values using one hot encoding procedure.

S2.3 Transform complex numbers to real: Some features calculated with Pyradiomics include complex numbers in which the imaginary part is always zero. To provide a uniform and suitable number format for ML predictors, features with complex numbers have been transformed to float point data types by taking only their real parts.

S2.4 Removing NaN columns: Not-a-Number (NaN) columns are numeric columns with missing values and cannot be employed in an ML predictor. Although some missing values can be imputed, in this case, we have decided not to do so to avoid introducing bias and additional uncertainty in the predictors.

S2.5 Feature Normalization: Once NaN columns have been removed, the features have been normalized using the standard scale method, which scales all the features to the same range and thus avoids biases from scale differences between features.

S3: Exploratory Data Analysis (EDA):
The objective of the exploratory data analysis is to identify the most relevant characteristics and the structure of the data set to select the most appropriate predictors for each task. In EDA, a correlation between the features and the target variable has been calculated to identify possible linear relationships between the input features and the target variable. Additionally, correlation analysis has been performed between the features to identify correlation and collinearity. Finally, a distribution analysis has been applied to the target variables, showing a considerable imbalance between the classes.

S4: Data split: To maintain reproducible experimentation and fair comparison between the different predictors, the data set was split into 80% for training and 20% for testing using the same random seed and stratified sampling. Additionally, to this data partition, we also employ the HECKTOR challenge partition, where the test set is composed of centers MDA, USZ, and CHB.

S5: Feature selection and model tuning:
Due to the high number of features after the preprocessing step, ~ 2427 for task T1 and ~ 2035 for task T2, a feature selection process has been applied

to reduce the complexity, avoid the curse of dimensionality [29] and focus on those features strongly related to the target variable. The feature selection process has been executed over the train set.

The *feature selection* (FS) process encompasses the following steps:

   FS1 For each input feature, calculate the univariate correlation coefficient with the target variable.

   FS2 Rank features based on the absolute value of the correlation coefficients calculated in step FS1.

   FS3 Choose the highest top 20 features based on the rank.

   FS4 Filter the train and test parts with selected features.

Once the features were selected, we trained four models (M1-M4) in the same setup, as described below.

   M1 Support Vector Machine (SVM) with a linear kernel and a C parameter of C=10, and a *class_weight* parameter set in *weighted*.

   M2 Decision Tree (DT) with a maximum depth of 50 levels and with impurity Ginny metric as bifurcation measure.

   M3 Random forest (RF) model with 100 estimators.

   M4 DL predictor is a Feed-forward neural network in which hyperparameters have been selected employing 5-fold cross-validation and three candidate architectures, with hidden layers ranging from 2 to 6. The number of neurons in each hidden layer varies from 2^2 to 2^8 on a logarithmic scale.

These models have been chosen due to their similar performance and interpretability.

S6: Rule extraction process: To extract logic rules from a pre-trained DL predictor, we have employed the algorithm DEXiRE [9]. Such an algorithm extracts boolean rules from a DL predictor by binarizing network activations and then inducing logical rules using the binary activations and the rule inductor algorithms. Finally, the inducted rules are recombined (from local to global) and expressed in terms of input features (see Fig. 2 – DEXiRE pipeline).

S7: Metrics evaluation:

To measure the predictors' performance, we have employed the following performance metrics (PM):

   PM1 Cohen-Kappa (CK-score): Cohen-kappa score measures the agreement level between the conclusions of two experts. Its value oscillates between -0,20 to 1.0. Negative or low CK-score values indicate not or slight agreement, whereas high values indicate total or strong agreement [32].

   PM2 F1-score: F1-score belongs to a family of metrics (F-measure) or (F-score). Its value oscillates between 0.0 to 1.0. Higher F1-score values indicate high performance. The F1-score calculates the harmonic mean between the precision and recall measures, combining specificity and sensibility measures. For the case of the multiclass F1-score reported, we employed the weighted mean to consider the imbalance of the dataset.

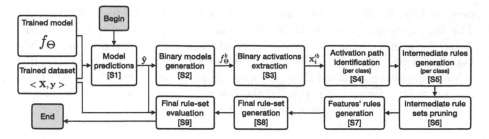

Fig. 2. DEXiRE pipeline to extract rules from DL predictor [9].

Additionally, to measure the ability of the rule set to explain the original model, we have used the following measures:

PM3 Fidelity: Fidelity measures how similar are the predictions from the rule set to the prediction of the original DL predictor.

PM4 Rule length: Rule length measures the number of atomic unique Boolean terms in the rule set.

All the models have been trained and evaluated on the same train and test partitions to make analysis and comparison of results possible. In addition, DNNs have been tested and compared against a set of baseline models with similar capabilities, i.e., Support Vector Machine (SVM), Decision Tree, and Random Forest.

4 Results and Analysis

It is worth recalling that the overall objective of this study is to explain a DL predictor's behavior (IIPV status prediction or tumor staging in H&N cancer) through rule sets extracted employing the decompositional rule extraction algorithm DEXiRE. Table 1 presents the rule sets' average performance for task T1 (HPV diagnosis) and T2 (Cancer staging), employing the metrics described in step S7 (see Sect. 3).

Datasets for tasks T1 and T2 are highly unbalanced, which can affect the predictors' and rule sets' performance. We have executed three experiments with different balancing and partitions for each task to test the possible effect of high dataset imbalance on the rule generation process and the rule set's performance.

Table 1 summarizes DEXiRE's rule sets for task T1 (HPV diagnosis) and T2 (Tumor staging) in three different dataset balance configurations. First, the imbalance dataset, in which partitions have been randomly selected while maintaining the proportions of the target variables. The dataset has been balanced with an oversampling technique (SMOTE) in the second configuration. In the third configuration, the dataset follows HECKTOR's challenge partitions, which are focused on medical centers' generalization. The rest of the Table 1 is organized as follows, the third column summarizes the average and standard deviation of the rule length (number of features involved in the rule), with values ranging from 7.6 to 14.8 terms for task T1 and 11.2 to 13.6 terms for task T2. The fourth column presents the fidelity measure, which describes the similarity degree between the rule sets' predictions and those from the original model. The highest fidelity value for T1 is ≈ 80%, while for task T2 is ≈ 73%. The fifth column shows the obtained F1-score, with values above ≈ 80% for task T1 and around ≈ 70% for task T2. Finally, the last column apprises the Cohen-kappa score ≈ 12% for T1 and above 20% for T2.

Table 1. DEXiRE's rule set performance on Task T1 (HPV diagnosis) and task T2 (cancer staging) for different dataset partition and balancing conditions. Numerical results are reported with *average value* ± *standard deviation*. The best results in each task are highlighted in bold.

Task	Data set balancing	Rule length	Fidelity	F1-score	CK-score
T1	Imbalanced	8.8 ± 0.9797	0.7622 ± 0.2490	0.8507 ± 0.0097	0.1207 ± 0.0315
	Balanced	14.8 ± 0.9797	**0.8075 ± 0.0237**	**0.8833 ± 0.0162**	**0.1366 ± 0.0682**
	Hecktor partitioning	**7.6 ± 0.7999**	0.7304 ± 0.0836	0.8230 ± 0.0724	0.0827 ± 0.0702
T2	Imbalanced	**11.2 ± 0.9797**	**0.7343 ± 0.0450**	0.7226 ± 0.0330	**0.2740 ± 0.0811**
	Balanced	13.6 ± 0.7999	0.7312 ± 0.0429	**0.7354 ± 0.0203**	0.2620 ± 0.0455
	Hecktor partitioning	**11.2 ± 0.9797**	0.6459 ± 0.0261	0.7168 ± 0.0294	0.2085 ± 0.0412

Appendix A shows the rule set recording the highest performance for each experiment for both T1 and T2.

4.1 Task T1 HPV Diagnosis

Task T1 performs a binary classification employing the radiomics features to predict whether a given patient is HPV positive or not. Results obtained for different datasets' configuration are described in the following subsections.

Experiment with Imbalanced Dataset. The dataset has not been modified in this setting, retaining its natural imbalance of 90% positive and 10% negative samples. Table 2 shows the results obtained by the baseline models, the DL predictor, and the DEXiRE's rule set concerning performance metrics PM1 to PM4 (Sect. 3 – step S7). The first column shows the F1-score is reported with all the values over approximately 80%, and the DL predictor obtained the best score (91%). The second column shows the Cohen-Kappa score (CK-score), whose values range from 12% to 28%. Once again, the DL predictor obtains the maximum score. The rule length and fidelity metrics concern only the rule set. The average rule length for this experiment is 8.8 boolean terms, and the fidelity is 76%.

Table 2. Results for baseline models, DL predictor and extracted rule set in task T1 with the imbalanced dataset. Numerical results are reported with *average value ± standard deviation*. The highest results in each column are highlighted in bold.

Model	F1-score	CK-score	Rule length	Fidelity
SVM	0.8743 ± 0.0104	0.2157 ± 0.1434	NA	NA
Decision tree	0.8647 ± 0.0224	0.2699 ± 0.1164	NA	NA
Random forest	0.8689 ± 0.0146	0.1933 ± 0.0957	NA	NA
Neural Network	**0.9153 ± 0.0123**	**0.2872 ± 0.0557**	NA	NA
DEXiRE's rule set	0.8507 ± 0.0097	0.1207 ± 0.0315	8.8 ± 0.9797	0.7622 ± 0.2490

Experiment with Balance Dataset. To test the effect of an artificial balancing dataset technique, we have executed an experiment with a balanced training set employing the oversampling method SMOTE, which allows the drawing of new samples from the minority class based on the neighbors. Table 3 shows the obtained results. In particular, the first column shows the F1-score (all the values are over approximately 80%, and the DL predictor got the best score of 91%). The second column shows the CK-scores, ranging from 13% to 63%. Again, the SVM has obtained the maximum score with ≈ 63%. The average rule length for this experiment is 14.8 boolean terms, and the fidelity is ≈ 80%.

Table 3. Results for baseline models, DL predictor and extracted rule set in task T1 with the balanced dataset using SMOTE. Numerical results are reported with *average value±standard deviation*. The highest results in each column are highlighted in bold.

Model	F1-score	CK-score	Rule length	Fidelity
SVM	0.8170 ± 0.0082	**0.6358 ± 0.0162**	NA	NA
Decision tree	0.8368 ± 0.0219	0.1996 ± 0.0888	NA	NA
Random forest	0.8824 ± 0.0142	0.2781 ± 0.0883	NA	NA
Neural Network	**0.9161 ± 0.0092**	0.2425 ± 0.0409	NA	NA
DEXiRE's rule set	0.8833 ± 0.0162	0.1366 ± 0.0682	14.8 ± 0.9797	0.8075 ± 0.0237

Experiment with the HECKTOR Partition. An essential task within PET-CT medical image analysis is the ability to generalize the results of the prediction models to different medical centers with equipment from different manufacturers and slightly different protocols. This challenge still demands further research and more flexible and robust techniques. To test the rule sets' generalization ability to various centers, we have used the partitions employed in the HECKTOR 2022 challenge, which provides a reproducible inter-center generalization scenario. Table 4 summarizes the results, recording the highest F1-score of 0,9724, obtained by the random forest predictor, followed by the SVM with an F1-score of 0,9432, the decision tree with a value of 0,9329, the

DL predictor with a value of 0,9025, and the rule set with a value of 0,8840. Concerning the CK-score, the highest result is obtained by the SVM predictor with a value of 0.2732, followed by the decision tree with a value of 0.2004, the random forest with a value of 0.1447, the neural network with a value of 0.1259, and the rule set with a value of 0.0241. The F1-score has shown variations of up to 11%. Similarly, the CK-score shows variations of up to 11%.

Table 4. Results for baseline models, DL predictor and extracted rule set in task T1 with HECKTOR partition. Numerical results are reported with *average value* ± *standard deviation*. The highest results in each column are highlighted in bold.

Model	F1-score	CK-score	Rule length	Fidelity
SVM	0.9432 ± 0.0094	**0.2732 ± 0.1135**	NA	NA
Decision tree	0.9329 ± 0.0087	0.2004 ± 0.0535	NA	NA
Random forest	**0.9724 ± 0.0029**	0.1447 ± 0.0893	NA	NA
Neural Network	0.9399 ± 0.0169	0.1376 ± 0.0826	NA	NA
DEXiRE's rule set	0.8230 ± 0.0724	0.0827 ± 0.0702	7.6 ± 0.7999	0.7304 ± 0.0836

4.2   Task T2 Cancer Staging

Task T2 performs a multiclass classification to stage and grade tumor. The cancer stage scale is progressive, ranging from a minimum value of 1 to a maximum of 4. In this dataset, the imbalance between the target classes is enormous, but the case of class 1, which has a total of 4 samples over 640, is particularly noteworthy. With such a few samples, applying any effective technique to balance the dataset without introducing bias and errors is very difficult. For this reason, we have decided to remove class 1 from the dataset and perform the subsequent experiments with three categories that, although still imbalanced, provide enough information to apply balance techniques and train the predictors effectively.

Experiments with Imbalanced Dataset. Table 5 presents the results obtained by the baseline models, the DL predictor, and the rule set concerning performance metrics PM1 to PM4 (Sect. 3 step S7) with the imbalanced dataset. The first column shows the F1-score with all the values over approximately 70%, and the DL predictor obtained the best score (76%). The second column reports the CK-score, whose values range from 8% to 27%. Here again, the DL predictor obtains the maximum score. The average rule length for this experiment is 11.2 boolean terms, and the fidelity is 73%.

Table 5. Results for baseline models, DL predictor and extracted rule set in task T2 with imbalanced dataset 3 classes. Numerical results are reported with *average value±standard deviation*. The highest results in each column are highlighted in bold.

Model	F1-score	CK-score	Rule length	Fidelity
SVM	0.7212 ± 0.0208	0.1374 ± 0.0641	NA	NA
Decision tree	0.7169 ± 0.0161	0.0892 ± 0.0699	NA	NA
Random forest	0.7380 ± 0.0147	0.1016 ± 0.0640	NA	NA
Neural Network	**0.7654 ± 0.0262**	**0.2776 ± 0.0661**	NA	NA
DEXiRE's rule set	0.7226 ± 0.0330	0.2740 ± 0.0811	11.2 ± 0.9797	0.7345 ± 0.0450

Experiments with Balanced Dataset. Table 6 presents the results obtained by the baseline models, the DL predictor, and the rule set. The first column shows the F1-score with all the values over ≈ 70%, and the random forest predictor obtained 84% as the best score. The second column reports the CK-score, whose values range from 12% to 35%. Again, the random forest has obtained the maximum score with ≈ 34%. The average rule length for this experiment is 13.6 boolean terms, and the fidelity is ≈ 72%.

Table 6. Results for baseline models, DL predictor, and extracted rule set in task T2 with the balanced 3 classes dataset using SMOTE. Numerical results are reported with *average value±standard deviation*. The highest results in each column are highlighted in bold.

Model	F1-score	CK-score	Rule length	Fidelity
SVM	0.7810 ± 0.1012	0.1274 ± 0.0340	NA	NA
Decision tree	0.7413 ± 0.0135	0.2186 ± 0.0368	NA	NA
Random forest	**0.8480 ± 0.0050**	**0.3469 ± 0.0307**	NA	NA
Neural Network	0.8058 ± 0.0140	0.2085 ± 0.5079	NA	NA
DEXiRE's rule set	0.7760 ± 0.0291	0.2359 ± 0.0423	13.6 ± 0.9797	0.7189 ± 0.0353

Experiments with HECKTOR Partition. Table 7 summarizes the results obtained using the imbalanced dataset with the HECKTOR partition. The highest reported F1-score is 84%, obtained by the random forest predictor, followed by the DL predictor with an F1-score of 80%, the SVM with 78%, the rule set with 77%, and the decision tree with 74%. The random forest predictor obtained the highest CK-score with a value of 0.3469, followed by the rule set with 0.2359, the decision tree with 0.2186, the neural network with a value of 0.2085, and the SVM with 0.1274. The F1-score shows variations up to 10%. Similarly, the CK-score shows variations up to 50%.

Table 7. Results for baseline models, DL predictor and extracted rule set in task T2 with the imbalanced (3 classes dataset) using HECKTOR partitions. Numerical results are reported with *average value ± standard deviation*. The highest results in each column are highlighted in bold.

Model	F1-score	CK-score	Rule length	Fidelity
SVM	0.7763 ± 0.0197	0.0851 ± 0.0763	NA	NA
Decision tree	0.7887 ± 0.0170	0.1548 ± 0.0370	NA	NA
Random forest	**0.8374 ± 0.0168**	0.1282 ± 0.0860	NA	NA
Neural Network	0.8021 ± 0.0761	0.1773 ± 0.2856	NA	NA
DEXiRE's rule set	0.7168 ± 0.0142	**0.2085 ± 0.0412**	11.2 ± 0.9797	0.6459 ± 0.0261

5   Discussion

This section elaborates on the results and performance obtained during the explanation of DL predictors (i.e., HPV diagnosis and tumor staging).

5.1   On Rules and Metrics for Task T1

Looking at the results obtained in the three experimental setups for task T1, the balanced dataset generated the rule set with the overall best performance, yet having the highest number of terms. Thus, the obtained results suggest a correlation between rule sets' length and performance. However, more extended rule sets are challenging to be understood, reducing their quality as explainers. Balancing the rule set's predictive ability and complexity (number of terms) is necessary yet not immediate.

5.2   On Rules and Metrics for Task T2

Looking at the results obtained in the three experimental setups, the average rule set length in this task is higher w.r.t. T1, reflecting the increased complexity of this task. Moreover, the overall average rule sets' performance of fidelity and F1-score are sensibly below T1's results. However, T2's CK-score is higher than T1's. This difference can be attributed to the different rule induction methods employed by DEXiRE for the binary and multiclass cases. While the former uses one-rule learning, the latter uses decision trees, which produce more robust and flexible rules.

5.3   Good Explanations, but What About the Predictors?

The results allow inferring that rule sets are good explainers, since they mimic the behavior of the original DL predictor on the training set with high quality. However, the results obtained in the test set and the CK-score for most experiments show results below other predictors. Indicating a limit to the generalization ability of the rule sets concerning more robust structures such as DL models and kernel methods.

5.4   Beyond Metrics

Using more than one metric to evaluate the rule sets generation is a good practice. However, it is possible to observe discrepancies between the consistently high F1-score and the consistently low CK-score. Such discrepancy is due to the data set imbalance that affects only the F1-score.

Indeed, Table 2 (imbalanced dataset) shows the rule set might not be the best, but from the F1-score perspective, it competes with the other predictors — although the CK-score is relatively poor. The performance differences between the rule set's efficacy measured with the F1-score and the one measured with the CK-score can be explained because the F1-score is the harmonic average of precision and recall. Therefore, in an imbalanced dataset, high precision and recall values on the majority class could produce a high F1-score, even if the class imbalance biases the metric. However, this is not the case for the CK-score. Indeed, it is based on the agreement between two experts, discounting the random influence. Proof of this explanation can be found in Table 3, where the results reported are obtained after balancing the dataset. In this table, the rule set's F1-score performance and CK-score are similar to the ones obtained by other models, including decision trees, SVM, and random forest.

5.5   The Influence of Data Partition on Rule Sets

In task T1, the rule set performance metrics are similar to those obtained by decision trees, SVM, and random forest — except for the HECKTOR partition. Thus, we can infer that the partitions' selection affects the models' performance and the rule extraction process. Moreover, the generalization ability of rule sets is limited to the samples in the training set.

The selection of data splits in tasks T1 and T2 can influence the performance evaluation because of the disparity between sample distribution in different centers. Despite the generalization ability of ML models and the regularization terms, overfitting for a particular center is an issue to be considered. In particular, the rule sets are less flexible, and they tend to overfit the training set to approximate the behavior of the original model on the train set as much as possible.

5.6   Imbalanced Datasets in Medical Domain and Bias Predictors

Imbalanced datasets are common in the medical domain. Such a condition is exacerbated in clinical studies, mainly because of the study of rare diseases or because screening trials focus on ill individuals. Indeed, this is the case for the datasets employed in task T1 HPV diagnosis and T2 cancer staging. Figures 3 and 4 show the sample counting for each target class.

A significant imbalance in the dataset can cause poor performance on the predictors, overfitting, and biases. This is because many optimization algorithms in ML/DL predictors privilege majority class and global accuracy over minority classes. As mentioned above, even the rule sets are affected by this phenomenon.

During rule extraction (step S6), some rule sets have been generated biased to predict only the majority class. Over the years, different solutions have been proposed to solve the imbalance in medical datasets. Rahman and Davis [44] proposed to balance medical datasets employing SMOTE. Although this approach works well, it can only be applied in some cases. For example, in task T2, this approach could not be employed in class 1 because there are not enough samples (4) to perform the interpolation.

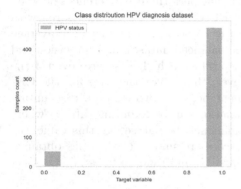 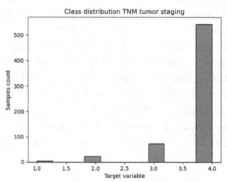

Fig. 3. Class distribution histogran for target variable HPV status dataset employed for task T1.

Fig. 4. Class distribution histogran for target variable TNM staging from dataset employed for task T2.

Rule sets provide domain-contextualized explanations, a logical language that human and artificial agents can understand. Logical language constitutes an advantage in safe-critical domains like medical diagnoses and prognoses because clinicians can validate this knowledge based on their expertise and through a rigorous reasoning process and extract conclusions that can be employed to support their daily work. DL predictors are not extensively used in clinical diagnosis due to the need for more transparency and domain-contextualized explanation of their internal behaviors that enable clinicians to trust their predictions. However, with the introduction of logic and semantic explanations, trust in DL predictors is expected to increase, and they could become part of daily clinical workflows, improving efficiency and effectiveness and helping clinicians and patients to understand their diagnoses.

5.7   Decompositional Rule Extraction Advantages

Decompositional rule extraction methods have several advantages over other post-hoc XAI methods. In particular, the DEXiRE rule extraction algorithm has the following advantages.

- Logical rule sets can be understood by human and artificial agents. More-over, they simplify the knowledge exchange between agents in heterogeneous environments.
- Rule sets, as symbolic objects, can be easily shaped into other symbolic objects like arguments or natural language explanations.
- Decompositional rule extraction algorithms generate rules by inspecting every neuron activation and better reflecting the internal behavior of the DL pre-dictor.
- Rule sets can be formally verified to assess their correctness.
- Besides extracted rule sets, DEXiRE can provide intermediate rule sets that describe the logical behavior in hidden layers, enabling model refinement and better understanding.
- Alongside the rule sets, DEXiRE can provide activation paths that describe the most frequent neural activation patterns to a given input, identifying the neurons that contribute more to the predictors' final decision.
- Rule sets can also be employed to perform inference and reasoning.

5.8   Limitations and Shortcomings

Despite the significant advancement in XAI in recent years, several challenges still need to be solved to apply XAI methods in safe-critical domains like medical diagnosis. The following briefly describes some limitations and shortcomings when using the DEXiRE algorithm in the medical domain.

- It is not possible to extract rule sets from every DL predictor. This is due to the non-linearity and complexity of the DL predictors' decision functions, which boolean rules cannot accurately approximate in all cases.
- DEXiRE algorithm is a very flexible algorithm able to extract rule sets from a wide range of DL architectures. However, currently, DEXiRE can only be applied to classification tasks. More research is required to extend DEXiRE to other machine learning tasks like regression or reinforcement learning.
- Rule sets depend on the models' architecture and data set partitions, making them less flexible in responding to never-before-seen cases or outliers. For this reason, we propose to use rule sets to understand and validate DL models rather than to perform large-scale inference processes.

To overcome these limitations, we have proposed several research paths, described at the end of the Conclusions and Future Work section.

6   Conclusions and Future Work

This study can conclude that the DEXiRE method enables the extraction of rule sets from DL predictors, aiming to make data-driven classifiers more transparent and facilitating the understanding of the motivations behind models' predictions to researchers and clinicians. In particular, it extracted rules from DL predic-tors trained on HPV diagnosis (T1) and TNM staging (T2) for H&N cancer, employing the decompositional rule extraction tool, namely DEXiRE. For both analyzed tasks T1 and T2 (HPV status and TNM staging), we conducted three experiments with imbalanced (original), balanced (SMOTE), and HECKTOR (inter-center) data partitioning. Finally, the rule sets and their performance metrics have been compared with baseline predictors to test their generalization, prediction, and explaining abilities. Elaborating on the obtained results and analysis, we can summarize the following:

- Concerning the F1-score metric, the extracted rule sets have shown similar performance among the predictors (i.e., SVM, decision tree, and random forest) and slightly lower performance of those obtained from the DL predictors.
- Concerning the CK-score, the extracted rule sets performance has shown better results in the multiclass task (T2) than in the binary classification (T1). This is because DEXiRE uses explainable layers (ExpL) and one-rule learning as rule induction methods for binary classification and decision trees for the multiclass case, inducing more robust rule sets.
- The rule sets are less flexible than ML/DL predictors. Therefore, they have a limited generalization capability and are more useful for providing post-hoc explanations than for making large-scale inferences.
- The decompositional rule extraction algorithm DEXiRE is affected by data partitions and dataset imbalance — since they impact the entropy, frequency of neuron activation, and terms' thresholds.
- Longer rule sets have shown better predictive performance and fidelity. However, it is harder to comprehend longer rule sets. A balance between performance and explainability is necessary for an optimal rule set.

Finally, we envision the following future works:

(i) To conduct further experiments focused on the inter-center rule set generalization, extracting rule sets based on the data from certain medical centers and applying them to other medical centers, (ii) Tumor staging task can also be analyzed using regression models. Thus we intend to extend DEXiRE enabling the explanation of regression DL predictors, (iii) To reduce the effect of unbalanced datasets in DEXiRE, we have proposed to extend DEXiRE to include sample and class weight to deal with imbalanced datasets, and (iv) To make DEXiRE more flexible and robust, we intend to extend it using fuzzy logic, which would allow a better approximation of the DL predictors' decision function.

Acknowledgments. This work is supported by the Chist-Era grant CHIST-ERA19-XAI-005, and by *(i)* the Swiss National Science Foundation (G.A. 20CH21_195530), *(ii)* the Italian Ministry for Universities and Research, *(iii)* the Luxembourg National Research Fund (G.A. INTER/CHIST/19/14589586), *(iv)* the Scientific, and Research Council of Turkey (TÜBİTAK, G.A. 120N680).

A   Appendix Rule Sets

In this appendix, examples of the rule sets extracted from DL predictors, in each experiment are presented.

A.1   Rule Sets for Task T1

Table 8 presents the best rule set extracted from DL predictor trained with imbalanced (original) dataset.

Table 8. Rule set extracted from DL predictor using DEXiRE and the imbalanced dataset.

Rule
IF [(everything_mergeddilat2mm_PT_gldm_DependenceVariance $\leq$ -0.6998) $\wedge$ (everything_mergedshell2mm_CT_firstorder_Skewness $>$ -0.6944)] $\vee$ [(everything_mergedBBox_PT_firstorder_Minimum $\leq$ -0.802) $\wedge$ (everything_mergedBBox_PT_glszm_SmallAreaLowGrayLevelEmphasis $\leq$ -0.2344) $\wedge$ (everything_mergeddilat2mm_PT_gldm_DependenceVariance > -0.6998)] $THEN$ 0
IF [(everything_mergedBBox_PT_firstorder_Minimum $\leq$ -0.802) $\wedge$ (everything_mergedBBox_PT_glszm_SmallAreaLowGrayLevelEmphasis $>$ -0.2344) $\wedge$ (everything_mergeddilat2mm_PT_gldm_DependenceVariance $>$ -0.6998)] $\vee$ [(everything_mergedBBox_PT_firstorder_Minimum $>$ -0.802) $\wedge$ (everything_mergeddilat2mm_PT_gldm_DependenceVariance $>$ -0.6998)] $\vee$ [(everything_mergeddilat2mm_PT_gldm_DependenceVariance $\leq$ -0.6998) $\wedge$ (everything_mergedshell2mm_CT_firstorder_Skewness ≤ -0.6944)] $THEN$ 1

Table 9 presents the best rule set extracted from DL predictor trained with SMOTE balanced dataset using.

Table 9. Rule set extracted from DL predictor using DEXiRE and the balanced dataset.

Rule
IF[(everything_mergedBBox_CT_glcm_MaximumProbability $\leq$ 0.21) $\wedge$ (everything_merged_PT_glcm_Imc1 $>$ 0.5062) $\wedge$ (everything_mergeddilat8mm_CT_firstorder_Maximum $\leq$ -1.0149) $\wedge$ (everything_mergedshell2mm_CT_gldm_LowGrayLevelEmphasis $>$ -0.2934)] $\vee$ [(everything_mergedBBox_CT_glcm_MaximumProbability $>$ 0.21) $\wedge$ (everything_merged_PT_glcm_Imc1 $>$ 0.5062)] $\vee$ [(everything_mergedBBox_CT_glcm_MaximumProbability $\leq$ 0.21) $\wedge$ (everything_merged_PT_glcm_Imc1 $>$ 0.5062) $\wedge$ (everything_mergedshell2mm_CT_gldm_LowGrayLevelEmphasis $\leq$ -0.2934)] $\vee$ [(everything_mergedBBox_PT_firstorder_10Percentile $>$ -1.4104) $\wedge$ (everything_merged_PT_glcm_Imc1 $\leq$ 0.5062) $\wedge$ (everything_mergeddilat4mm_CT_glcm_DifferenceEntropy $\leq$ 1.4793) $\wedge$ (everything_mergedshell4mm_PT_glrlm_GrayLevelNonUniformity $\leq$ 1.5895)] $THEN$ 1
IF[(everything_mergedBBox_PT_firstorder_10Percentile $>$ -1.4104) $\wedge$ (everything_merged_PT_glcm_Imc1 $\leq$ 0.5062) $\wedge$ (everything_mergeddilat4mm_CT_glcm_DifferenceEntropy $\leq$ 1.4793) $\wedge$ (everything_mergedshell4mm_PT_glrlm_GrayLevelNonUniformity $>$ 1.5895)] $\vee$ [(everything_mergedBBox_PT_firstorder_10Percentile $>$ -1.4104) $\wedge$ (everything_merged_PT_glcm_Imc1 $\leq$ 0.5062) $\wedge$ (everything_mergeddilat4mm_CT_glcm_DifferenceEntropy $>$ 1.4793)] $\vee$ [(everything_mergedBBox_PT_firstorder_10Percentile $\leq$ -1.4104) $\wedge$ (everything_merged_PT_glcm_Imc1 $\leq$ 0.5062)] $\vee$ [(everything_mergedBBox_CT_glcm_MaximumProbability $\leq$ 0.21) $\wedge$ (everything_merged_PT_glcm_Imc1 $>$ 0.5062) $\wedge$ (everything_mergeddilat8mm_CT_firstorder_Maximum $>$ -1.0149) $\wedge$ (everything_mergedshell2mm_CT_gldm_LowGrayLevelEmphasis $>$ -0.2934)] $THEN$ 0

Table 10 presents the best rule set extracted from DL predictor trained with the HECKTOR partition.

Table 10. Rule set extracted from DL predictor using DEXiRE and the HECKTOR partition.

Rule
$IF[[(everything\_merged40\%\_CT\_firstorder\_Median >$ $0.0167) \wedge (everything\_mergeddilat8mm\_PT\_glcm\_Idn \leq$ $0.526) \wedge (everything\_mergedshell2mm\_shape\_Flatness \leq -0.4546) \wedge$ $(everything\_mergedshell4mm\_CT\_gldm\_SmallDependenceHighGrayLevelEmphasis \leq$ $0.6285)] \vee [(everything\_mergedshell2mm\_shape\_Flatness > -0.4546) \wedge$ $(everything\_mergedshell4mm\_CT\_gldm\_SmallDependenceHighGrayLevelEmphasis \leq$ $0.6285)] \vee [(everything\_mergeddilat8mm\_PT\_glcm\_Idn >$ $0.526) \wedge (everything\_mergedshell2mm\_shape\_Flatness \leq -0.4546) \wedge$ $(everything\_mergedshell4mm\_CT\_gldm\_SmallDependenceHighGrayLevelEmphasis \leq$ $0.6285)] \ THEN \ 1$
$IF[[(everything\_merged40\%\_CT\_firstorder\_Median \leq 0.0167)$ $\wedge(everything\_mergeddilat8mm\_PT\_glcm\_Idn \leq$ $0.526) \wedge (everything\_mergedshell2mm\_shape\_Flatness \leq -0.4546)\wedge$ $(everything\_mergedshell4mm\_CT\_gldm\_SmallDependenceHighGrayLevelEmphasis \leq$ $0.6285)]\vee$ $(everything\_mergedshell4mm\_CT\_gldm\_SmallDependencHighGrayLevelEmphasis >$ $0.6285)] \ THEN \ 0$

A.2 Rule Sets for Task T2

Table 11 presents the best rule set extracted from DL predictor trained with the imbalanced dataset.

Table 11. Rule set extracted from DL predictor using DEXiRE and the imbalanced dataset.

Rule
$IF[[(everything\_mergeddilat16mm\_PT\_firstorder\_Minimum \leq$ $1.1895) \wedge (everything\_mergeddilat8mm\_PT\_glrlm\_GrayLevelNonUniformity >$ $-0.5826) \wedge (everything\_mergedshell8mm\_PT\_glszm\_SmallAreaEmphasis >$ $1.3477)] \vee [(everything\_mergeddilat16mm\_PT\_firstorder\_Minimum >$ $1.1895) \wedge (everything\_mergeddilat1mm\_shape\_Flatness \leq 0.2752)] \vee$ $[(everything\_mergeddilat16mm\_CT\_glszm\_SizeZoneNonUniformityNormalized >$ $-0.5867) \wedge (everything\_mergeddilat16mm\_PT\_firstorder\_Minimum \leq 1.1895) \wedge$ $(everything\_mergeddilat8mm\_PT\_glrlm\_GrayLevelNonUniformity \leq -0.5826)] \ THEN \ 1$
$IF[[(everything\_mergeddilat16mm\_PT\_firstorder\_Minimum >$ $1.1895) \wedge (everything\_mergeddilat1mm\_shape\_Flatness > 0.2752)] \ THEN \ 0$
$IF[[(everything\_mergeddilat16mm\_PT\_firstorder\_Minimum \leq 1.1895)\wedge$ $(everything\_mergeddilat8mm\_PT\_glrlm\_GrayLevelNonUniformity >$ $-0.5826) \wedge (everything\_mergedshell8mm\_PT\_glszm\_SmallAreaEmphasis \leq 1.3477)]\vee$ $[(everything\_mergeddilat16mm\_CT\_glszm\_SizeZoneNonUniformityNormalized \leq$ $-0.5867) \wedge (everything\_mergeddilat16mm\_PT\_firstorder\_Minimum \leq 1.1895) \wedge$ $(everything\_mergeddilat8mm\_PT\_glrlm\_GrayLevelNonUniformity \leq -0.5826)] \ THEN \ 2$

Table 12 presents the best rule set extracted from DL predictor trained with the balanced dataset.

Table 12. Rule set extracted from DL predictor using DEXiRE and the imbalanced dataset.

Rule
$IF[(Chemotherapy > 0.3396) \wedge$
$(everything\_mergedBBox\_CT\_gldm\_GrayLevelNonUniformity > -1.0517) \wedge (everything\_mergeddilat8mm\_PT\_glrlm\_GrayLevelNonUniformity > -0.3239) \wedge (everything\_mergedshell2mm\_CT\_gldm\_DependenceNonUniformity \leq -0.5275)] \vee [(everything\_merged40\%\_shape\_Maximum2DDiameterSlice \leq 0.123) \wedge (everything\_mergedBBox\_CT\_gldm\_GrayLevelNonUniformity > -1.0517) \wedge$
$(everything\_mergeddilat16mm\_CT\_glszm\_SizeZoneNonUniformityNormalized > -0.7827) \wedge (everything\_mergeddilat16mm\_PT\_firstorder\_Minimum \leq -0.7768) \wedge (everything\_mergeddilat2mm\_CT\_glcm\_JointAverage \leq 0.6604) \wedge (everything\_mergeddilat8mm\_PT\_glrlm\_GrayLevelNonUniformity \leq -0.3239)] \vee [(everything\_merged40\%\_shape\_Maximum2DDiameterSlice \leq 0.123) \wedge (everything\_mergedBBox\_CT\_gldm\_GrayLevelNonUniformity > -1.0517) \wedge$
$(everything\_mergeddilat16mm\_CT\_glszm\_SizeZoneNonUniformityNormalized \leq -0.7827) \wedge (everything\_mergeddilat2mm\_CT\_glcm\_JointAverage \leq 0.6604) \wedge (everything\_mergeddilat8mm\_PT\_glrlm\_GrayLevelNonUniformity \leq -0.3239)] \vee [(everything\_merged40\%\_shape\_Maximum2DDiameterSlice > 0.123) \wedge (everything\_mergedBBox\_CT\_gldm\_GrayLevelNonUniformity > -1.0517) \wedge$
$(everything\_mergeddilat8mm\_PT\_glrlm\_GrayLevelNonUniformity \leq -0.3239)] \vee [(everything\_merged40\%\_shape\_Maximum2DDiameterSlice \leq 0.123) \wedge (everything\_mergedBBox\_CT\_gldm\_GrayLevelNonUniformity > -1.0517) \wedge$
$(everything\_mergeddilat2mm\_CT\_glcm\_JointAverage > 0.6604) \wedge (everything\_mergeddilat8mm\_PT\_glrlm\_GrayLevelNonUniformity \leq -0.3239)] \vee [(Chemotherapy > 0.3396) \wedge$
$(everything\_mergedBBox\_CT\_gldm\_GrayLevelNonUniformity > -1.0517) \wedge (everything\_mergeddilat8mm\_PT\_glrlm\_GrayLevelNonUniformity > -0.3239) \wedge (everything\_mergedshell2mm\_CT\_gldm\_DependenceNonUniformity > -0.5275)] \ THEN \ 2$
$IF[(Chemotherapy \leq 0.3396) \wedge$
$(everything\_mergedBBox\_CT\_gldm\_GrayLevelNonUniformity > -1.0517) \wedge$
$(everything\_mergeddilat8mm\_PT\_glrlm\_GrayLevelNonUniformity > -0.3239)] \vee [(Age > -0.9538) \wedge (Chemotherapy \leq 0.3285) \wedge (everything\_mergedBBox\_CT\_gldm\_GrayLevelNonUniformity \leq -1.0517) \wedge (everything\_mergedshell8mm\_CT\_glcm\_Contrast \leq 2.4235)] \ THEN \ 0$
$IF[(Chemotherapy > 0.3285) \wedge$
$(everything\_mergedBBox\_CT\_gldm\_GrayLevelNonUniformity \leq -1.0517)] \vee [(everything\_merged40\%\_shape\_Maximum2DDiameterSlice \leq 0.123) \wedge (everything\_mergedBBox\_CT\_gldm\_GrayLevelNonUniformity > -1.0517) \wedge (everything\_mergeddilat16mm\_CT\_glszm\_SizeZoneNonUniformityNormalized > -0.7827) \wedge (everything\_mergeddilat16mm\_PT\_firstorder\_Minimum > -0.7768) \wedge (everything\_mergeddilat2mm\_CT\_glcm\_JointAverage \leq 0.6604) \wedge (everything\_mergeddilat8mm\_PT\_glrlm\_GrayLevelNonUniformity \leq -0.3239)] \vee [(Age > -0.9538) \wedge (Chemotherapy \leq 0.3285) \wedge (everything\_mergedBBox\_CT\_gldm\_GrayLevelNonUniformity \leq -1.0517) \wedge (everything\_mergedshell8mm\_CT\_glcm\_Contrast > 2.4235)] \vee [(Age \leq -0.9538) \wedge (Chemotherapy \leq 0.3285) \wedge (everything\_mergedBBox\_CT\_gldm\_GrayLevelNonUniformity \leq -1.0517)] \ THEN \ 1$

Table 13 presents the best rule set extracted from DL predictor trained with the balanced dataset.

Table 13. Rule set extracted from DL predictor using DEXiRE and the imbalanced dataset.

Rule
$IF[(Chemotherapy \leq -1.0322) \wedge$
$(everything\_mergedshell2mm\_CT\_gldm\_SmallDependenceEmphasis \leq 0.1661)] \vee [(Chemotherapy > -1.0322) \wedge$
$(everything\_merged40\%\_CT\_glcm\_JointEnergy \leq 0.3687) \wedge (everything\_mergeddilat16mm\_shape\_Maximum2DDiameterSlice \leq -0.2465) \wedge (everything\_mergeddilat4mm\_CT\_glszm\_SmallAreaEmphasis > -0.6568)]\ THEN\ 1$
$IF[(Chemotherapy > -1.0322) \wedge$
$(everything\_merged40\%\_CT\_glcm\_JointEnergy > 0.3687) \wedge (everything\_mergeddilat16mm\_shape\_Maximum2DDiameterSlice \leq -0.2465) \wedge (everything\_mergeddilat4mm\_CT\_glszm\_SmallAreaEmphasis > -0.6568)] \vee [(Chemotherapy > -1.0322) \wedge (everything\_mergeddilat16mm\_shape\_Maximum2DDiameterSlice > -0.2465)] \vee [(Chemotherapy > -1.0322) \wedge (everything\_mergeddilat16mm\_shape\_Maximum2DDiameterSlice \leq -0.2465) \wedge (everything\_mergeddilat4mm\_CT\_glszm\_SmallAreaEmphasis \leq -0.6568)]\ THEN\ 2$
$IF[(Chemotherapy \leq -1.0322) \wedge$
$(everything\_mergedshell2mm\_CT\_gldm\_SmallDependenceEmphasis > 0.1661)]\ THEN\ 0$

References

1. Özel et al., H.: Use of pet in head and neck cancers (2015). https://doi.org/10.5152/tao.2015.863
2. Andrearczyk, V., Oreiller, V., Boughdad, S., Rest, C.C.L., Elhalawani, H., Jreige, M., Prior, J.O., Vallières, M., Visvikis, D., Hatt, M., et al.: Overview of the hecktor challenge at miccai 2021: automatic head and neck tumor segmentation and outcome prediction in pet/ct images. In: Head and Neck Tumor Segmentation and Outcome Prediction: Second Challenge, HECKTOR 2021, Held in Conjunction with MICCAI 2021, Strasbourg, France, September 27, 2021, Proceedings, pp. 1–37. Springer (2022)
3. Arrieta, A.B., et al.: Explainable artificial intelligence (xai): concepts, taxonomies, opportunities and challenges toward responsible ai. Inform. Fusion **58**, 82–115 (2020)
4. Augasta, M.G., Kathirvalavakumar, T.: Rule extraction from neural networks-a comparative study. In: International Conference on Pattern Recognition, Informatics and Medical Engineering (PRIME-2012), pp. 404–408. IEEE (2012)

5. Augustin, J.G., et al.: Hpv detection in head and neck squamous cell carcinomas: What is the issue? 10 (2020). https://doi.org/10.3389/fonc.2020.01751
6. Bagher-Ebadian, H., et al.: Application of radiomics for the prediction of hpv status for patients with head and neck cancers. Med. Phys. **47**(2), 563–575 (2020). https://doi.org/10.1002/mp.13977
7. Bray, F., et al.: Global cancer statistics 2018 (2018). https://doi.org/10.3322/caac.21492
8. Chaturvedi, A.K., et al.: Human papillomavirus and rising oropharyngeal cancer incidence in the united states. J. Clin. Oncol. **29**(32), 4294–4301 (2011). https://doi.org/10.1200/JCO.2011.36.4596
9. Contreras, V., et al.: A dexire for extracting propositional rules from neural networks via binarization. Electronics **11**(24) (2022). https://doi.org/10.3390/electronics11244171, https://www.mdpi.com/2079-9292/11/24/4171
10. Craven, M.W., Shavlik, J.W.: Understanding time-series networks: a case study in rule extraction. Int. J. Neural Syst. **8**(04), 373–384 (1997)
11. Das, A., Rad, P.: Opportunities and challenges in explainable artificial intelligence (xai): A survey. arXiv preprint arXiv:2006.11371 (2020)
12. Exarchos, K.P., Goletsis, Y., Fotiadis, D.I.: Multiparametric decision support system for the prediction of oral cancer reoccurrence. IEEE Trans. Inf Technol. Biomed. **16**(6), 1127–1134 (2012). https://doi.org/10.1109/TITB.2011.2165076
13. Fitzmaurice, C., et al.: The global burden of cancer 2013. JAMA Oncol. **1**(4), 505–527 (2015)
14. Främling12, K.: Contextual importance and utility in r: the 'ciu'package (2021)
15. Galati, L., et al.: Hpv and head and neck cancers: Towards early diagnosis and prevention. Tumour Virus Research p. 200245 (2022)
16. Gillies, R.J., Schabath, M.B.: Radiomics improves cancer screening and early detection. Cancer Epidemiol., Biomarkers Prevent. **29**(12), 2556–2567 (2020). https://doi.org/10.1158/1055-9965.EPI-20-0075
17. Gillison, M.L., Chaturvedi, A.K., Anderson, W.F., Fakhry, C.: Epidemiology of human papillomavirus-positive head and neck squamous cell carcinoma. J. Clin. Oncol. **33**(29), 3235–3242 (2015). https://doi.org/10.1200/JCO.2015.61.6995
18. Graziani, M., et al.: A global taxonomy of interpretable AI: unifying the terminology for the technical and social sciences. Artif. Intell. Rev. **56**(4), 3473–3504 (2023)
19. Guidotti, R., Monreale, A., Ruggieri, S., Turini, F., Giannotti, F., Pedreschi, D.: A survey of methods for explaining black box models. ACM Comput. Surv. (CSUR) **51**(5), 1–42 (2018)
20. Gupta, P., et al.: Explain your move: Understanding agent actions using specific and relevant feature attribution. In: International Conference on Learning Representations (ICLR) (2020)
21. Gupta, P., Kaur Malhi, A.: Using deep learning to enhance head and neck cancer diagnosis and classification, pp. 1–6 (2018). https://doi.org/10.1109/ICSCAN.2018.8541142
22. Halicek, M., et al.: Deep convolutional neural networks for classifying head and neck cancer using hyperspectral imaging. J. Biomed. Opt. **22**(6), 060503 (2017). https://doi.org/10.1117/1.JBO.22.6.060503
23. Hayashi, Y., Yukita, S.: Rule extraction using recursive-rule extraction algorithm with j48graft combined with sampling selection techniques for the diagnosis of type 2 diabetes mellitus in the pima indian dataset. Inform. Med. Unlocked **2**, 92–104 (2016)

24. Huang, S.H., O'Sullivan, B.: Overview of the 8th edition tnm classification for head and neck cancer. Current Treatment Options in Oncology (2017). https://doi.org/10.1007/s11864-017-0484-y

25. Institute, N.C.: Cancer staging (2022). https://www.cancer.gov/about-cancer/diagnosis-staging/staging

26. Institute, N.C.: Cancer staging (2022). https://www.cancer.gov/about-cancer/diagnosis-staging/staging

27. Junn, J.C., Soderlund, K.A., Glastonbury, C.M.: Imaging of head and neck cancer with ct, mri, and us. Seminars in Nuclear Medicine $51(1)$, 3–12 (2021). https://doi.org/10.1053/j.semnuclmed.2020.07.005, https://www.sciencedirect.com/science/article/pii/S0001299820300763 imaging Options for Head and Neck Cancer

28. Knapič, S., Malhi, A., Saluja, R., Främling, K.: Explainable artificial intelligence for human decision support system in the medical domain. Mach. Learn. Knowl. Extract. $3(3)$, 740–770 (2021)

29. Köppen, M.: The curse of dimensionality. In: 5th Online World Conference on Soft Computing in Industrial Applications (WSC5). vol. 1, pp. 4–8 (2000)

30. Lechner, M., Liu, J., Masterson, L., et al.: Hpv-associated oropharyngeal cancer: epidemiology, molecular biology and clinical management. Nat. Rev. Clin. Oncol. $19(3)$, 306-327 (2022). https://doi.org/10.1038/s41571-022-00603-7

31. Leijenaar, R.T., et al.: Development and validation of a radiomic signature to predict HPV (p16) status from standard CT imaging: a multicenter study. Br. J. Radiol. $91(1086)$, 20170498 (2018). https://doi.org/10.1259/bjr.20170498

32. McHugh, M.L.: Interrater reliability: the kappa statistic. Biochemia medica $22(3)$, 276–282 (2012)

33. Mayerhoefer, M.E. et al.: Introduction to radiomics. J. Nuclear Med. $61(4)$, 488–495 (2020)

34. Mishra, S., Sturm, B.L., Dixon, S.: Local interpretable model-agnostic explanations for music content analysis. In: ISMIR. vol. 53, pp. 537–543 (2017)

35. Mody, M., Rocco, J.W., Yom, S.S., Haddad, R.I., Saba, N.F.: Head and neck cancer: high-end technology is no guarantee of high-quality care (2022). https://doi.org/10.1016/S0140-6736(22)00426-3

36. Molnar, C.: Interpretable machine learning. Lulu. com (2020)

37. Moradi, M., Samwald, M.: Post-hoc explanation of black-box classifiers using confident itemsets. Expert Syst. Appl. **165**, 113941 (2021)

38. Muddamsetty, S.M., Jahromi, M.N., Moeslund, T.B.: Expert level evaluations for explainable ai (xai) methods in the medical domain. In: Pattern Recognition. ICPR International Workshops and Challenges: Virtual Event, January 10–15, 2021, Proceedings, Part III. pp. 35–46. Springer (2021)

39. Myronenko, A., Siddiquee, M.M.R., Yang, D., He, Y., Xu, D.: Automated head and neck tumor segmentation from 3D pet/CT (2022). arXiv:2209.10809

40. Nóbrega, C., Marinho, L.: Towards explaining recommendations through local surrogate models. In: Proceedings of the 34th ACM/SIGAPP Symposium on Applied Computing, pp. 1671–1678 (2019)

41. Oreiller, V., et al.: Head and neck tumor segmentation in pet/CT: the hecktor challenge. Med. Image Anal. **77**, 102336 (2022)

42. of Otolaryngology. Head, A.A., Foundation, N.S.: Tnm staging of head and neck cancer and neck dissection classification (2014)

43. Puri, N., et al.: Explain your move: Understanding agent actions using specific and relevant feature attribution. arXiv preprint arXiv:1912.12191 (2019)

44. Rahman, M.M., Davis, D.N.: Addressing the class imbalance problem in medical datasets. Int. J. Mach. Learn. Comput. **3**(2), 224 (2013)
45. Raji, I.D., Yang, J.: About ml: Annotation and benchmarking on understanding and transparency of machine learning lifecycles. arXiv preprint arXiv:1912.06166 (2019)
46. Rebaud, L., Escobar, T., Khalid, F., Girum, K.B., Buvat, I.: Simplicity is all you need: Out-of-the-box nnunet followed by binary-weighted radiomic model for segmentation and outcome prediction in head and neck pet/CT (09 2022). https://doi.org/10.13140/RG.2.2.30709.04328/1
47. Sabbatini, F., Ciatto, G., Calegari, R., Omicini, A.: On the design of psyke: A platform for symbolic knowledge extraction. In: WOA, pp. 29–48 (2021)
48. van der Schroeff, M.P., de Jong, R.J.B.: Staging and prognosis in head and neck cancer. Oral Oncol. **45**(4–5), 356–360 (2009)
49. Setiono, R., Leow, W.K.: Fernn: an algorithm for fast extraction of rules from neural networks. Appl. Intell. **12**(1–2), 15–25 (2000)
50. Slack, D., Hilgard, A., Singh, S., Lakkaraju, H.: Reliable post hoc explanations: Modeling uncertainty in explainability. Adv. Neural. Inf. Process. Syst. **34**, 9391–9404 (2021)
51. Song, B., et al.: Radiomic features associated with hpv status on pretreatment computed tomography in oropharyngeal squamous cell carcinoma inform clinical prognosis. Front. Oncol. **11**, 744250 (2021). https://doi.org/10.3389/fonc.2021.744250
52. Strobel, M.: Aspects of transparency in machine learning. In: Proceedings of the 18th International Conference on Autonomous Agents and MultiAgent Systems, pp. 2449–2451 (2019)
53. Suh, C., Lee, K., Choi, Y.: Oropharyngeal squamous cell carcinoma: radiomic machine-learning classifiers from multiparametric MR images for determination of HPV infection status (2020). https://doi.org/10.1038/s41598-020-74479-x
54. Takes, R.P., et al.: Future of the TNM classification and staging system in head and neck cancer. Head & neck **32**(12), 1693–1711 (2010)
55. Van Griethuysen, J.J., et al.: Computational radiomics system to decode the radiographic phenotype. Can. Res. **77**(21), e104–e107 (2017)
56. Zarlenga, M.E., Shams, Z., Jamnik, M.: Efficient decompositional rule extraction for deep neural networks. arXiv preprint arXiv:2111.12628 (2021)
57. Zhu, X., Wang, D., Pedrycz, W., Li, Z.: Fuzzy rule-based local surrogate models for black-box model explanation. IEEE Trans. Fuzzy Syst. (2022)

Metrics for Evaluating Explainable Recommender Systems

Joris Hulstijn[1](✉)(iD), Igor Tchappi[1](iD), Amro Najjar[1,2](iD), and Reyhan Aydoğan[3,4](iD)

[1] University of Luxembourg, Esch-sur-Alzette, Luxembourg
joris.hulstijn@uni.lu
[2] Luxembourg Institute of Science and Technology (LIST), Esch-sur-Alzette, Luxembourg
[3] Computer Science, Özyeğin University, Istanbul, Turkey
[4] Interactive Intelligence, Delft University of Technology, Delft, Netherlands

Abstract. Recommender systems aim to support their users by reducing information overload so that they can make better decisions. Recommender systems must be transparent, so users can form mental models about the system's goals, internal state, and capabilities, that are in line with their actual design. Explanations and transparent behaviour of the system should inspire trust and, ultimately, lead to more persuasive recommendations. Here, explanations convey reasons why a recommendation is given or how the system forms its recommendations. This paper focuses on the question how such claims about effectiveness of explanations can be evaluated. Accordingly, we investigate various models that are used to assess the effects of explanations and recommendations. We discuss objective and subjective measurement and argue that both are needed. We define a set of metrics for measuring the effectiveness of explanations and recommendations. The feasibility of using these metrics is discussed in the context of a specific explainable recommender system in the food and health domain.

Keywords: Evaluation · Metrics · Explainable AI · Recommender systems

1 Introduction

Artificial intelligence is becoming more and more pervasive. However, there are also concerns about bias in the algorithm, bias in the data set, or stereotyping users, to name a few examples [9]. In particular, if autonomous systems take decisions, how can they justify and explain these decisions, to those who are affected? These concerns have led to an increasing interest in *responsible* AI [9] and specifically in *explainable* AI (XAI) [2], witness the special issue [27].

An important application of explainable AI is found in *recommender systems* [1,35]. A recommender system is "any system that guides a user in a personalized way to interesting or useful objects in a large space of possible options or that produces such objects as output" [5, p. 2]. Increasingly, recommender systems also provide explanations [35]. There are two types: explanations that motivate the system's choice of recommendations, and explanations that clarify how the system works, to derive the recommendations. In this paper, we focus on the former type of explanations.

D. Calvaresi et al. (Eds.): EXTRAAMAS 2023, LNAI 14127, pp. 212–230, 2023.
https://doi.org/10.1007/978-3-031-40878-6_12

If the purpose is to persuade users to change their behaviour, only predicting which recommendation best fits a user profile, is not enough. Users expect reasons that motivate why this recommendation was given, and not another. That is why researchers now aim to build systems that provide an explanation, personalized to the user's preferences and to the context. In addition, recommender systems are becoming more interactive. A recommendation must be followed by an opportunity for feedback [16]. This allows users to correct misunderstandings and ask follow-up questions.

Designing interactive systems is a complex task. Unlike graphical user interfaces, dialogue systems have no visible menu-structure to display next possible moves [32]. The expectations that users do have about artificial intelligence are often wrong [17]. So the design should guide the user on how to control the interaction. The aim is to make the system transparent to the user. A system is called *transparent* when the user's mental model of the system's intent (purpose), beliefs (current internal state) and capabilities (way of working), corresponds to its actual purpose, state and capabilities [22]. A transparent system should inspire trust [39]. Note that transparency is part of many frameworks for ethical AI e.g. [14] and of the AI legislation proposed by the EU [10].

Consider for example an interactive recommender system in the food and health domain: it selects a recipe on the basis of user preferences and general knowledge about food and health. After that, the system allows feedback and provides explanations in an interactive manner [6]. Explainable recommender systems such as these, need to be *evaluated*. Claims about the usefulness of the recommendations and about the relevance and comprehension of explanations, and the overall effect on the transparency of the system and ultimately on the trust that users have in the system and its recommendations, must be measured. That suggests the following research question:

Can we define a system of metrics to evaluate explainable recommender systems, in the food and health domain?

The research method for this paper is conceptual and is mostly based on a literature study. Currently, there is no consensus in the literature on how to evaluate effectiveness of explainable AI [15, 36]. For instance, there is a debate whether one should use *subjective measures* [8, 37], or *objective measures*. There is not even consensus on the main concepts, such as explainability, transparency, or trust [43]. So before we can define metrics for evaluation, we must first analyze these concepts and how they relate. So we will discuss several conceptual models and define metrics for the main concepts.

The intended research contribution of this paper, is twofold: (i) to provide clarity on the main concepts used in explainable recommender systems, in particular explainability, transparency, and trust, and (ii) to define metrics, that can precisely and reliable measure these concepts, so explainable recommender systems can be evaluated.

We realize that the context in which a system is used, determines the way a system must be evaluated. In order to illustrate and guide our definitions for a specific context, we will use an example of a specific explainable food recommender system [6].

The remainder of the paper is structured as follows. Section 2 starts with a review of evaluation methods of interactive systems in general, and about explainable recommender systems in particular. After that, we will briefly detail the case in Sect. 3. In Sect. 4, we specify a series of conceptual models, and define the required set of metrics. The paper ends with a list of challenges and a number of recommendations.

Table 1. Comparing objective and subjective system evaluation

	Objective measurement	Subjective measurement
Purpose	measure *task success* of interaction with the system on the basis of observation and log-files	measure *perceived success* of interaction with the system, on the basis of user studies and questionnaires
Way of working	Annotators assess interaction behaviour according to definitions.	Users fill in questionnaires with Likert scales, open or closed questions or card sorting tasks
Metrics	task completion rate, comprehension, duration, misunderstandings	perceived usefulness, perceived ease of use, user satisfaction, trust, transparency

2 Overview

In the following sections, we review some of the literature on evaluating interactive systems in general, and explainable recommender systems in particular. We discuss a number of issues and dilemmas. The argument is largely based on Hoffman et al. [15], and Vorm and Combs [39]. The Q-methodology [29] is also discussed.

2.1 Subjective or Objective Evaluation

Suppose we want to evaluate the effectiveness of a system design in context. To evaluate effectiveness, we first need to define the objectives of the system. Given the objectives, there are two ways in which we can collect evidence of the effectiveness of a system: *subjective*, by asking end-users about their experiences in interviews or questionnaires [8,37], or *objective*, by having developers observing functionalities and system behaviour directly, or from log-files [12,42]. Table 1 lists examples of both. Observe that objective measures test specific features in development, whereas subjective measures look at the total impression of the system on the user.

In general, we believe that we need both perspectives: they have separate purposes and strengthen each other. For example, suppose subjective evaluation reveals that most users like the system (high user satisfaction), but some group of users do not. Analysis of the log-files may show, that the dialogue duration for the users who did not like the system, is longer than for those who liked it. In that case we found that satisfaction (subjective) depends on duration (objective). We can even go further and analyze the longer dialogues in detail. Perhaps, a specific type of misunderstanding causes delays. In that case, the system can be re-designed to avoid such misunderstandings. It is also possible that the objective and subjective measures diverge. In that case, it depends on the purpose of the evaluation, which type of measure takes precedence. For testing individual system features, objective measures remain useful, even if end-users do not perceive the differences. But for over-all system success, it is user satisfaction that counts. For more on this discussion in the context of evaluating explainable AI, see [15] and [36, p3].

2.2 Technology Acceptance

One of the most influential models for evaluating information systems is the Technology Acceptance Model (TAM) [8], and later adjustments [37]. Note that often, the terms 'adoption', 'acceptance' and 'use', are used interchangeably, although they are not completely equivalent. We start from a simple model of psychology: when people make a decision to perform some action, they first form an attitude towards that action. If that attitude is positive, or more positive than for alternative actions, they will form the intention to pursue the action. That intention will then produce that action.

Which attitudes affect the intention to use a system? The original idea is very simple, which also explains its attractiveness (Fig. 1(a)). In deciding to use or adopt a system, people make a trade-off between expected benefits and expected costs or efforts in using it. If the system is helpful in performing the task (useful), the user is more likely to consider using it. However, using the system will also take some effort. One has to learn to use the system and there may be misunderstandings and delays (ease of use). However, when considering to use a system, the user does not yet know the system. That means, that the intention to adopt a system is usually based on a system description. Therefore, the model uses 'perceived ease of use' and 'perceived usefulness' as the main variables, and not the actual usefulness, or actual ease of use. Subjective judgements like 'perceived usefulness' can be measured using Likert-scales. Well-tested and practical questionnaires exist, and results can be statistically analyzed, for example using regression models. There are no time-series, for example, or feedback loops. This simplicity partly explains the popularity of TAM models.

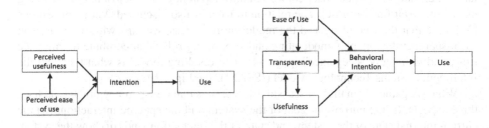

Fig. 1. (a) Technology Acceptance Model [8] and (b) ISTAM Model [39]

The Technology Acceptance Model also has clear disadvantages. The model only looks at the individual user, not at the corporate or social environment in which the system will be used. The model is about the decision to use a system, beforehand. It does not evaluate actual usage, afterwards. Moreover, the model suggests that intentions always lead to successful action; it doesn't look at feasibility. In the TAM model, technology is seen as a black-box. There is no evaluation of the effect of design choices and specific functionalities. Furthermore, the model is psychologically too simple. For example, it does not cover learning effects, habits, or previous experience.

Some of these disadvantages have been addressed in later adjustments and improvements to the model. In particular, the unified model of Venkatesh et al. [37] adds variables for social influence, and facilitating conditions. In addition, control variables for gender, age, experience and voluntariness of use, are taken into account.

It is relatively easy to add additional variables to TAM models. For example, trust has been added in the context of e-commerce systems [30]. System Usability Scale (SUS) [20] is a well-known alternative for TAM-models. It measures usability, which combines both ease of use, and usefulness in a single scale of 10 questions.

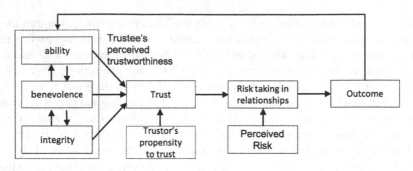

Fig. 2. Trust and Feedback [25]

In the same research tradition, Vorm and Combs [39] extend the TAM model, but now for evaluating intelligent systems (Fig. 1b). The notion of transparency is added as the key intermediate variable, that influences the behavioural intention. Vorm and Combs discuss various conceptions of transparency. Based on earlier work [4], they make a distinction between transparency for monitoring (i.e., what is happening?), transparency for process visibility (i.e., how does it work?), transparency for surveillance (i.e., interactivity and user control) and transparency for disclosure (i.e., opening up secrets regarding purpose). The relation to trust is also discussed. Vorm and Combs [39] state that the role of trust in the model more or less overlaps with transparency: "Transparency factors play moderating and supporting roles that combine to influence trust, and ultimately acceptance" (page 14). The resulting model is what they call the Intelligent Systems Technology Model (ISTAM), see Fig. 1(b).

What we gather from these discussions [22, 23], is that transparency involves at least three aspects: (i) the purpose or goal of the system and the specific interaction, (ii) the current internal state of the system, and state of the interaction, and (iii) how the system works, and ways for the user to control the interaction.

2.3 Trust

Trust has been discussed in many disciplines and fields. Here, we will follow the tradition in economics, that relates trust to the willingness to take risk in collaborating with another person, but without additional guarantees or controls over that other person's behaviour. The propensity to take risks, is part of the character of the trustor. We can also look at trustworthiness, the properties that are needed for the trustee to be trusted. Mayer et al. [25] define three properties of trustworthiness: (i) ability (or competence): can the trustee perform the task, (ii) benevolence: does the trustee want to do good to the trustor, and (iii) integrity: does the trustee follows a set of personal principles?

Trust is a relationship, so it depends both on aspects of the trustor and the trustee. In general, the likelihood that a trustor will trust a trustee will depend on (i) the trustor's propensity to trust, and (ii) the trustee's perceived trustworthiness (ability, benevolence and integrity). This is a nice definition, but it doesn't tell us how trust is won or lost. Which signals inspire trust in a person? What is the effect of repeated interactions? Mayer et al. [25] show an interactive model, that allows feedback (Fig. 2). The outcome of a (repeated) event, will influence the future trustor's assessment of the trustee' trustworthiness. In general, when the outcome is positive, this will increase trust; when the outcome is negative, this will reduce trust, for the next time around.

Lewicki and Bunker [18] study trust in work relationships. Based on older models of trust in personal relationships, they conclude that trust develops in three stages: calculus-based trust, knowledge-based trust, and identification-based trust (Table 2).

Now we need to map these models of inter-personal trust, to trust in machines. The regularity that underlies calculus-based trust is also the main source for trust in a machine [38]. For example, I trust a coffee machine to give me coffee, based on previous experiences, or on testimonies from other people. Such trust based on testimonies of a group of people is often called *reputation* [11]. It is also possible to use knowledge in trusting a machine. For example, I have a naive mental model: the weight of the coin will tip a leaver, that triggers release of a paper cup, coffee powder and hot water. True or not, that mental model allows me to operate the machine. I also have knowledge about the purpose. I trust the machine will give me coffee, because I know that is the vendor's business model. Moreover, I trust that some regulator has put safety regulations into place. We do not believe it is possible to use identification-based trust in the case of machines, at least not with current state of the art in artificial intelligence.

This example shows that theories about trust, especially calculus-based trust (regularities) and knowledge-based trust (mental model), are similar to theories about transparency [22]. Previous experience, as well as knowledge about the design, about the internal state, and about the purpose of the machine will induce trust in the machine.

That ends our discussion of trust. We may conclude that trust is an important factor that influences the intention to use a system or continue to use a system. We distinguish trust in the machine, mediated by knowledge of the design, the internal state, and the purpose of the machine, and institutional trust in the organizations that developed the machine, and that now operate the machine. We can also conclude that there are many parallels between trust and transparency, and that trust in a machine, depends on transparency of the system design. However, unlike Vorm and Combs [39], we do not believe we can reduce trust to transparency. Transparency is a system property, a requirement, that can be designed and tested for, whereas trust is a user attitude, but also an objective to achieve by designing the system in a certain way. That means, that to evaluate effectiveness of the design, these variables should be measured independently.

Table 2. Trust develops in stages [18]

	Calculus-based trust	Knowledge-based trust	Identification-based trust
based on	consistency of behaviour; repeated observations	knowledge of beliefs and goals that underlie behaviour	identification with values and background
example	coffee-machine	chess opponent	former classmate
usage	allow users to inspect what happened (trace)	help users build a mental model; explain	build a relationship

2.4   On Evaluation

We discussed models of trust and transparency, and ideas about evaluation. How can all of this be put together? In this section we discuss the model of Hoffman et al. [15], that was influential in the discussion on evaluation of explainable AI systems (Fig. 3).

In yellow, the model shows a flow. The user receives an initial instruction. This affects the user's initial mental model. The instruction is followed by an explanation, which revises the user's mental model, and subsequently enables better performance. Can we adjust the model Hoffman et al. [15] for recommendation? Yes. As we have seen, explainable recommendation dialogues proceed in three stages: (i) collection of user preferences, (ii) recommendation, and (iii) feedback and explanation. Therefore we have added a step, shown in dark yellow, for recommendation. That means that implicitly, the evaluation of the first step, elicitation of user preferences, is part of the evaluation of the second step, recommendation.

In green, the model shows how to measure these variables. In particular, effectiveness of the explanation is tested by *goodness criteria*, which can be assessed by developers, on the basis of log-files, and a *test of satisfaction*, a subjective measure, asking end-users whether they are satisfied with the explanation. The effect of an explanation on a user's mental model is tested by a *test of comprehension*, similar to an exam question: is the user's mental model in line with reality? Finally, the effect on performance can be a tested by a *test of performance*, related to the task. A recommendation is also evaluated by a goodness criteria, and by user satisfaction, shown here in dark green.

In grey, the model shows the expected effect of explanations on trust. Initially, the user may have inappropriate trust or mistrust, based on previous conceptions. After recommendation and explanation, the user's mental model changes, and leads to more appropriate trust, which enables more appropriate use of the system.

Goodness criteria measure the success conditions, in this case of an explanation. For example: the given explanation must match the type of explanation that was asked for. If the user wants to know how the system works, it should not the purpose. These criteria can be assessed relatively objectively by comparing specified functionality with the behaviour shown on the log-files. At least two coders should verify inter-coder

Fig. 3. Evaluating explainable AI in various stages, adjusted from [15]. Components in dark yellow and dark green are added here for explainable recommendation (Color figure online)

Table 3. Goodness criteria for evaluating an explanation [15]

The explanation helps me **understand** how the [software, algorithm, tool] works	Y/N
The explanation of how the [software, algorithm, tool] works is **satisfying**	Y/N
The explanation of the [software, algorithm, tool] sufficiently **detailed**	Y/N
The explanation of how the [software, algorithm, tool] works is sufficiently **complete**	Y/N
The explanation is **actionable**, that is, it helps me know how to use the [software, algorithm, tool]	Y/N
The explanation lets me know how **accurate** or **reliable** the [software, algorithm] is	Y/N
The explanation lets me know how **trustworthy** the [software, algorithm, tool] is	Y/N

agreement on the scores [42]. Hoffman et al. provide a list of goodness criteria for explanations (Table 3). This is meant as an objective measure. However, we can see the list is written from the perspective of an end-user, so it looks like a subjective measure. What to do? First, these criteria can be re-used in user satisfaction test. Second, we can in fact define objective criteria. We will show that in Sect. 4.

A *test of satisfaction* for an explanation aims to test "the degree to which users feel that they understand the AI system or process being explained to them." According to Hoffman et al., user satisfaction is measured by a series of Likert scales for key attributes of explanations: understandability, feeling of satisfaction, sufficiency of detail, completeness, usefulness, accuracy, and trustworthiness. As discussed above, satisfaction seems to overlap with the goodness criteria. Hoffman et al. explain the difference as follows. Relative to goodness, satisfaction is contextualized. It is measured after the interaction, all factors included. The measurements are meant for a different audience. The goodness test is meant for developers and the satisfaction test is for end-users.

A *test of comprehension* aims to test the effectiveness of the explanation on the mental model. Similar to an exam question: is the user able to remember and reproduce elements of the explanation? For example, users can be asked to reproduce how a particular part of the system works, reflect on the task, or be asked to make predictions, for which knowledge of the system is needed. There are many ways in which mental models can be elicited [15, Table 4]. Consider methods like think aloud protocols, task reflection (how did it go, what went wrong?), card sorting tasks (which questions are most relevant at this stage?), selection tasks (identifying the best representation of the mental model), glitch detector tasks (identifying what is wrong with an explanation), prediction tasks, diagramming tasks (drawing a diagram of processes, events and concepts), and a shadow box task (users compare their understanding to that of a domain expert). Various methods have to be combined, to make tests more reliable.

Finally, a *test of performance* aims to objectively test over-all effectiveness of a system. One could take the *success rate*: count the number of successfully completed dialogues, relative to the total number of dialogues. For goal-directed dialogue, progress towards the goal can be measured objectively. Consider for example a system applied in retail [33]. Here, the conversion rate is a measure of success: how many potential customers end up buying a product. We can also try to evaluate *communicative success*. The effectiveness of an explanation is inversely proportional to the number of misunderstandings. Thus, one could identify indicators of misunderstanding (e.g. overly long duration, signs of frustration, aborted dialogues), and count the relative number of such

Table 4. Trigger questions [15]

	Triggers	User/Learner's Goal
1.	How do I use it?	Achieve the primary ask goals
2.	How does it work?	Feeling of satisfaction at having achieved an understanding of the system, in general (global understanding)
3.	What did it just do?	Feeling of satisfaction at having achieved an understanding of the system, in general (local understanding)
4.	What does it achieve?	Understanding of the system's functions and uses
5.	What will it do next?	Feeling of trust based on the observability and predictability of the system
6.	How much effort will this take?	Feeling of effectiveness and achievement of primary task
7.	What do I do if it gets it wrong?	Desire to avoid mistakes
8.	How do I avoid the failure modes?	Desire to mitigate errors
9.	What would it have done if x were different?	Resolution of curiosity at having achieved an understanding of the system
10.	Why didn't it do z?	Resolution of curiosity at having achieved an understanding of the local decision

misunderstandings. The over-all purpose of a recommendation system is to *convince* the user, and perhaps even to induce them to change behaviour. Objectively establishing such a change of behaviour is the ultimate test of success. That concludes our discussion of Hoffman et al. [15]. It serves as a good basis for designing evaluation experiments.

In a more recent paper, Van der Waa et al [36] discuss how to evaluate explainable AI. They conduct experiments, comparing two types of explanations: rule-based and example-based. These explanations are compared on user satisfaction and general system performance. They also discuss the advantages of combining subjective measures with more detailed behavioural analysis, on the basis of observable behaviour.

Another technique for subjective measurement is the *Q-methodology*, from HCI [29]. Using the trigger-questions in Table 4 from [15], Vorm and Miller [40] suggest to evaluate explainable systems by having the user select the question, which they would like to ask at that point in the interaction. Users are asked to sort 36 cards with questions. Vorm and Miller carefully developed the question bank. For example: "How current is the data used in making this recommendation?" or "Precisely what information about me does the system know?". Factor analysis determines specific groups of users with similar preferences. In this way, four groups of users are found [40]: 1: Interested and Independent, 2: Cautious and Reluctant, 3: Socially Influenced, and 4: Egocentric. This shows that different types of users have various needs for explanations. A system should be flexible enough to handle these needs.

3 Application

In this section, we discuss a specific system, that is currently being developed [6]. The system is an explainable recommendation system, for the food and health domain [35]. The system is interactive and provides personalized recommendations and explanations. The system is developed in two versions: a web-based platform allowing the users to experience both the explanation-based interactive recommender and its replica without the explanations and critiques component (i.e., a regular recommender). This allows us to assess the effectiveness of explanation and of interaction.

A user interaction involves three stages, with the following success conditions.

- Stage 1. *User preference elicitation.* Ask user about preferences. Afterwards, the system must know enough user preferences to select a recipe; preferences are consistent and are correctly understood.
- stage 2. *Recommendation of a recipe.* The recipe must fit the user preferences, and follow from knowledge about food, recipes, and healthy lifestyles.
- Stage 3. *Explanation and interaction.* The explanation must fit the user's request. The explanation must be personalized to the user's preferences and be relevant in context, and the subsequent interaction must be coherent.

If we classify the system, we can say that the *application domain* is food and health. The *task* is recommendation, but it also involves elements of *persuasion*, as we intend users to follow a healthy lifestyle. In some cases, that means convincing the user and making them change behaviour. In other words, the system is intended as a nutrition virtual coach (NVC) [35]. In order to persuade the user, trust and transparency are crucial.

Persuasion is often about breaking a habit. What seems to work, based on conversations with nutritionists, is to set personal goals, and help users attain those goals, by measuring the current state, the distance to the goal, and suggesting ways of getting closer. Measuring weight can be used to quantify progress towards the goal, and calories are used to quantify the required energy intake of a meal. Long-term relationship building, as required for a nutrition virtual coach, is out of scope for this research prototype, but it does play a role in the over-all design of the system, and in future research.

In this domain, generally we find that explanations are of two types: preference related explanations, which are based on the user preferences which were inferred or stated just before, or health related explanations, which are based on general knowledge about food an health [6]. Here we show an example of each.

- *Health-related*: Protein amount covers user needs for a meal.
  "This recipe contains X grams of protein, which is about Y % of your daily requirement. Your body needs proteins. Consuming the necessary amount is important!"
- *Preference-related*: User's chosen cuisine matches the recipe.
  "This recipe is a typical part of the cuisine you like: Z."

4   Towards Metrics

In this section, we specify the metrics to evaluate claims about effectiveness of an explainable recommender system, in the context of the food and health domain, as detailed in Sect. 3. Consider the research model in Fig. 4.

On the right, interpret effectiveness as the direct effect of an interaction on the user, in terms of user satisfaction (performance), transparency and trust. That aspect refers to the recommendation part of the task, and also the explanation. In addition, repeated interactions should have an indirect effect on the user, in terms of a change of behaviour, for instance a healthier choice of food. That aspect refers to the persuasion task.

On the left, the system design is detailed. We see a system as a white-box, with separate functionalities. Each of the modules may have an effect on the user interaction.

These modules are: the algorithm for generating a recommendation, the knowledge base about health and food, the goals and plans of the system during interaction, the user interface, the user model that represents user preferences, and the data set with all the recipes that can be recommended. Each of these modules must be evaluated separately and as part of the system (unit test; integration test).

In the middle, we discuss two moderator variables, that may strengthen or weaken the effect of the system design on the success variables. First, *explanation*, whether the system is able to provide explanations about its recommendations. Second, *interaction*, whether the systems allows feedback and interaction about recommendations and explanations that fit the context. These are the interventions that we want to study.

This model can test the over-all effect of recommendations, and the specific effect of explanations and interactive dialogues on user satisfaction, transparency and trust, and ultimately, on behavioural change. However, the model also has some disadvantages. The model disregards several variables that are familiar from TAM, in particular perceived usefulness and perceived ease of use and the intention to use a system. Model (1) focuses not so much on the decision to start using a system (as in TAM), but rather on evaluating actual use of a system. In addition, the 'effectiveness' variables (user satisfaction, transparency, trust) need to be worked out in more detail.

Therefore, we developed the model in Fig. 5. On the left, we see the various modules that make up the recommender system design. If these modules function effectively, they have a positive influence on transparency. In addition, the system design has a direct effect on the use of the system (long red arrow). Transparency in turn has an effect on the perceived usefulness and perceived ease of use, which in turn affect the intention to use, and usage itself, as in the ISTAM model [39]. It is also possible that the system design has a direct effect on perceived usefulness and perceived ease of use. Transparency is expected to have an effect on trust. After all, the perceived competence, benevolence and integrity of the system and organization that deploys the system, are mediated by the interface. Trust, in turn, has an effect on the intention to use, and ultimately, on the suggested behaviour change. Finally, we also consider a feedback loop, back from usage to trust. However, such feedback loops are difficult to test for.

Like in Model (1) we test for two moderator variables: explanation and interaction. These features are expected to affect transparency, and indirectly affect perceived ease of use and perceived usefulness, as well as trust. Moreover, they are expected to have a direct effect on use (success rate and failure rate).

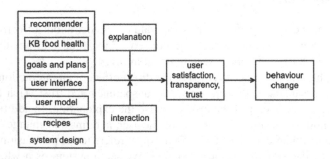

Fig. 4. Towards a Model for Evaluation (1)

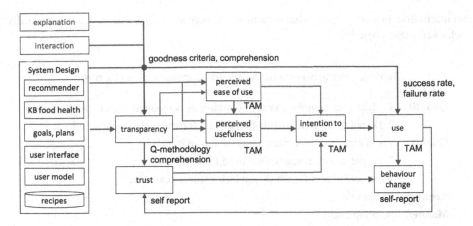

Fig. 5. Towards a Model for Evaluation (2)

Accordingly, we propose a table of metrics for each these variables (Table 5). Symbol '–' means that the variable is inversely related to success. 'Goodness criteria' refers to the tables with specific goodness criteria per functionality, as discussed in Sect. 4.1. TAM instruments refers to established questionnaires.

Table 5. Metrics to measure variables in Model for Evaluation (2)

Variable	Measures
system design	goodness criteria, comprehension
	ease of use: – duration, – number of misunderstandings, – time to learn
	usefulness: success rate, – failure rate
transparency	Q-methodology
	comprehension
	goodness criteria: shows purpose, internal state, and how it works
trust	self-report, willingness to recommend
perceived ease of use	TAM instruments
perceived usefulness	TAM instruments
intention to use	TAM instruments
use	counting, TAM instruments
behavioural change	self-report

4.1 Goodness Criteria for Recommendation and Explanation

An important part of the evaluation methods depend on requirements or goodness criteria for the various components and functionalities. The most important functionalities are the ability to provide a recommendation, and the ability to provide an explanation, in an interactive manner. Under what conditions can we say that a system has successfully achieved these objectives?

Table 6. Gricean maxims for cooperative communication [13, p 46]

Quantity	1. Make your contribution as informative as is required (for current purposes)
	2. Do not make your contribution more informative than is required
Quality	Try to make your contribution one that is true
	1. Do not say what you believe to be false
	2. Do not say that for which you lack adequate evidence
Relation	Be relevant
Manner	Be perspicuous
	1. Avoid obscurity of expression
	2. Avoid ambiguity
	3. Be brief (avoid unnecessary prolixity)
	4. Be orderly

The following example shows the type of functionality that we develop. How suitable is the explanation in this dialogue?

U. I'd like some toasted white bread for breakfast.
S. You should eat whole meal bread.
U. Why?
S. Because you said you wanted to lose weight, and eating whole meal bread instead of white bread is a good way to reduce the number of quick calories per meal, and it is well known that reducing the number of quick calories per meal will help you lose weight.

We start by clarifying the relation between explanations and transparency. Transparency is a property of a system. The system must reveal its purpose, inner state, and how it works [22]. Transparency is not a property of an explanation, except in the sense of 'clarity' or 'being based on evidence'. Instead, part of the purpose of having explanations, is for the system to be more transparent. There are other methods to make a system more transparent too, such as a user manual, a suitable persona, etc.

4.2 Good Explanation

What makes a good explanation? An explanation is a form of assertion. That means, that we can follow Grice's maxims for cooperative communication [13, p. 46]: quantity, quality, relation and manner (see Table 6). The point about manner, specifically to be brief, is also made by Mualla et al [28], who advocate parsimonious explanations.

There is a lot of research on what makes a good explanation, in various fields. Properties of everyday explanations are summarized in a survey paper by Miller [26]:

1. Explanations are *contrastive*: distinguish an outcome from counterfactual outcomes.
2. Explanations are *selected* from a range of possible reasons.
3. Explanations are not necessarily based on probabilities, but rather on *narratives*.
4. Explanations are social, and are usually part of *interactions*.

These four characteristics can be summarized by stating that explanations are inherently *contextual*. We will discuss them one by one. Ad 1. An explanation must not only be generic (e.g. based on laws of nature), but also involve specific facts of the case of the user, that show why other alternative advice is not given [41]. Ad 2. What are reasons? For natural events, they are causal histories built from facts and natural laws [19]. For human behaviour, they are goals which can be inferred by abduction, because they most likely motivate those actions in that context [24]. In our case, the reasons are ingredients which match user preferences. Other reasons are natural laws of nutrition (vegetables have low calories; pasta has high calories), and motivational goals of the user, to maintain a certain weight, for example. Ad 3. Miller [26] criticizes some technical research on intelligible algorithms, which focuses scientific explanations. A doctor would justify a treatment to a colleague using probabilities, but for lay people, stories often work better. Ad 4. Interactive dialogues with explanations are preferred, because they give the user more control. In case of a problem, the user can just ask.

Generally, there are several levels or successive rounds of explanation.

- Level 1. Why this recommendation? Because of *facts* (user preferences, ingredients, recipes) and a *rule* (knowledge about food and health)
- Level 2. Why that rule? Because the rule is true and relevant relative to a *goal*. Why those facts? Because the procedure for selecting these facts is *valid*.
- Level 3. Why that goal? Because the goal (promote healthy choices) helps to promote social values (health), which represent who we are (virtual nutritionist).

This example of explanation levels is based on value-based argumentation [3], which also has three levels: (1) actions, facts and rules, (2) goals, and (3) social values.

The properties of explanations discussed so far, are relatively abstract. How can they be built into algorithms? Rosenfeld [31] presents four metrics for evaluating explainable artificial intelligent systems: D, R, F, and S. Here D stands for the performance difference between the black-box model and a transparent model, R considers the size of the explanation (i.e., number of rules involved in given explanations), F takes the relative complexity into consideration, by counting the number of features to construct an explanation, and D measures the stability of the explanations (i.e., ability to handle noise perturbations) [31]. In the context of explaining recommendations, we can measure the following aspects:

- *Improvement Effect:* Test the system with and without explanations and observe the effect of explanations on system performance. We can list a number of performance metrics such as acceptance rate, average acceptance duration, and average number of interactions spent for acceptance of the recommendation.
- *Simplicity of Explanations:* This can be measured with the length of the explanations and to what extent that can be grasped by the user (comprehension).

- *Cognitive Effort Required:* An explanation may focus on a single decision criterion (e.g., only nutrition levels) to reduce the user's cognitive effort. Some explanations may point out several criteria (e.g., nutrition levels, user's goals such as losing weights, and their preferences on ingredients) at one time, which may increase the cognitive load. We can count the number of criteria captured in a given explanation.
- *Accuracy of Explanations:* The system generates recommendations based on its objectives and its beliefs about the user's goal and preferences. What if it is wrong? Then, the system may generate explanations which conflict with actual preferences. In such case, users give feedback on the explanations by pointing out their mistakes. We can analyze the given feedback to determine the accuracy of the explanations.

Table 7. Goodness criteria for recommendation, adjusted from criteria in Table 3 [15].

The recommendation **helps me to decide** [what action to do/which recipe to cook]	Y/N
The recommendation [what action to do/which recipe to cook] is **satisfying**	Y/N
The recommendation [what action to do/which recipe to cook] is **sufficiently detailed**	Y/N
The recommendation [what action to do/which recipe to cook] is **sufficiently complete**	Y/N
The recommendation is **actionable**, that is, it helps me to carry out my decision	Y/N
The recommendation lets me know how **accurate** or **reliable** the [action/recipe] is	Y/N
The recommendation lets me know how **trustworthy** the [action/recipe] is	Y/N

4.3 Good Recommendation

In recommendation systems, performance metrics are often borrowed from information retrieval: *precision* (fraction of given recommendations that are relevant) and *recall* (fraction of potentially relevant recommendations that are given). One can balance precision and recall, by means of the F-measure. Alternatively, people use the area under the Receiver Operator Characteristic (ROC) curve, to measure how well the algorithm scores on this trade-off between precsion and recall, see [34].

What makes a good recommendation? We can see a recommendation as a response to a request for advice. The same Gricean maxims apply (Table 6). In the context of our application that suggest the following requirements.

- *Quality*: the recipe must be an existing recipe, and fit the agreed dietary goals.
- *Quantity*: the recipe must be detailed enough to be able to make it. All ingredients and quantities must be listed and clear.
- *Relation*: the recipe must respond to the request of the user and fit the context. Specifically, the recipe must match the user preferences, if such recipes exist. If no such recipes exist, a clear no-message must be given.
- *Manner*: the recipe is presented clearly and with diagrams or photographs to illustrate. The recipe must not be too long or detailed [28].

The goodness criteria for recommendations are similar to those for explanations in Table 3. For comparison, we have adjusted them to fit recommendations (Table 7).

5   Discussion

Building explainable recommender systems in the food and health domain, has ethical consequences. This is why we care about explainability, transparency and trust. In a previous paper, we have given a survey of ethical considerations for nutritional virtual coaches [7]. Here, we will discuss a few examples.

First, the factual *information* about food and ingredients must be true, informative, and relevant (Gricean Maxims). The data set and knowledge bases used must be fair and present a representative coverage of foods and tastes. This is not trivial, as food is related to culture and identity. The system will collect personal data from the user, namely food preferences and health related data. These data are sensitive, and must be adequately protected. We observe a trade-off between privacy and relevance. If more detailed personal data is collected, a better recommendation can be made. A metric to test this balance, is to check how many requested data items, are actually used.

Second, the system makes recommendations and provides explanations. A related ethical issue is *control*: in case of a conflict between user preferences and healthy choices, who determines the final recommendation? Suppose the user asks for a hamburger? Suggesting a more healthy alternative may be seen as patronizing. Here, we believe the solution is to be transparent: whenever a requested unhealthy choice is *not* recommended, this must always be explained. The explanation is contrastive. Moreover, the recommendation must be in line with the stated purpose and 'persona' of the system: chef (good food) or nutritionist (advice).

Another ethical issue is *sincerity*. A recommendation or explanation must be trusted not to have a hidden purpose, like commercial gain [21]. If a system provides a clear explanation, a user can verify that reason. Moreover, if the explanation is contrastive [26], indicating why some recipes are not shown, and interactive, allowing users to vary the request to see how that affects the recommendation, this will make the reasoning mechanism transparent, and make it easier to detect a hidden purpose [41].

Third, the systems aims to *persuade* the user, for instance to make healthier choices for food. To do so, the system makes use of argumentation techniques. An ethical issue is how much persuasion we accept from a machine. Again, this depends on the stated purpose and persona of the system (e.g. chef or nutritionist). Who is ultimately responsible? Here, the answer is to develop the system as a tool to support a nutritionist in coaching a large group of clients. After deployment a qualified nutritionist should remain as *human-in-the-loop*, with meaningful control over the persuasion process.

To summarize, an explainable recommender system offers many opportunities for manipulation [21]. Manipulation is harder to achieve, if the system is transparent: the user can verify and compare the stated purpose with actual behavior.

6   Conclusions

In this paper, we have discussed models and metrics for evaluation interactive explainable recommender systems. We pointed out the debate between subjective measurement (perceived ease of use, perceived usefulness, user satisfaction) and objective measures (goodness criteria, task success, misunderstandings). We argue that subjective and objective evaluation strengthen each other. For example, consider the following design principle: users who experience a misunderstanding are less likely to be satisfied. Misunderstandings might be clear from the log-files. Satisfaction depends on users. So, in order to test this design principle, one needs to compare both objective and subjective evaluation metrics.

The model of [15] forms a good basis to develop evaluation metrics for explainable recommender systems, except that the notion of 'goodness criteria' needs to be worked out, and must be more clearly separated from user satisfaction. For acceptance testing and user evaluations, the famous TAM model is relevant [8,37]. Trust can be added. Following Vorm and Combs [39] we believe that transparency is crucial, and that trust is largely influenced by transparency. However, unlike [39] we believe trust is a separate notion, that can be measured, by subjective measures.

So, our evaluation model is based on three components: (i) the ISTAM model [39], which combines TAM and Transparency, (ii) trust [25] and specifically trust in machines [38], and (iii) various objective measures, such as goodness criteria and fit to the context, success rate, number of misunderstandings, and over-all performance [15].

Especially for applications in the food and health domain, building an explainable recommender system has important ethical considerations [7]. An important part of the solution is to provide explanations and be transparent about the system's purpose and way of working. This should allow the user to verify the behaviour of the system and decide if this forms a basis to trust the system and the recommendations it makes.

Acknowledgments. This work has been supported by CHIST-ERA grant CHIST-ERA19-XAI-005, and by (i) the Swiss National Science Foundation (G.A. 20CH21_195530), (ii) the Italian Ministry for Universities and Research, (iii) the Luxembourg National Research Fund (G.A. INTER/CHIST/19/14589586), (iv) the Scientific and Research Council of Turkey (TÜBİTAK, G.A. 120N680).

References

1. Adomavicius, G., Tuzhilin, A.: Toward the next generation of recommender systems: a survey of the state-of-the-art and possible extensions. IEEE Trans. Knowl. Data Eng. **17**, 734–749 (2005)
2. Anjomshoae, S., Calvaresi, D., Najjar, A., Främling, K.: Explainable agents and robots: results from a systematic literature review. In: Autonomous Agents and Multi Agent Systems (AAMAS 2019), pp. 1078–1088 (2019)
3. Atkinson, K., Bench-Capon, T., McBurney, P.: Computational representation of practical argument. Synthese **152**(2), 157–206 (2006)
4. Bernstein, E.: Making transparency transparent: the evolution of observation in management theory. Acad. Manag. Ann. **11**(1), 217–266 (2017)
5. Burke, R., Felfernig, A., Göker, M.H.: Recommender systems: an overview. AI Mag. **32**, 13–18 (2011)
6. Buzcu, B., Varadhajaran, V., Tchappi, I.H., Najjar, A., Calvaresi, D., Aydoğan, R.: Explanation-based negotiation protocol for nutrition virtual coaching. In: PRIMA 2022. LNCS, vol. 13753, pp. 20–36. Springer (2022). https://doi.org/10.1007/978-3-031-21203-1_2

7. Calvaresi, D.: Ethical and legal considerations for nutrition virtual coaches. In: AI and Ethics, pp. 1–28 (2022)
8. Davis, F.D.: Perceived usefulness, perceived ease of use, and user acceptance of information technology. MIS Q. **13**(3), 319–340 (1989)
9. V. Dignum. Responsible Artificial Intelligence: How to Develop and Use AI in a Responsible Way. Springer (2019). https://doi.org/10.1007/978-3-030-30371-6
10. European Commission. Proposal for a Regulation of the European Parliament and of the Council laying down harmonised rules on Artificial Intelligence (Artificial Intelligence Act) and amending certain union legislative acts (2021)
11. Falcone, R., Castelfranchi, C.: Trust and relational capital. Comput. Math. Organ. Theory **17**(2), 179–195 (2011)
12. Goodhue, D.L.: Understanding user evaluations of information systems. Manage. Sci. **41**(12), 1827–1844 (1995)
13. Grice, H.P.: Logic and conversation. In: Cole, P., Morgan, J.L. (eds.) Syntax and Semantics, vol. 3, pp. 41–58. Academic Press, New York (1975)
14. HLEG. Ethics guidelines for trustworthy AI (2019)
15. Hoffman, R.R., Mueller, S.T., Klein, G., Litman, O.: Metrics for explainable ai: challenges and prospects. arXiv:1812.04608 [cs.AI] (2018)
16. Jannach, D., Pearl, P., Ricci, F., Zanker, M.: Recommender systems: past, present, future. AI Mag. **42**, 3–6 (2021)
17. Kriz, S., Ferro, T.D., Damera, P., Porter, J.R.: Fictional Robots as a Data Source in HRI Research, pp. 458–463. IEEE (2010)
18. Lewicki, R.J., Bunker, B.B.: Developing and maintaining trust in work relationships. In: Trust in Organizations, pp. 114–139. Sage Publications (1996)
19. Lewis, D.: Causal explanation, pp. 214–240. Oxford University Press, Oxford (1986)
20. Lewis, J.R., Sauro, J.: Item benchmarks for the system usability scale. J. Usability Stud. **13**(3), 158–167 (2018)
21. Lima, G., Grgić-Hlača, N., Jeong, J.K., Cha, M.: The conflict between explainable and accountable decision-making algorithms. In: FACCT, pp. 2103–2113. ACM, Seoul, Republic of Korea (2022)
22. Lyons, J.B.: Being transparent about transparency: A model for human-robot interaction, pp. 48–53. AAAI (2013)
23. Lyons, J.B., Havig, P.R.: Transparency in a human-machine context: approaches for fostering shared awareness/intent. In: Shumaker, R., Lackey, S. (eds.) VAMR 2014. LNCS, vol. 8525, pp. 181–190. Springer, Cham (2014). https://doi.org/10.1007/978-3-319-07458-0_18
24. Malle, B.F.: How people explain behavior: a new theoretical framework. Pers. Soc. Psychol. Rev. **3**(1), 23–48 (1999)
25. Mayer, R.C., Davis, J.H., Schoorman, F.D.: An integrative model of organizational trust. Acad. Manag. Rev. **20**(3), 709–734 (1995)
26. Miller, T.: Explanation in artificial intelligence: insights from the social sciences. Artif. Intell. **267**, 1–38 (2019)
27. Miller, T., Hoffman, R., Amir, O., Holzinger, A.: Special issue on explainable artificial intelligence. Artif. Intell. **307**, 103705 (2022)
28. Mualla, Y., et al.: The quest of parsimonious XAI: a human-agent architecture for explanation formulation. Artif. Intell. **302**, 103573 (2022)
29. O'Leary, K., Wobbrock, J.O., Riskin, E.A.: Q-methodology as a research and design tool for HCI, pp. 1941–1950. ACM, Paris (2013)
30. Pavlou, P.A., Gefen, D.: Building effective online marketplaces with institution-based trust. Inf. Syst. Res. **15**(1), 37–59 (2004)
31. Rosenfeld, A.: Better metrics for evaluating explainable artificial intelligence. In: AAMAS, pp. 45–50, Richland, SC (2021)

32. Smith, R.W., Hipp, D.R.: Spoken Language Dialog Systems: A Practical Approach. Oxford University Press, Oxford (1994)
33. Christina Soyoung Song and Youn-Kyung Kim: The role of the human-robot interaction in consumers' acceptance of humanoid retail service robots. J. Bus. Res. **146**, 489–503 (2022)
34. Tintarev, N., Masthoff, J.: Explaining recommendations: design and evaluation. In: Ricci, F., Rokach, L., Shapira, B. (eds.) Recommender Systems Handbook, pp. 353–382. Springer, Boston, MA (2015). https://doi.org/10.1007/978-1-4899-7637-6_10
35. Trang Tran, T.N., Atas, M., Felfernig, A., Stettinger, M.: An overview of recommender systems in the healthy food domain. J. Intell. Inform. Syst. **50**(3), 501–526 (2018)
36. van der Waa, J., Nieuwburg, E., Cremers, A., Neerincx, M.: Evaluating XAI: A comparison of rule-based and example-based explanations. Artif. Intell. **291**, 103404 (2023)
37. Venkatesh, V., Morris, M.G., Davis, G.B., Davis, F.D.: User acceptance of information technology: toward a unified view. MIS Q. **27**(3), 425–478 (2003)
38. Vermaas, P.E., Tan, Y.-H., van den Hoven, J., Burgemeestre, B., Hulstijn, J.: Designing for trust: a case of value-sensitive design. Knowl. Technol. Policy **23**(3–4), 491–505 (2010)
39. Vorm, E.S., Combs, D.J.Y.: Integrating transparency, trust, and acceptance: The intelligent systems technology model (ISTAM). Int. J. Hum.-Comput. Interact., 1–19 (2022)
40. Vorm, E.S., Miller, A.D.: Modeling user information needs to enable successful human-machine teams: designing transparency for autonomous systems. In: Schmorrow, D.D., Fidopiastis, C.M. (eds.) HCII 2020. LNCS (LNAI), vol. 12197, pp. 445–465. Springer, Cham (2020). https://doi.org/10.1007/978-3-030-50439-7_31
41. Wachter, S., Mittelstadt, B., Russell, C.: Counterfactual explanations without opening the black box: automated decisions and the GDPR. Harvard J. Law Technol. **31**(2), 841–887 (2018)
42. Walker, M.A., Litman, D.J., Kamm, A., Abella, A.: PARADISE: A framework for evaluating spoken dialogue agents. In: Proceedings of the 35th Annual meeting of the ACL/EACL, pp. 271–280, Madrid (1997)
43. Wallkötter, S., Tulli, S., Castellano, G., Paiva, A., Chetouani, M.: Explainable embodied agents through social cues: a review. ACM Trans. Hum.-Robot Interact. **10**(3), 27:2–27:24 (2021)

Leveraging Imperfect Explanations
for Plan Recognition Problems

Ahmad Alelaimat(✉), Aditya Ghose, and Hoa Khanh Dam

Decision Systems Lab, School of Computing and Information Technology,
University of Wollongong, Wollongong, NSW 2522, Australia
{aama963,aditya,hoa}@uow.edu.au

Abstract. Open environments require dynamic execution of plans
where agents must engage in settings that include, for example, replanning, plan reusing, plan repair, etc. Hence, real-life Plan Recognition (PR) approaches are required to deal with different classes of
observations (e.g., exogenous actions, switching between activities, and
missing observations). Many approaches to PR consider these classes
of observations, but none have dealt with them as deliberated events.
Actually, using existing PR methods to explain such classes of observations may generate only so-called imperfect explanations (plans that
partially explain a sequence of observations). Our overall approach is to
leverage (in the sense of plan editing) imperfect explanations by exploiting new classes of observations. We use the notation of capabilities in
the well-known Belief-Desire-Intention (BDI) agents programming as an
ideal platform to discuss our work. To validate our approach, we show
the implementation of our approach using practical examples from the
Monroe Plan Corpus.

Keywords: Plan recognition · Plan updating · BDI agents

1 Introduction

It is generally accepted that an agent system with practical extensions can carry
out tasks that would not otherwise be achievable by its basic reactive system.
Often when the environment is highly dynamic and/or the task is complicated for
the basic reactive behaviour of the agent system, developers resort to extending
or adding new modules to the agent system. A typical example of extending agent
models can be found far and wide in the state-of-the-art Belief-Desire-Intention
(BDI) paradigm [1], including, but not limited to, extending the architecture
with self-awareness [2], automated planning [3], and reconfigurability [4].

Much of the work done on PR involves abductive reasoning (e.g., [5,6], and [7]),
which seeks to abductively infer plans by mapping the observed actions to plan
libraries. A major drawback to Abductive Plan Recognition (APR) is that target

---

A. Ghose—Passed away prior to the submission of the manuscript. This is one of the
last contributions by Aditya Ghose.

plans are usually not from the plan library. This can be due to several reasons, some of which are related to extending the observed agent system with different modules, such as the works presented in [2–4]. Actually, appealing to APR approaches to explain such observed actions would only generate imperfect explanations. An imperfect explanation is one that partially explains a sequence of actions. Another notion of PR is discovering plans by executing action models (in the sense of classical automated planning) to best explain the observed actions. Nevertheless, this can take a great deal of time in complex problems. Our approach is a third way which does not align itself with either notion. Still, it can potentially improve the explanatory power of plan libraries without the need for action model execution.

In this paper, we address the problem of *leveraging imperfect explanations*, the process of modifying existing hypotheses to explain an observed sequence of actions. We expect leveraging imperfect explanations to help answer the following questions:

- *What was the agent's original plan based on its new (or unusual) observed actions?*
- *Were the observed changes in a plan execution intentional?*

Answering these questions can be useful in understanding the capabilities of the observed agent. Consider, for example, a system composed of heterogeneous autonomous agents, some of which are equipped with a form of societal control (e.g., such as the one described in [8]). In such systems, leveraging imperfect explanations could be used to gain insight into the actual norm-modification operators of the observed agents compared to non-normative agents.

We show that when the observed agent operates in a domain model known to the observer, imperfect explanations can be a valuable guide to explain unknown plans that involve new classes of observations. Hence, our approach can be seen as a post-processing stage for various single-agent plan library-based PR techniques. To avoid arbitrary modification of hypotheses, we also introduce a classification model that can determine the settings (e.g., noisy or explanatory) in which an unknown plan has been observed. We demonstrate the performance of the proposed approach using the Monroe Plan Corpus.

The remainder of this work is organized as follows. Related work is discussed in Sect. 2. In Sect. 3, we introduce some preliminaries on the notion of capabilities and BDI agents programming, which are the two main ingredients of our work. Section 4 presents our running example with pointers to different scenarios. We formalize the problem of this work in Sect. 5. Section 6 describes our approach to leveraging imperfect explanations for PR problems. Empirical evaluation is described in Sect. 7 before we conclude and outline future work in Sect. 8.

2   Related Work

Real-life PR systems are required to deal with domains in which new classes of observations are frequently observed. Roughly speaking, there are three noticeable domains regarding new classes of observations in PR problems: (1) Exploratory domains - the observed behaviour is a subject of exploratory and discovery learning (e.g., mistakes, exogenous and repeating activities), (2) Noisy domains - the observed behaviour is characterized by imperfect observability (e.g., extraneous, mislabeled and missing activities), and (3) Open domains - the observed behaviour is characterized by new, deliberated classes of observations (e.g., reconfigured plans). We concentrate our review on how existing PR approaches viewed/handled new classes of observations. We then use these classes in later sections where we describe our approach to leveraging imperfect explanations.

We first describe works that assume exploratory domains. With the intention of inferring students' plans, Mirsky et al. [7] proposed a heuristic plan recognition algorithm (called CRADLE) that incrementally prunes the set of possible explanations by reasoning about new observations and by updating plan arguments, in which explanations stay consistent with new observations. Uzan et al. [9] introduced an off-line PR algorithm (called PRISM) to recognize students' plans by traversing the plan tree in a way that is consistent with the temporal order of students' activities. Amir et al. [10] proposed an algorithm (called BUILDPLAN) based on recursive grammar to heuristically generate students' problem-solving strategies.

Many prior approaches to PR focused on dealing with noisy domains. Massardi et al. [11] classified noise in PR problems into three types: missing observations, mislabeled observations and extraneous actions and proposed a particle filter algorithm to provide robust-to-noise solutions to PR problems. Sohrabi et al. [12] transformed the PR problem into an AI planning problem that allows noisy and missing observations. Ramírez and Geffner [13] used classical planners to produce plans for a given goal G and compare these plans to the observed behaviour O. The probability distribution $P(G|O)$ can be computed by how the produced plans are close to the observed behaviour. Sukthankar and Sycara [14] proposed an approach for pruning and ranking hypotheses using temporal ordering constraints and agent resource dependencies.

PR systems are also required to deal with open domains, where the observed plans are usually not from the used plan library. Avrahami-Zilberbrand and Kaminka [15] describe two processes for anomalous and suspicious activity recognition: one using symbolic behaviour recognizer (SBR) and one using utility-based plan recognizer (UPR), respectively. Mainly, SBR filters inconsistent and ranking hypotheses, while UPR allows the observer to incorporate his preferences as a utility function. Zhuo et al. [16] also address the problem of PR in open domains using two approaches: one using an expectation-maximization algorithm and one using deep neural networks. A notable difference from other approaches is that the work of Zhuo et al. [16] is able to discover unobserved actions by constantly sampling actions and optimizing the probability of hypotheses.

Our work is not competing to, but complementing most of the previous works on PR, where it can be seen as a post-processing task for various single-agent plan library-based PR techniques. More precisely, leveraging imperfect explanations for PR problems can be viewed as an activity that occurs just before ruling out imperfect explanations or considering an observation as exploratory or noisy. In contrast with the literature, we focus on improving the explanatory power of plan libraries by leveraging imperfect explanations and exploiting new classes of observations.

3 Preliminaries

This section reviews the prior research used in the remainder of this work. First, we clarify the link between the notion of capability and BDI programming, and then we describe the representation we use for capabilities and plans.

3.1 BDI Programming and Capabilities

An agent system with BDI architecture [1] commonly consists of a belief base (what the agent knows about the environment), a set of events (desires that the agent would like to bring about), a plan library (a set of predefined operational procedures), and a base of intentions (plans that the agent is committed to executing).

Fundamentally, the reactive behaviour of BDI agent systems includes the agent system handling events by selecting an event to address from the set of pending events, selecting a suitable plan from the plan library, and stacking its program into the intention base. A plan in the plan library is a rule of the form $\epsilon : \nu \leftarrow \varrho$, where the program ϱ is a predefined strategy to handle the event ϵ whenever the context condition ν is believed to be true by the agent. A plan can be selected for handling an event ϵ if it is relevant and applicable, i.e., designed with respect to the event ϵ, and the agent believes that the context of the plan ν is a logical consequence of its belief base, respectively. A program ϱ often can be presented as a set of actions that result in changes in the environment state. For the purpose of this work, we ignore other elements (e.g., trigger events or guards) in the plan body. Note that using BDI programming languages such as Jason [17] and 2APL [18], information on action pre- and post-conditions can only be defined in a separate file (called simulated environment), making this information invisible to the agent.

To reason about actions and their specifications, we need to access information about the preconditions and postconditions of all available actions. We shall refer to *capability* as an explicit specification of action preconditions and postconditions. A capability has been understood in intelligent agent studies as having at least one way to achieve some state of affairs, where it can be used only if its preconditions are believed to be true [2,19]. For the purposes of this work, we shall concentrate mostly on the plan library. We do not, therefore, discuss other issues related to integrating the notion of capabilities into the BDI paradigm. For a detailed introduction to integrating the notion of capabilities into the BDI system, the reader is referred to [2,19].

3.2 Capability and Plan Representation

Our representation of agent capabilities is closely related to [2,19], but significantly influenced by the action theory found in classical automated planning, such as what has been presented in situation calculus [20] and STRIPS reasoning [21]. Following this representation, we use a language of propositional logic $\mathcal{L}$ over a finite set of literals $L = \{l_1, \ldots, l_n\}$ to represent the set of states of the environment S in which the agent is situated, such that each state of the environment $s \in S$ is a subset of L, i.e., $l_i \in s$ defines that the propositional literal l_i holds at the state s. As mentioned before, a capability specification describes an action that a BDI agent can carry out along with its pre- and post-conditions. Notationally, capability specification is triple subsets of L, which can be written as a rule of the form $\{pre(c)\}c\{post(c)\}$, where

- $\{pre(c)\}$ is a set of predicates whose satisfiability determines the applicability of the capability,
- c is the capability, and
- $\{post(c)\}$ is a set of predicates that materialize with respect to the execution of the capability.

It is not hard to see that, by sequentially grouping capabilities that are dedicated to bringing about some state of affairs, the sequence $C = \langle c_1, \ldots, c_n \rangle$ can be seen as an operational procedure to resolve that state. Consider the plan p, we use

- $\{pre(p)\}$ as the conditions under which the plan is applicable,
- $C = \langle c_1, \ldots, c_n \rangle$ as the plan body, and
- $\{post(p)\}$ as the conditions associated with the end event of the plan,

As such, we use the notation $p = \{pre(p)\}C\{post(p)\}$ to represent plan p specifications.

For the purpose of reasoning about the execution of agent capabilities, we work with a simple representation of the possible ways the plan can be executed, which we term normative plan traces. A normative plan trace is one such that (1) the specification of capabilities completely determines the transition on states in S, i.e., if $s \in S$ and c is applicable to s, then it produces another state $s' \in S$, (2) the capabilities are guaranteed to execute sequentially, e.g., knowing that capability c_2 immediately follows capability c_1, then c_2 cannot be executed until $post(c_1)$ holds, and (3) a plan execution can not be interleaved with other plans.

4 Running Example

As a running example, we consider variants of Monroe County Corpus for emergency response domain [22].

Example 1. As shown in Fig. 1, the agent aims to provide medical attention to patients. It just receives requests from medical personnel, drives to the patient's location (loc), loads the patient (pt) into the ambulance (amb), drives back to the hospital (h), and takes the patient out of the ambulance. Moreover, the agent is also equipped with capabilities related to the emergency response domain problem.

```
@h1 +provide_medical_attention(patient)
    :   available(ambulance)
    <- drive_to(ambulance,patient);
       get_in(patient,ambulance);
       drive_to(ambulance,hospital);
       get_out(patient,hospital).
@h10 +!at(ambulance,loc)
    :   +at(ambulance,loc)
    <- true.
@h11 +!at(ambulance,loc)
    : not +at(ambulance,loc).
    <- drive_to(P);
    !at(ambulance,loc).
```

```
{not at(abm,loc)} drive_to(loc) {at(abm,loc)}
{at(abm,loc)} get_in(pt,amb) {in(pt,abm)}
{in(pt,abm),at(amb,h)} get_out(pt,amb) {out(pt,abm),at(pt,h)}
{not reachable(pt)} call(air_amb) {at(pt,h)}
{not breathing(pt),has(pt,pulse)} cpr(pt) {breathing(pt)}
```

Fig. 1. Providing medical attention plan and capabilities.

We argue that existing APR applications are inadequate in settings where the observed agent is characterized by extensibility and deliberation. To illustrate this, let us consider the following example.

Example 2. Consider the following scenarios that may arise in this emergency response domain:

i After arriving at the scene, the agent performed CardioPulmonary Resuscitation (CPR) on the patient and loaded the patient into the ambulance.
ii The agent called an air ambulance and drove back to the hospital without loading the patient into the ambulance due to rough terrain, poor weather, etc.
iii The agent drove back to the hospital without loading the patient into the ambulance as the patient went missing.

Although simple, Example 2 is far from trivial. First of all, it is not difficult to recognize that the basic reactive behaviour of the BDI agent system (described in the previous section) cannot produce the behaviours depicted in these scenarios on its own, since it does not have those plans in its plan library.

Arguably, there is at least an extension to the system that enabled such edits in the agent behaviour. Indeed, the state-of-the-art BDI agent framework exhibits a large number of extensions to the reactive behaviour of the BDI agent system. For example, the behaviour illustrated in scenario (i) involves an insertion edit (where the agent added CPR execution into the plan). It is possible for the agent to add an action(s) to a plan by incorporating automated planning into its system, such as in [3]. Scenario (ii) involves a substitution edit (where the agent replaced loading the patient into the ambulance with calling an air ambulance). It is possible for the agent to replace a capability with another one by leveraging an extension such as reconfigurability [4]. Finally, scenario (iii) represents a deletion edit (where the agent dropped loading and taking the patient into and out of the ambulance from the plan), which is doable if the agent is equipped with a task aborting mechanism, such as the one presented in [23].

5   Problem Formulation

A plan library H is a set of BDI plans, each of which contains a sequence of capabilities $\langle c_1, \ldots, c_n \rangle$ as its body, where each c_i, $1 \leq i \leq n$, is the capability name and a list of typed parameters. We assume the presence of a capability library, denoted by $\mathscr{A}$, comprises the set of all available capabilities specifications related to the domain problem. An observation of an *unknown plan* $\bar{p}$ is denoted by $O = \langle o_1, \ldots, o_n \rangle$, where $o_i \in \mathscr{A} \cup \{\varnothing\}$, i.e., the observation o_i is either a capability in $\mathscr{A}$ or an empty capability $\varnothing$ that has not been observed. Note that the plan $\bar{p}$ is not necessarily in H, and thus mapping O to the H may generate only imperfect explanations.

When reasoning about new classes of observation, one can classify an observed capability by four types: (1) match, when the capability is correctly observed, (2) insertion edit, when the observed capability is added to a normative plan trace, (3) deletion edit, when the observed capability is dropped from a normative plan trace, (4) substitution edit, when a capability that is to be executed as part of a normative plan trace is replaced with another one. We propose to describe these three edits in plan execution using operations as follows.

Definition 1 (Abductive edit operation, sequence). Let p with $C = \langle c_1, \ldots, c_n \rangle$ as its body be an imperfect explanation for the observation $O = \langle o_1, \ldots, o_n \rangle$. An abductive edit operation is the insertion, deletion, or substitution of capabilities in C according to the observations in O. An abductive insertion of an observed capability o_i is denoted by $(\varnothing \rightarrow o_i)$, deletion of c_i is denoted by $(c_i \rightarrow \varnothing)$ and substitution of c_i with o_i is denoted by $(c_i \rightarrow o_i)$. An abductive edit sequence $AES = \langle ae_1, \ldots, ae_n \rangle$ is a sequence of abductive edit operations. An AES derivation from C to O is a sequence of sequences $C_0, \ldots, C_n$, such that $C_0 = C$ and $C_n = O$ and for all $1 \leq i \leq n$, $C_{i-1} \rightarrow C_i$ via ae_i.

Definition 2 (Extended plan library). An Extended plan library is a couple $EPL = (H, AES)$, where

1. H is a plan library, and

2. AES is a sequence of abductive edit operations.

Definition 3 (APR problem). Considering our settings, the APR problem can be defined by a 4-tuple APR = (EPL, O, explain, $\mathscr{A}$), where:

1. EPL is an extended plan library,
2. O is an observed trace of capabilities,
3. explain is a map from plans and sub-plans of H to subset of O, and
4. $\mathscr{A}$ is a library of capability specifications.

As such, the solution to APR is to discover an unknown plan $\bar{p}$, which is a plan with an edited sequence of capabilities as its body that best explains O given EPL and $\mathscr{A}$. Again, this can be challenging since the plan $\bar{p}$ is not necessarily in H, and thus mapping O to the H may generate only imperfect explanations.

6   Approach

Our approach to leveraging imperfect explanations consists of four phases, as shown in Fig. 2. These phases are (1) Classification of unknown plans, (2) Abductive editing of imperfect explanations, (3) Reasoning about the validity of the edited plan, and (4) Abductive updating of imperfect explanations. We describe each of these steps in greater detail in the following sub-sections. Note that solid arrows refer to leveraging imperfect explanations phases and dotted arrows refer to required inputs.

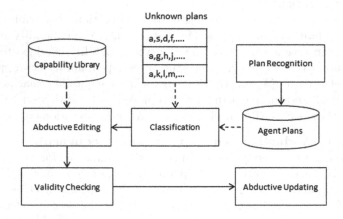

Fig. 2. Overview of the approach.

As illustrated in Fig. 2, leveraging imperfect explanations takes as inputs

1. An unknown plan,
2. An approximation of the plan(s) that have been used to generate input (1), and

3. A set of all available capabilities specifications related to the domain problem.

While inputs (1) and (2) are used in the classification and updating phases, inputs (2) and (3) are used in the editing and validation phases. Note also that our approach still requires an external plan recognition mechanism to provide imperfect explanations (hence the dotted arrow from agent plans to the classification phase).

6.1 Classification

Given an unknown plan, before any decision can be made concerning leveraging imperfect explanations, it is first necessary to determine the characteristic of the environment in which the unknown plan has been carried out (i.e., we do not want to build on noisy or exploratory observations). To that end, we use decision tree learning to classify unknown plans. Although decision trees are not the only means of classification, they are highly interpretable models [24]. Following the classification described in Sect. 2, the following taxonomy for classification is proposed:

EE Exploratory environment - observed behaviour is a subject of exploratory or discovery learning. Much of the work done on PR for exploratory domains considers trial-and-error, activity repeating and interleaving as features of exploratory behaviours [7,9,10].

NE Noisy environment - observed behaviour is characterized by imperfect observability. Previous studies (e.g., [12] and [11]) reported noisy observations as those that cannot be explained by the actions of any plan for any given goal (computing all possible plans for a given goal is fully described in [6]).

OE Open environment - observed behaviour is characterized by extensibility. Many studies on intelligent agents (e.g., [4,25,26]) consider rational changes in plan execution (see Sect. 6.3 for how rational changes are validated) as a feature of engaging the agent with open environments.

A given unknown plan is classified into one of the classes {**EE, NE, OE**}, each representing the settings in which the unknown plan has been executed. An unknown plan is assigned to a class membership based on its characteristic features by comparing it to its imperfect explanation (i.e., approximation of the unknown plan). For example, unknown plans that contain actions that any plan for any given goal cannot explain are labelled as **NE**, whilst unknown plans that contain rational edits compared to their imperfect explanations are labelled as **OE**. Historical instances are labelled manually while the test data is not labelled, so the decision tree can classify whether the unknown plan is a result of **EE**, **NE**, or **OE**.

According to state-of-the-art PR and intelligent agents [4,6,7,9–12,25,26], we initially extracted a number of features related to **EE**, **NE**, and **OE**.

1. **Unreliable action:** This binary feature represents whether an unknown plan contains action(s) that any plan for any given goal cannot explain.

2. **Trial-and-error**: This binary feature represents whether an unknown plan contains multiple attempts to achieve a desirable effect using different activities.
3. **Action repeating**: This binary feature represents whether an unknown plan contains multiple attempts to achieve a desirable outcome using the same activity with different parameters.
4. **Activity interleaving**: This binary feature represents whether an unknown plan contains the execution of an activity while waiting for the results of the current activity.
5. **Rational editing**: This binary feature represents whether an unknown plan contains rational edit(s) compared to its normative plan traces.

Classification takes place as a supervised multi-class classification making use of the features described above. Classification of unknown plans is applied before the actual process of leveraging imperfect explanations due to avoid useless wait (i.e., we do not want to build on noisy or exploratory observations). If an unknown plan is classified as **OE**, the plan will be taken as input. Continuing with our running example, both scenarios (i) and (ii) were classified as **OE**. This is due to containing the feature of rational changes, which strongly correlates to **OE**. Whiles scenario (iii) was classified as a **NE** because any plan of the given goal could not explain it.

In this section, we have seen one possible way of classifying changes in plan execution by attending to the settings in which it has been carried out (i.e., noisy, explanatory and open). However, in fact, many other features can be adopted to classify these changes hence a better understanding of what the target agent is actually doing and why. For example, these changes can be divided into mistakes and exploratory activities in exploratory domains. Another example, noisy observations can be classified as sensor failure or programming errors (e.g., inappropriate belief revision). However, such classifications are outside the scope of the present work, since this is, in general, an intractable problem.

6.2   Abductive Editing

Our guiding intuition here is that a plan that serves as an imperfect explanation for an observed behaviour could possibly be edited (modified) according to that observation, thus improving the explanatory power of the plan library. We realize a computational solution to leveraging imperfect explanations by appealing to the optimal edit distance [27] between an unknown plan and its imperfect explanation and using its corresponding edit sequence.

Let plan p with body $C = \langle c_1, \ldots, c_n \rangle$ be an imperfect explanation of the observations $O = \langle o_1, \ldots, o_m \rangle$, with the former having length n and the latter length m. Recall that turning the plan body C into O requires a sequence of edit operations. Each of these operations can be weighted by a cost function, denoted by $w(ae)$. For example, one can set the cost function to return 0 when the capability is correctly observed and to return 1 otherwise.

With a cost function in hand, the abductive plan edit distance between C and O is given by a matrix d of size $n \times m$, defined by the recurrence.

$$d[i, 0] = i$$
$$d[0, j] = j$$

$$d[i, j] = \begin{cases} d[i-1, j-1] & \text{(correctly observed)} \\ \\ \min \begin{cases} d[i-1, j] + w(\varnothing, o_i) & \text{(insertion edit)} \\ d[i, j-1] + w(c_i, \varnothing) & \text{(deletion edit)} \\ d[i-1, j-1] + w(c_i, o_i) & \text{(substitution edit)} \end{cases} \end{cases}$$

After filling the matrix, the value in the bottom-right cell of the d, or $d[m, n]$, will represent the minimum cost to turn the plan body C into the observation sequence O, and this cost is the abductive plan edit distance. For the corresponding AES to be computed, we need to traceback the choices that led to the minimum edit cost in the above recurrence. Hence, turning the imperfect explanation p into an unknown plan $\bar{p}$ that best explains the observations in O can be seen as applying an AES that corresponds to the abductive plan edit distance of p.

Example 3. Continuing with our running example, assume the imperfect explanations h1 with a body as shown below:

```
drive_to(loc),get_in(pt,amb),drive_to(loc),get_out(pt,amb).
```

An observation sequence O from scenario (ii) as shown below:

```
drive_to(pt),call(air_amb),drive_to(h)
```

And let $w(ae) = 0$ when the capability is correctly observed and $w(ae) = 1$ otherwise. Based on the above inputs, the abductive edit operations needed to turn the body part of plan h1 into O are:

1. $get\_in(pt, amb) \rightarrow call(air\_amb)$, and
2. $get\_out(pt, amb) \rightarrow \varnothing$.

Example 3 illustrates how imperfect explanations can be leveraged merely by editing. Although edits in Example 3 sound rational, they may be invalid in other scenarios. Hence, there are two important questions yet to be discussed: how to ensure (1) that the edited plan is consistent and (2) at the end of its execution, the goal is achieved. We will address these two questions in the following subsection.

6.3   Validity Checking

We build on the approach of monitoring plan validity proposed by [28]. Our theory of edited plan validity uses the accumulative effects (denoted as accum) of [29] and ensures consistency during the process of plan editing, given the capability library $\mathscr{A}$. To monitor an edited plan validity, two plans are generated for every PR problem - one with the minimum edit cost, i.e., the imperfect explanation, and one that explains O completely, i.e., $\mathsf{explain}(\bar{\mathsf{p}}) = \mathsf{O}$. Assuming an idealised execution environment, plan validity can be defined as follows.

Definition 4 (Valid Plan). Consider the capability library $\mathscr{A}$ and a plan specification $\mathsf{p} = \{\mathsf{pre}(\mathsf{p})\}\mathsf{C}\{\mathsf{post}(\mathsf{p})\}$ for plan p with body $\mathsf{C} = \langle \mathsf{c}_1, \ldots, \mathsf{c}_n \rangle$, we say that the plan p is valid in the state s if

1. $\mathscr{A} \models (\langle \mathsf{pre}(\mathsf{c}_1), \ldots, \mathsf{pre}(\mathsf{c}_n) \rangle, \mathsf{s})$, and
2. $\mathsf{accum}(\mathsf{p}) \models \mathsf{post}(\mathsf{p})$.

As such, with respect to $\mathscr{A}$ and the current state of the world s, the preconditions of each capability in the plan body of p will be satisfied, and the final effect scenario $\mathsf{accum}(\mathsf{p})$ associated with the end state event of the plan p execution entails its post-condition specifications. An important detail of this definition is that a single edit can impact the consistency of the plan. Also, it can change the final effect scenario associated with the end state of the plan. We now identify a valid edited plan.

Definition 5 (Valid edited plan). Consider the plan $\mathsf{p} = \{\mathsf{pre}(\mathsf{p})\}\mathsf{C}\{\mathsf{post}(\mathsf{p})\}$ as an imperfect explanation of O. Let $\bar{\mathsf{p}}$ be an edited plan that has identical pre- and post-conditions to p except that $\bar{\mathsf{p}}$ has an edited sequence of capabilities $\bar{\mathsf{C}}$. We say that the edited plan $\bar{\mathsf{p}}$ preserves the validity of p if $\bar{\mathsf{p}} = \{\mathsf{pre}(\mathsf{p})\}\bar{\mathsf{C}}\{\mathsf{post}(\mathsf{p})\}$ is valid.

Where it is possible for the plan recognition model to find two or more different imperfect explanations that have the same edit cost and are valid, further post-processing may be required for more reliable results. We overtake this problem by seeking stronger goal entailment and consistency conditions, such as the plan internal analysis described in [30].

6.4   Abductive Updating

Now, we consider the problem of what needs to be done to improve the explanatory power of the plan library when an edited plan is found valid according to the checks described above. An easy solution is to create a new plan that has identical triggering and context parts to the imperfect explanation, except that $\bar{\mathsf{p}}$ has an edited body $\bar{\mathsf{C}}$ that explains O, i.e., $\mathsf{explain}(\bar{\mathsf{p}}) = \mathsf{O}$. However, this may increase the complexity of determining applicable plans at run-time. More interestingly, we offer a semi-automated solution for merging an edited plan with its imperfect explanation.

Procedure 1 (Abductive Updating). Consider the imperfect explanation p and its edited plan $\bar{p}$ and let $AES = \langle ae_1, \ldots, ae_n \rangle$ be a derivation from p to $\bar{p}$.

```
 1: For each aeᵢ in AES:
 2:   Replace cᵢ with subgoal sgᵢ
 3:   Create two subplans p1 and p2, and let
 4:   triggering(p1) = triggering(p2) = sgᵢ
 5:   switch(aeᵢ):
 6:     case(cᵢ → oᵢ)
 7:       context(p1) = pre(cᵢ)
 8:       context(p2) = pre(oᵢ)
 9:       body(p1) = cᵢ
10:       body(p2) = oᵢ
11:     case(cᵢ → ∅)
12:       context(p1) = pre(cᵢ)
13:       context(p2) = not pre(cᵢ)
14:       body(p1) = cᵢ
15:       body(p2) = true
16:     case(∅ → oᵢ)
17:       context(p1) = not pre(oᵢ)
18:       context(p2) = pre(oᵢ)
19:       body(p1) = true
20:       body(p1) = oᵢ
```

Procedure 1 facilitates the merging between an imperfect explanation, i.e., one with the minimum edit cost, and its edited plan, i.e., one that explains O completely. Fundamentally, for each ae_i in AES, the procedure replaces the corresponding capability with a sub-goal sg_i (line 2), for which two sub-plans p_1 and p_2 are created (line 3-4). Our guiding intuition behind creating two sub-plans is to improve the explanatory power of the plan library while maintaining its construction.

Recall that there are three ways in which a plan can be edited: deletion, insertion, and substitution. Hence, there are three ways in which sub-plans can be created (line 5). For example, let us consider the substitution edit (line 6): In this case, plan p_1 takes the corresponding capability c_i as its body and $\text{pre}(c_i)$ as its context part (lines 7 and 9). While p_1 takes the corresponding observation o_i as its body and $\text{pre}(o_i)$ as its context part (lines 8 and 10). However, there are situations where valid edited plans could possibly be merged with their imperfect explanations, but they should not be merged. For example, avoiding negative interactions between goals. For readers interested in how we can deal with the feasibility of plans merging, we refer to [31].

7   Evaluation

In this section, we present the evaluation of our approach. First, we present the setup for evaluation. Next, evaluation results are described.

7.1 Evaluation Setup

We implemented our approach as a plugin for our Toolkit XPlaM[1] [32] and evaluated it using Monroe Plan Corpus [22]. For the purposes of this work, the corpus has been rewritten using AgentSpeak(L) programming language and executed using Jason interpreter [17]. Observation traces[2] have been collected using a debugging tool called Mind Inspector [17]. Unknown plans have been classified using C4.5 decision tree algorithms [33]. Abductive plan editing is implemented using the well-known Levenshtein distance. For deriving accumulative effects, we implemented a state update operator similar to the one described in [29] using the NanoByte SAT Solver package[3]. For testing, experiments have been implemented on an Intel Core i5-6200 CPU (with 8 GB RAM).

7.2 Performance Results

We ran a number of experiments to test the scalability of our approach with respect to the number of new classes of observations. The corpus used in these experiments has been executed to consider all possible traces (the number of executed actions is 80). New classes of observations were artificially added to the observation traces. The number of capabilities in $\mathscr{A}$ is set to 20. Figure 3 compares the number of explanations generated by our approach for a different number of new classes of observations.

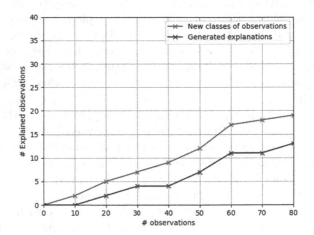

Fig. 3. Performance results.

---

[1] The source code for XPlaM Toolkit (including the code for the approach presented here) has been published online at https://github.com/dsl-uow/xplam.

[2] We published the datasets supporting the conclusions of this work online at https://www.kaggle.com/datasets/alelaimat/xplam.

[3] https://github.com/nano-byte/sat-solver.

Our first study shows that with a moderate number of new classes of observations and a reasonable number of capabilities in $\mathscr{A}$, abductive plan editing would possibly improve the explanatory power of plan libraries in open environment settings. Note that the scalability of the abductive plan editing should be tied to the performance of the classification task. Nevertheless, the results depicted in Fig. 3 are partially independent of the used classifier (C4.5 decision tree). Actually, we allowed for unknown plans that contain unreliable actions to be considered as inputs. We argue that additional features are required for a more accurate classification of unknown plans.

7.3   Speed

Aiming to find out how fast abductive plan editing would take to explain unknown plans, we ran a number of experiments with Monroe Plan Corpus. Recall that executing action models (i.e., planning) is another technique to explain unknown plans. Hence, we use diverse planning [34] as a benchmark to assess the complexity of our approach. Diverse planning aims at discovering a set of plans that are within a certain distance from each other. The discovered set is then used to compute the closest plan to the observation sequence. Figure 4 shows the time required to discover a valid explanation for variate numbers of new classes of observations using diverse planning (as used by LPG-d planner [34]) to the time required by our plan editing approach.

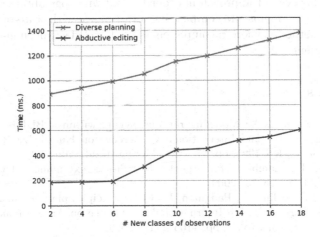

Fig. 4. Explaining time required for plan editing and diverse planning.

Figure 4 shows that, unlike diverse planning, for different number of new classes of observations, plan editing is a relatively faster approach to explaining unknown plans. However, the reader should keep two details in mind. First, the performance of plan editing is tied to the performance of the used dynamic programming algorithm (in our case, Levenshtein distance). A more fine-grained evaluation, therefore, should include different dynamic programming algorithms (e.g., Hamming distance). Secondly, during our experiments, we noticed that the size of $\mathscr{A}$ can highly affect the performance of diverse planning and plan editing. We will investigate these two details as part of our future work.

8   Conclusion

Much of the work done on APR requires a plan library to infer the top-level plans of the observed agent. Nevertheless, in open environment settings, target plans are usually not from plan libraries due to reusing plans, replanning and agent self-awareness, etc. This work builds on a more sophisticated notion of APR, which seeks to improve the explanatory power of plan libraries by way of leveraging imperfect explanations and exploiting new classes of observations. In this work, we proposed a classification of unknown plans based on the characteristics of the environment in which they have been carried out. As far as we know, this has been absent in PR research. Furthermore, we presented a theory based on capabilities and plans and introduced the notion of abductive plan editing. Finally, we described how imperfect explanations could be updated with new classes of observations in a rational fashion.

A number of extensions of this work are of interest, including applications of plan library reconfigurability [4], plan editing in online settings, and dealing with incomplete action models (i.e., learning unknown activities). Furthermore, we plan to improve our approach in order to deal with logs obtained from noisy and exploratory domains and compare its performance with state-of-the-art plan recognition methods, where incompleteness of knowledge and non-determinism might be present.

References

1. Georgeff, M., Rao, A.: Modeling rational agents within a BDI-architecture. In: Proceedings of the 2nd International Conference on Knowledge Representation and Reasoning, KR 1991, pp. 473–484. Morgan Kaufmann (1991)
2. Padgham, L., Lambrix, P.: Agent capabilities: extending BDI theory. In: AAAI/IAAI, pp. 68–73 (2000)
3. De Silva, L., Sardina, S., Padgham, L.: First principles planning in BDI systems. In: Proceedings of the 8th International Conference on Autonomous Agents and Multiagent Systems, vol. 2, pp. 1105–1112 (2009)
4. Cardoso, R.C., Dennis, L.A., Fisher, M.: Plan library reconfigurability in BDI agents. In: Dennis, L.A., Bordini, R.H., Lespérance, Y. (eds.) EMAS 2019. LNCS (LNAI), vol. 12058, pp. 195–212. Springer, Cham (2020). https://doi.org/10.1007/978-3-030-51417-4_10
5. Singla, P., Mooney, R.J.: Abductive Markov logic for plan recognition. In: 25th AAAI Conference on Artificial Intelligence (2011)
6. Geib, C.W., Goldman, R.P.: A probabilistic plan recognition algorithm based on plan tree grammars. Artif. Intell. **173**(11), 1101–1132 (2009)

7. Mirsky, R., Gal, Y., Shieber, S.M.: CRADLE: an online plan recognition algorithm for exploratory domains. ACM Trans. Intell. Syst. Technol. (TIST) **8**(3), 1–22 (2017)

8. Meneguzzi, F.R., Luck, M.: Norm-based behaviour modification in BDI agents. In: AAMAS, vol. 1, pp. 177–184 (2009)

9. Uzan, O., Dekel, R., Seri, O., et al.: Plan recognition for exploratory learning environments using interleaved temporal search. AI Mag. **36**(2), 10–21 (2015)

10. Amir, O., et al.: Plan recognition in virtual laboratories. In: 22nd International Joint Conference on Artificial Intelligence (2011)

11. Massardi, J., Gravel, M., Beaudry, E.: Error-tolerant anytime approach to plan recognition using a particle filter. In: Proceedings of the International Conference on Automated Planning and Scheduling, vol. 29, pp. 284–291 (2019)

12. Sohrabi, S., Riabov, A.V., Udrea, O.: Plan recognition as planning revisited. In: IJCAI, pp. 3258–3264 (2016)

13. Ramírez, M., Geffner, H.: Probabilistic plan recognition using off-the-shelf classical planners. In: 24th AAAI Conference on Artificial Intelligence (2010)

14. Sukthankar, G., Sycara, K.P.: Hypothesis pruning and ranking for large plan recognition problems. In: AAAI, vol. 8, pp. 998–1003 (2008)

15. Avrahami-Zilberbrand, D., Kaminka, G.A.: Keyhole adversarial plan recognition for recognition of suspicious and anomalous behavior. In: Plan, Activity, and Intent Recognition, pp. 87–121 (2014)

16. Zhuo, H.H., Zha, Y., Kambhampati, S., Tian, X.: Discovering underlying plans based on shallow models. ACM Trans. Intell. Syst. Technol. (TIST) **11**(2), 1–30 (2020)

17. Bordini, R.H., Fred Hübner, J., Wooldridge, M.: Programming Multi-agent Systems in AgentSpeak Using Jason, vol. 8. Wiley (2007)

18. Dastani, M.: 2APL: a practical agent programming language. Auton. Agent. Multi-Agent Syst. **16**, 214–248 (2008)

19. Padgham, L., Lambrix, P.: Formalisations of capabilities for BDI-agents. Auton. Agent. Multi-Agent Syst. **10**(3), 249–271 (2005). https://doi.org/10.1007/s10458-004-4345-2

20. Reiter, R.: The frame problem in the situation calculus: a simple solution (sometimes) and a completeness result for goal regression. In: Artificial and Mathematical Theory of Computation, pp. 359–380. Citeseer (1991)

21. Fikes, R.E., Nilsson, N.J.: Strips: a new approach to the application of theorem proving to problem solving. Artif. Intell. **2**(3–4), 189–208 (1971)

22. Blaylock, N., Allen, J.: Generating artificial corpora for plan recognition. In: Ardissono, L., Brna, P., Mitrovic, A. (eds.) UM 2005. LNCS (LNAI), vol. 3538, pp. 179–188. Springer, Heidelberg (2005). https://doi.org/10.1007/11527886_24

23. Thangarajah, J., Harland, J., Morley, D., Yorke-Smith, N.: Aborting tasks in BDI agents. In: Proceedings of the 6th International Joint Conference on Autonomous Agents and Multiagent Systems, pp. 1–8 (2007)

24. Gilpin, L.H., Bau, D., Yuan, B.Z., Bajwa, A., Specter, M., Kagal, L.: Explaining explanations: an overview of interpretability of machine learning. In: 2018 IEEE 5th International Conference on Data Science and Advanced Analytics (DSAA), pp. 80–89. IEEE (2018)

25. Stringer, P., Cardoso, R.C., Huang, X., Dennis, L.A.: Adaptable and verifiable BDI reasoning. arXiv preprint arXiv:2007.11743 (2020)

26. Dennis, L.A., Fisher, M.: Verifiable self-aware agent-based autonomous systems. Proc. IEEE **108**(7), 1011–1026 (2020)

27. Navarro, G.: A guided tour to approximate string matching. ACM Comput. Surv. (CSUR) **33**(1), 31–88 (2001)
28. Fritz, C., McIlraith, S.A.: Monitoring plan optimality during execution. In: ICAPS, pp. 144–151 (2007)
29. Ghose, A., Koliadis, G.: Auditing business process compliance. In: Krämer, B.J., Lin, K.-J., Narasimhan, P. (eds.) ICSOC 2007. LNCS, vol. 4749, pp. 169–180. Springer, Heidelberg (2007). https://doi.org/10.1007/978-3-540-74974-5_14
30. Gou, Y.: A computational framework for behaviour adaptation: the case for agents and business processes (2018)
31. Thangarajah, J., Padgham, L., Winikoff, M.: Detecting & exploiting positive goal interaction in intelligent agents. In: Proceedings of the Second International Joint Conference on Autonomous Agents and Multiagent Systems, pp. 401–408 (2003)
32. Alelaimat, A., Ghose, A., Dam, H.K.: XPlaM: a toolkit for automating the acquisition of BDI agent-based digital twins of organizations. Comput. Ind. **145**, 103805 (2023)
33. Quinlan, J.R.: C4.5: Programs for Machine Learning. Elsevier (2014)
34. Nguyen, T.A., Do, M., Gerevini, A.E., Serina, I., Srivastava, B., Kambhampati, S.: Generating diverse plans to handle unknown and partially known user preferences. Artif. Intell. **190**, 1–31 (2012)

Reinterpreting Vulnerability to Tackle Deception in Principles-Based XAI for Human-Computer Interaction

Rachele Carli[1,2]([✉])[iD] and Davide Calvaresi[3][iD]

[1] Alma Mater Research Institute for Human-Centered AI, University of Bologna,
Bologna, Italy
rachele.carli2@unibo.it
[2] CLAIM Group and AI RoboLab, University of Luxembourg, Luxembourg,
Luxembourg
[3] University of Applied Sciences Western Switzerland, Delémont, Switzerland
davide.calvaresi@hevs.ch

Abstract. Artificial intelligence (AI) systems have been increasingly
adopted for decision support, behavioral change purposes, assistance,
and aid in daily activities and decisions. Thus, focusing on design and
interaction that, in addition to being functional, foster users' acceptance
and trust is increasingly necessary. Human-computer interaction (HCI)
and human-robot interaction (HRI) studies focused more and more on
the exploitation of communication means and interfaces to possibly enact
deception. Despite the literal meaning often attributed to the term,
deception does not always denote a merely manipulative intent. The
expression "banal deception" has been theorized to specifically refer
to design strategies that aim to facilitate the interaction. Advances in
explainable AI (XAI) could serve as technical means to minimize the risk
of distortive effects on people's perceptions and will. However, this paper
argues that how the provided explanations and their content can exacerbate the deceptive dynamics or even manipulate the end user. Therefore,
in order to avoid similar consequences, this analysis suggests legal principles to which the explanation must conform to mitigate the side effects
of deception in HCI/HRI. Such principles will be made enforceable by
assessing the impact of deception on the end users based on the concept
of vulnerability – understood here as the rationalization of the inviolable
right of human dignity – and control measures implemented in the given
systems.

Keywords: XAI · Deception · Vulnerability

1 Introduction

Interactive AI systems are now used for many purposes that require a constant
exchange of information with the end user. Some of the main tasks performed

D. Calvaresi et al. (Eds.): EXTRAAMAS 2023, LNAI 14127, pp. 249–269, 2023.
https://doi.org/10.1007/978-3-031-40878-6_14

by such applications include: e-health goals [16], decision-making activities, support and guide in behavioral changes [11], e-administration proceedings [31], assistance for e-services [19].

The quality and frequency of interaction in many of these cases are crucial for two fundamental reasons. First, this allows the application to refine its outcomes and, consequently, to pursue the purpose for which it was developed more effectively and efficiently. Second, an interaction that is not only technically satisfying but also pleasant, at times "familiar", enables users to be more consistent in their engagement, to adhere better to the recommendations provided, and to rely on them to develop trust [68].

In light of the above, there has been a growing interest of researchers in the fields of HCI, HRI, and XAI in those dynamics and elements that, if correctly implemented and elicited, could foster interaction by acting on the psychological, cognitive, and emotional mechanisms of the human interlocutors. This relied on research in neuroscience, behavioral psychology, cognitive science, communication science, and interdisciplinary working groups [33]. One of the main results achieved by scholars led back to an aspect already dear to computer science: the theme of deception. It has entered the context of human-AI interaction since the Turing Test, demonstrating how the very concept of AI is based on the ability a system has to emulate the capabilities of human beings, regardless of whether they are objectively present or not [66]. The credibility of this appearance has a far more impactful influence on the perceived quality of the interaction than pure technical efficiency. For this reason, efforts have been made to implement AI systems more and more with design characteristics and communication features capable of targeting the brain areas that are involved in the perception of positive emotions such as cuteness, trustworthiness, sympathy, tenderness, and empathy. It is worth emphasizing that the primary purpose of such implementations is to ensure the good functionality of the application and, as a direct consequence, the possibility for the user to benefit from the resulting beneficial and supportive effects.

Thus, the concept of "banal deception" has recently been theorized [53]. In particular, it aims to delimit the difference between deception, understood as a mechanism of mendacity, manipulation, and the phenomenon described above, which identifies an expedient that may contribute to the user's best interest.

According to such premises, the crucial role of explanation clearly emerges in this context. It may have the capacity to make many processes performed by the AI system clearer to the non-specialist interlocutor, redefining the boundaries between appearance and reality, technology and humanity, functionality and emotionality [27]. In all those cases in which – or within the limits in which – banal deception is essential for the success of the interaction, XAI can counterbalance the emotional and unconscious scope of the user's reactions with accurate, punctual explanations, which leverage the more logical-rational part of the brain in return.

This paper argues that the explanation risks becoming an element of deception due to how it is produced, communicated (e.g., styles and tones), and its content. In this sense, it could elicit deceptive effects, leading to real manipulative consequences.

Therefore, this paper suggests a framework – based on enforceable legal principles – paving the way for structuring two possible tools to mitigate potential adverse effects. The first proposal concerns structuring a Vulnerability Impact Assessment, aiming at establishing the different degrees of banal deception that can be considered admissible, depending on the level with which they impact, exasperate, or assist the unavoidable human vulnerability. The second proposal concerns the formalization of a knowledge graph that identifies features and relationships between elements that make up vulnerability and implement them in the XAI system.

The rest of this paper is organized as follows.

Section 2 presents the state of the art, focusing on the original concept of deception, presenting the new theorization of 'banal deception' and emphasizing the role the XAI could play in either facilitating manipulative drifts or – conversely – protecting the user from them. To pursue the second objective, Sect. 3 elaborates on the basic legal principles that should be the framework to which the design of explanations should adhere. Section 4.1 and Sect. 4.2 present the dual approach through which this framework should be concerted, namely a new Impact Assessment vulnerability-based and a knowledge graph implemented in the system itself.

2   Background and State of the Art

This section presents the background and state of the art of disciplines intersecting explainability in HCI, such as deception and XAI.

2.1   Deception in HCI

The subject of deception has a long history in computer science and, more specifically, HCI and – more recently – HRI. However, researchers in these disciplines are often skeptical about describing the outcome of their work or their approach in terms of deception due to the predominantly negative connotation that this concept has inherited from the humanities, especially the legal sciences. However, while the legal semantics is often connected to an act aimed at circumventing, misleading, and inducing disbelief for the benefit of the deceiver and necessarily to the detriment of the deceived, the same is not always true from a more strictly technological point of view.

Deception has become part of AI since the Touring Test. In what was essentially considered an HCI experiment, the computer program will only be able to win the game if it can "fool" the human interlocutor [41]. This means to assume a very human-centric perspective, where it is assessed whether the illusion – of intelligence in this case – has been programmed accurately and allegedly enough to not only be plausible but to convince the individual [60]. In other words, we start by analyzing human beings, their communicative strategies, and connotative semantics to understand how to program a successful interaction [6].

Considering only this angle, it is still possible to interpret the deceptive dynamic as aimed at misleading the other party. Nevertheless, it should be noted that the Turing test was structured as a game in which individuals participated willingly, following the game's rules. The playful background conveys the deceptive interaction with an innocuous and ethically acceptable connotation [35]. This is precisely what modern voice assistants and much interaction software claim to have inherited from the test. However, even in that context, giving the application the ability to respond with jokes is a way of playfully engaging the interlocutors, pushing them to challenge the limits of the imitation of humanity [50].

In these contexts, deception is domesticated and does not carry the negative connotation that manipulative ends have in common with other misleading expedients. Therefore, the envisioned outcome, subsequent research, and experimentation were set to imagine a future in which deception, conceived in its functional and non-harmful form, would become a useful tool for developing successful interactions with new technologies used on a daily basis.

2.2   From Deception to the ELIZA Effect

To make it possible efficient and effective interactions for the benefit of the users, HCI developed as a field of research aimed at adapting system interfaces, design features, and functionality to the perceptual and cognitive abilities of human beings. Thus, deception became a proper method to deflect any element that could uncover the artificial and aseptic nature of the AI system. Said otherwise, the standard approach in the design of applications deputed to interact continuously and closely with users became to exploit the fallibility, the unconscious psychological and cognitive mechanisms inherent in human nature [12].

An example of this evolution is represented by ELIZA, the software that pioneered new perspectives on chatbots [43]. It shows that the correctness and appropriateness of the outputs, given a certain input, are not the only crucial factors for a successful HCI. The so-called "social value" is also important [38]. That is to say that the fact that the application can play a certain role in the interaction, which remains consistent in itself throughout the whole exchange, has to be considered central [28]. This stems from the fact that individuals by nature attribute to their interlocutors – even humans – a specific role, which could be defined as a "social role" [61]. This has little to do with the actual identity of the other (e.g., if they are the professional to whom we turn for a consultation, a family member, or a stranger). What plays an essential role is the – social and not – value that people attribute to those with whom they interact. It consists of projections, past experiences, and emotional resistance. Referring more specifically to the HCI domain, the subconscious tendency of humans to believe that AI systems and software have their own behavior and that this is similar to that of peers has been termed the "ELIZA effect" [63].

Upon closer analysis, it demonstrated how even relatively unsophisticated programs can deceive the user through AI, creating an appearance of intelligence and agency [23].

This was instrumental in suggesting that humans are naturally inclined to attribute human appearance, faculties, and destinies to inanimate objects [29], and that this inescapable characteristic can be exploited to create efficient interactions.

Hence, the theorization of the ELIZA effect was conducive to bringing to light something already clear since the Imitation Game: AI is the result of projection mechanisms which ineradicably characterize individuals – despite their level of practical knowledge [59]. For "projection mechanisms" is meant that universal psychic modality by which people transfer subjective ideational content outwards – into other people, animals, even objects [45].

This led to an in-depth exploration of the unconscious mechanisms that lead human beings to prefigure a kind of "computer metaphor", according to which machines and software could be comparable to human beings. Such an examination was conducted mainly through disciplines like neuroscience, behavioral psychology, cognitive science, and communication science [20, 24, 48].

What mentioned so far has led to the creation of the CASA model: Computers Are Social Actors too [51]. According to such a paradigm, people applied to computers social rules and expectations similar to those they have towards humans. This is possible because each component in interface design conveys social meaning, even if this end is not pre-determined by programmers or designers. Concurrently, if it is somehow possible to anticipate this meaning, it is also possible to direct it with the result of programming for more efficient HCIs [65].

2.3 Banal Deception

The investigation conducted so far supports the idea, now widespread in the literature, that deception in HCI and HRI is often implemented and addressed as an essential element for the best functionality of AI systems and the increase of the user's comfort. Ultimately, it could even be described as a constitutive element of AI, without which it would not be possible to define artificial intelligence itself [53]. However, this is not to disregard its possible manipulative drifts. Especially from a legal standing point, the concrete outcome of a harmful event must often be considered more relevant, rather than the benevolent but unrealized intent with which the event was preordained. For this reason, research in the field of human-AI interaction has recently proposed a new terminology, namely *Banal Deception* [53]. It is adopted to frame that type of deception that does not arise with the direct intent to mislead but rather to facilitate the use of the application and the efficiency of achieving the intended purpose. Doing so would contribute to integrating AI technologies into everyday life for decision support, entertainment, and guidance in behavioral changes. Simone Natale [53] identifies five elements that can guide in profiling the phenomenon of banal deception:

Ordinary Character: ELIZA may be a good example. It did not present anything particularly extraordinary, and the same can be said of Siri, Alexa, or many other modern chatbots. Non-specialized users seem to focus on communicative and interactive aspects that make them often curious about other media. However, the AI technologies mentioned above induce them to believe that the appearance of personality and agency is actually real. This denotes the inherent vulnerability to deception, on the one hand. On the other hand, it also underlines that banal deception is probably imperceptible, but not without consequences. It may not be intent on manipulating, but it has the predetermined purpose of making AI systems enter the core of individuals' mental structure, and – to some extent – identity [54]. In fact, thanks to the mechanisms of trivial deception, in fact, such technologies can target specific areas of the human mind, elicit trust and emotional attachment, influence habits and tastes, shape the perception of reality.

Functionality: an application capable of eliciting positive emotions, trust, and reliability in the user will be used more often and with less skepticism. This allows a more intense flow of data, which is indispensable for improving performance.

Obliviousness: being extremely subtle, as well as being a decisive part of the design, this deceptive phenomenon is not perceived by the user and is often lowered to the rank of mere technical expedient without further investigation. Nonetheless, overcoming the barriers of consciousness and awareness is also effective in physiologically balanced, well-informed subjects [5]. They can recognize the artificial nature of the application rationally, without being able to "resist" the mechanisms of anthropomorphism and personification proper of their primordial cognitive structure.

Low Definition: chatbots and other AI systems programmed according to the logic of the banal deception are neither necessarily very sophisticated from an aesthetic point of view nor particularly characterized in impersonating a single, fixed, communicative/social role. This is because human beings tend to attribute meaning to what they interact with in an intimate and/or continuous way [67]. Leaving (intentionally) the possibility to the user to exploit their imagination to fill the gaps left by programmers or designers allows customization that translates to an emotional level of familiarity, empathy, and attachment.

To Be Programmed: although banal deception relies on mechanisms inherent in human cognitive structures, it is voluntarily programmed by technical experts on the basis of studies aimed at investigating human perceptual mechanisms, with the precise purpose of targeting similar structures to pursue the "functionality" described above. In other words, banal deception needs the – unconscious – cooperation of the user to work, but it is *ex-ante* – consciously – pre-ordered by AI systems' developers.

2.4 Explainable AI

Interpretable and Explainable AI is a discipline that has not yet found a unanimous definition, as it lends itself to work among different disciplines [34]. Nevertheless, it can be detailed through its primary objective: to make data-driven recommendations, predictions/results, and data processing comprehensible to the final user [3,10].

This is necessary since human beings have a tendency to attribute mental states to artificial entities (a.k.a., agents) leveraging the evaluation of their objective behavior/outputs [7,42]. This in itself can lead to two possible inauspicious effects: (i) creating a false representation of the AI system and its capabilities and (ii) attributing an emotional-intentional valence to its answers/actions.

Starting from such assumptions, the explanation generated has been conceived by the scientific community as a valid aid so that the intentional stance [25] that the user will inevitably project onto the technology is as objective and realistic as possible. This should happen despite the prior knowledge possessed by the subject in question. Thus, according to XAI theorists, it would be possible to pursue a twofold result: to limit the negative effects of anthropomorphism and foster interaction.

Yet, this interpretation cannot in itself exhaust the complete analysis of a dynamic - that of HCI - which is multi-factorial.

This becomes clear considering the phenomenon called "Mindless behavior". Such an expression is commonly used to delineate the subconscious mechanism through which people apply social conventions to artificial agents. The reason why this evaluation seems to take place "mindlessly". Indeed, it stems from the fact that individuals reveal such a way of interpreting the interaction regardless of the level of awareness they have of the actual nature of the AI system [62].

This brings us back to the above discussion of the mechanisms of banal deception. In this context, making evident the mechanical and inanimate nature of the application, its lack of consciousness and intentionality, opening the black box by revealing the hidden mechanisms and rationals behind the processing of data would seem to have no bearing on the subconscious empathic dynamics that users are in any case naturally induced to enact.

2.5 XAI in the Realm of Banal Deception

Explaining is considered critical in making the operation of the system/robot and the nature of the output as transparent as possible. This facilitates its use and (most importantly) its trustworthiness, desirability, and pleasant interaction. Ultimately, it is conceived as an essential tool to shorten the distance often perceived between the technicality of AI and the unskilled user.

Nevertheless, depending on the characteristics of the explanation and the manner in which it is given, it may itself represent an element of banal deception – as described above. Furthermore, in some circumstances, it may reinforce the deceptive mechanisms already inherent in the application, crossing the line between deception and manipulation.

This discussion will not delve into the conceptual and semantic analysis of these two themes, for which we refer to, among others [18,46]. For the purposes of the analysis conducted here, we point out that the circumscription of the manipulation concept is still much debated in the scientific community. This also applies to the legal sphere, where it is relevant to determine cases and means by which to intervene to protect people's will and psychological integrity. Hence, reporting at least an overall conceptualization of manipulation is deemed appropriate. It is conceived as a dynamic that can circumvent individuals' critical thinking and logic [44], making them do something different from what they would have done or justified if they had not been subjected to the same manipulative techniques for the benefit of the manipulator. Thus, targeting self-awareness more and before even affecting rationality, manipulation can have deception as one of the means through which the purpose is pursued [22].

Acknowledging this brief examination, the explanation might be structured according to the logic of the banal deception, going to strengthen confidence and trust in the outcome of the application, to the detriment of the real interest and goal set by the user. For instance, some explanations, or the methodology/expedient by which it is provided, could be aimed at (i) making the interlocutors dependent on the use or feedback of the AI system, (ii) inducing them to pursue ends that merely benefit the producer, (iii) generating behavioral change that is harmful to the user, but still useful for general profiling purposes, (iv) eliciting the loss of significant social contacts (including the'second expert opinion' performed by a human specialist in the case of applications with potential impact on health).

To preserve the protective and positive purposes of XAI, to prevent it from becoming a tool of manipulation, and to make it a valuable aid in limiting the side effects of banal deception, it might be helpful to draw up a list of principles to which the explanation must conform. With this aim in mind, the principles suggested here are of a legal, rather than ethical, nature. This offers the benefit of making the framework below potentially enforceable with both *ex-ante* and *ex-post* logic. *Ex-ante*, ideally, it will have to be taken into account when programming and designing the AI system and its explainability. *Ex-post*, as it can be invoked in the event of a violation to require a forced adjustment or to correct any divergences that, through the interaction itself, the application will have developed.

3   Principles-Based Framework for Explanations

The European approach to AI seems to be delineated around the recurring concept of human-centered AI (HCAI) [2]. This entails aiming to create AI systems that support human capabilities rather than replacing or impoverishing them. Therefore, technological development should be oriented toward the benefit of human beings. From a European perspective, it is possible through the protection and enhancement of fundamental rights referred to in the European Charter of Human Rights. They reflect the constitutive values of European policies, are legally binding, and constitute the reference framework for the legal systems of the Member States - as well as often being used as a prototype for the legislation of other states at an international level.

Consequently, fundamental principles that must be considered essential for the design of human-centered explanations will be listed below. They have been identified starting with those most commonly referred to in the main regulations and guidelines issued by European Parliament and European Commission. A further skimming was carried out trying to identify those principles that were to be considered more directly involved in the dynamics analysed here – namely the possible manipulation of users' will, the potential distortion in the perception of reality, and the assessment of the risks attached to the interaction with AI systems. They could constitute the reference framework to make XAI a useful tool for mitigating the effects of banal deception on the end user, rather than exacerbating possible manipulative drifts.

3.1   Right to the Integrity of the Person

Article 3 of the European Chart of Human Rights protects individual physical and also psychological integrity [21]. In addition, the article refers to the value of free and informed consent in healthcare treatment. However, it is a commonly accepted interpretation that consent is conceived as the pivotal instrument of any act affecting a person or one of their available rights – as also demonstrated in the GDPR.

The reference to informed consent is certainly fundamental at a conceptual level. Indeed, it may be considered one of the reasons why a branch of legal experts sees the explanation as a valid tool for shortening – even removing – the information gap that recognizes the non-specialized user of AI as disadvantaged by default. In the scope of this study, on the contrary, this issue does not seem to be decisively relevant. Informing an individual of what is happening, why a given recommendation is being made, or how their data will be processed and stored is certainly essential. At the same time, if the primary purpose of the framework in question is to prevent explanation from becoming an instrument of manipulation, informing is a practice that is neither sufficient nor goal-oriented. This is mainly due to the phenomenon of mindless behavior described above and to the subliminal nature of banal deception. Furthermore, even if the user accepted the dynamic of banal deception, if the result implies the possible infringement of fundamental rights, this presumed acceptance would be considered null and void. This is because fundamental human rights are considered by law to be "unavailable", namely, not subjected to renunciation or negotiation by the holder.

It follows that, in pursuing the scope of the principle of individual integrity, an explanation should also guarantee the respect of the subsequent principles:

Physical Health: This is mainly affected by those AI systems that involve medical aspects or habits that impact health (e.g., quitting smoking or adopting a different diet). Here it is important to ensure that the interaction and the justifications provided for the recommendations offered follow the standards of care, medical guidelines, and principles of good medical practice that are also followed by human specialists [40,49]. More specifically, it will be important to give the user the most truthful and up-to-date view of their progress and of the appropriateness of their goal. In no way, for the pure purpose of incentivizing and increasing interaction, should the individual be induced to persist in use once the limits set by health standards have been reached (e.g., to continue to lose weight or to increase muscle mass once beyond good medical practice). The system must also be able to interrupt the flow of recommendations/explanations if the user gives signals that they want to use the service outside of safe standards (e.g., setting weight loss standards too low, unbalancing nutrient intake in an unhealthy way, persisting in refusing explanations to bring their goals closer to those set by medical standards). Moreover, according to *ex* Article 3, it is impossible to impose any treatment that the patient does not understand and accept. Likewise, the system cannot use an explanation that exploits means of subliminal and subtle persuasion such as those of banal deception, push to accept recommendations or outcomes that induce potentially health-impacting actions.

Physiological Health: Although the subject of mental and psychological health is becoming increasingly pervasive in the law, there is still no objective and uniquely accepted definition of it in doctrine. Among the first steps taken by jurisprudence was to decouple this concept from the occurrence of mental disorders in the clinical-pathological sense [57]. By interpretation, it could be useful to start from the very concept of integrity, which is brought back – by analogy with other areas – to the preservation of unity, of the compactness of the subject of analysis. Anything that interferes with this idea of integrity, causing a split in an individual's coherence with themselves, their beliefs, and their feelings, alters the integrity thus understood [13]. Therefore, the explanation must aim at mitigating those aspects of banal deception that may manipulate users' perception and will, thus leading them to prefigure a conception of reality, of themselves, and of their own needs. If, as briefly investigated above, manipulation is that phenomenon that goes beyond reasoning, the explanation must be structured in such a way as to counterbalance the functional effects of deception without leading to the distortion of individuals' self-awareness.

3.2   Respect for Private and Family Life

Article 7 of the European Chart of Human Rights protects the respect for private and family life. In this concept, the security and confidentiality of the home environment and correspondence are explicitly mentioned [21].

This article is often considered to have a rather broad semantic and applicative scope that is not easy to substantiate. Indeed, the concept of private life includes instances belonging to the aforementioned Article 3, encompassing aspects of physical and psychological integrity. However, its primary focus should be on aspects of identity and autonomy [1].

Identity: Personal identity from the perspective of pure private law is understood as the unique, personal recognition of an individual. However, nowadays, the law has also opened up in interpreting identity as the set not only of objective and verifiable data attributable to an individual. It also included the specialty of people's cognitive-psychological dimension, the way they perceive themselves, their beliefs, and their will. Thus, personal identity can be harmed by expedients aimed at inducing changes, habits, and desires that are not consistent with the idea that an individual has of themself and with the lifestyle and beliefs they have chosen for themselves. Manipulation is precisely able to target self-awareness and induce attitudes that are lucidly not justifiable or recognizable as proper by those who perform them. Therefore, in this context, the explanation must have the primary role of allowing the users at each stage of the interaction to realign with themselves, accepting only the recommendations they consider in line with their own convictions and goals. It must also always put the user in the position to question and challenge a recommendation/ motivation received with an exchange that includes acceptance and/ or rejection and a more active and argumentative understanding.

Inviolability of Private Space: Following the examination of the concept of identity above, arguing that the domestic environment – or more generally private – should be protected from external inferences means also to include elements that go beyond the mere concept of home or private property. In this *space*, it is also necessary to bring back habits and the deeper aspects of daily life [53]. For this reason, while banal deception has the primary purpose of facilitating the inclusion of AI systems in the most intimate contexts – both physically and cognitively – through the explanation, we should aim to ensure that this happens only to the extent that it is essential for more effective interaction.

Autonomy of Private Choice: The two satellite principles analyzed so far lead us to an incontrovertible conclusion: the protecting the individual autonomy, here to be understood as "freedom to choose for oneself" [47]. This implies the negative freedom to reject what one does not want or is not willing to accept. The explanation, this view, must be designed in such a way as to always ensure the possibility of rejecting both a given justification and a recommendation, as well as to go back on decisions previously taken in order to modify them potentially. XAI must become the main tool for the user to always keep in mind their ability to release themselves from the application and perceive that they can always choose for themselves in each phase of the interaction.

3.3 Human Dignity

The right to human dignity is presented last, but certainly not in importance. In fact, it is expressed in Article 1 of the European Charter of Human Rights [21], just as it is often the first right to be enunciated and guaranteed in most Constitutional Charters and international treaties [8]. The reason why it is the last to be analyzed here is due to the twofold approach with which it is addressed by the doctrine.

Human dignity is considered a "constellation principle", around which all others orbit and by reason of which all others find their justification and their – possible – balance [69]. This is why it is considered the founding element of freedom, justice, and peace [4], enforced as such by the United Nations General Assembly. Nonetheless, the difficulty in providing an objective and universally accepted empirical demonstration, its imperative character, and the lack of definition [30] have meant that – without discussing its legal value – its direct concrete application has been questioned.

Consequently, it might be useful to identify a still legally relevant concept that serves as an element of operability in the practice of the higher principle of human dignity – at least with regard to the human-AI interaction context.

Vulnerability: The principle of human dignity protects the intrinsic value that each individual possesses only as a human being [26] and, consequently, protects the individual's autonomy against forms of constraint. Both such aspects are the foundation of any comprehensive discussion of vulnerability [32]. The main difference is that any reference to human dignity often lends a universalist approach that has not always been easy to apply to concrete cases and disciplines. In other words, the reference to human dignity denotes a reference to a pivotal foundation of modern Constitutions, to a fundamental and inalienable human right that, as such, are easier to include in a principle-oriented argumentation that enshrines the theoretical frame of reference rather than an instrument directly applicable, without being translated into concepts of more immediate practical implementation [9]. On the contrary, vulnerability has already been used by the European Court of Human Rights as an indirect tool to evaluate the impact that some phenomena have on human dignity [64].

It follows that the concept of vulnerability can be used to substantiate the influence of banal deception at many levels. Depending on the result of such an investigation, it could become possible to draw up a range of possible repercussions. Depending on the range taken as a reference, it may be determined how to react – whether to correct, reevaluate, or stop the practice.

In particular, the objective and factual analysis using vulnerability as the materializing principle of the universal right of human dignity will allow the following satellite principles to be monitored and ensured:

Inclusion: An AI system capable of engaging the users at a psychological level, often going beyond their cognitive structures, is also able to reduce the level of socially relevant interactions, including those with healthcare specialists or psychotherapists (e.g., in the case of virtual nutritional coaches or behavioral changes applications [14,15]). The explanation can act as a pivotal element so that the phenomenon of banal deception does not result in induced or encouraged addiction and that the user always has the opportunity to interface with domain experts when the system recognizes the establishment of dynamics of dependence on interaction and loss of connection with reality (including the reality of one's physical or mental condition)

Humanisation of the Interaction: The above-mentioned dynamics could also lead to a phenomenon of dehumanization of individuals [56], conceived by the system as an aggregate of data and inputs, more than as human beings with their own weaknesses, their doubts, their inherent biases. The privileged, continuous, and often unique interaction with a responsive AI system that appears reliable and friendly can lead to conceiving that mode of interaction as the benchmark for evaluating all the others. This means creating – and consolidating – expectations of readiness in output, systematic argumentation, and acclimatization to errors that can generate two possible situations. On the one hand, the phenomenon of mechanomorphism [17], according to which the user becomes accustomed to the methods of communication, the timing, and the content provided by a given application, reshaping on it the expectations that are created on interactions with other human beings. In other words, technology becomes the model through which to navigate and act in the real world rather than the opposite. On the other hand, people can lower their expectations, their communicative level, and their complexity of thoughts to facilitate understanding of the system and its work. In this case, humans put themselves at the service of technology, anthropomorphizing its technical shortcomings instead of being its owners and users. In such a context, explanations can modulate interactive dynamics, ensure individuals always maintain control, specify technical dysfunctions, and modulate interaction times and pause from usage.

4   Principles in Action

The principles listed above represent the starting framework into which the concrete approach to deception in human-AI interaction can be inserted. We have realized that it is impossible to completely remove how banal deception impacts human experience with new technologies, both technically and functionally (i.e., from programming and psychological-perceptual reasons). Therefore, an approach that aims at realizing a true human-centered AI will have to address the issue, trying to modulate its impact, maximizing the benefits, and reducing the potential harmful effects.

This may be made possible through a two-phase approach: (i) a new method of assessing the impact of new technologies and their design and (ii) a control system to be implemented in the explanation and communications stages themselves.

For the sake of completeness, both levels will be presented. However, for the more specific purpose of this discussion, only the latter will be explored in depth, leaving a more detailed analysis of the former for future work.

4.1   Vulnerability Impact Assessment

The Vulnerability Impact Assessment (VuAI) aims to systematically identify, predict, and respond to the potential impacts of the technology used on human vulnerability. Moreover, in a broader sense, it could become crucial in assessing government policies at both European and Member States levels. It would be framed by international legal and ethical principles and fundamental human rights.

This could be an important instrument for mitigating possible harms occurred in, or because of, the interaction with AI systems designed in accordance with banal deception dynamics, while ensuring accountability. To this end, it could be relevant to make VuAI mandatory human rights due diligence for providers. Such an essential step can foster the achievement of the EU goals for the development and deployment of human-centered AI. It is also central for understanding and determining the levels of risk of AI systems, even when it is not immediate or objectively identifiable *ex-ante* the impact of the AI system on human rights, and even when there is little evidence and knowledge for detecting the risk level.

Once the reference structure is concretely developed, it will make it possible to divide the interaction models into classes, which consider (i) the interactive mode, (ii) the nature of the expected average user, and (iii) the ultimate goal of the interaction itself. Each of them will be linked to a range of impacts estimated on the profiles of human vulnerability, investigating whether it is respected and supported, exalted, or exploited for purposes not aligned with the right of human dignity (which we said underlies vulnerability and legitimizes enforceability). Each estimate of the impact on human vulnerability must correspond to a range of deception to be considered, not further reducible for reasons related to the functionality and acceptability of the AI system.

To this end, the Impact Assessment may consist of two elements: an assessment tool (e.g., a questionnaire or a semi-automatized feedback analysis tool) and an expert committee. The first is useful to define the features which may elicit or directly exploit vulnerability, thus inducing over-dependency and manipulating people's will. What would make hypothetical harms to vulnerability legally enforceable is the connection it has with human dignity. More precisely, the relation of direct derivation vulnerability has with this foundational right, as previously addressed.

The Expert Committee would analyze these aspects in the specific context of usage or with regard to the given technology under evaluation.

A more detailed and systematic definition of this new impact assessment and its scope will be further discussed in future works.

4.2   Principles-Based XAI

The framework described above represents the theoretical basis and justification element in a legal perspective of the modulation of deception in HCI.

To give substance to this new perspective, centered on a reassessment of the concept of vulnerability, it would be useful to formalize a knowledge graph. It would consist of a way to represent and structure the contextual (i.e. domain/application-related) concepts and information.

Its purpose would be to identify the characteristics of vulnerability, the nature of the relationships existing between its elements, and the influence of the context of use. In this way, systems leveraging XAI techniques would be able to identify and parse, with a good degree of approximation, any risk elements that might arise, even at run time, due to extensive interaction with the user. Once these "warnings" have been identified, the system will have to determine, select, and execute the ideal countermeasures.

The first approach would be to analyze/revise the explanations themselves, to counterbalance the otherwise exaggerated effects of banal deception. The main rebalancing effects of the interaction would consist in trying to engage the user as much as possible on a logical-rational level – both in the way the explanation is given and in its content. One way is to exploit design features that target areas of the brain antagonistic to those affected by the phenomenon of banal deception. Those same studies that have guided researchers in structuring the interaction "for deception" may provide insights into how to structure it to mitigate the same phenomenon (e.g., imposing semantic and thematic boundaries structured as logic rule sets). Moreover, contact with a second human opinion should be reiterated and encouraged, especially in the case of e-health applications or decision-making procedures. In doing so, it is important to reaffirm the individual's right to make autonomous choices and to ask for all necessary confirmation or information to form as critical and autonomous a thought as possible. In cases of intense risk to psychological integrity, mainly concerning aspects of presumed addiction, excessive dependence, isolation, and distortion of one's own initial will/goal, XAI-powered systems must suggest, even enforcing, a suspension of use and/or regular interaction (even periodically), until a decrease in the risk factors is registered, based on specific requests addressed to the user by the system itself. A possible strategy to enact such an intuition can be to periodically question the user's understanding and alignment with the necessary knowledge to safely use a given system and the "integrity" of their judgment/standing point.

If such an intervention might not be sufficient – or if the risk is high or difficult to assess by the application alone – the case should be handed over to a human domain expert that, for health/safety-critical applications, must always have the means to assess and intervene if necessary (e.g., a psychologist in stress-relieve personal assistants or a nutritionist/medical doctor in nutrition assistant scenarios).

5   Vulnerability as a Guiding Tool: Scepticism and Potentials

The application of the framework here proposed and the future theoretical investigations to be developed in this regard imply that vulnerability assumes a central role.

Especially in the European tradition, it might be natural to ask whether it is necessary to refer to a concept that is not purely legal in the strict sense. Indeed, it could be objected that a multiplication of legal principles does not benefit a clear, coherent, and streamlined application of the law, with the additional risk to become a mere exercise in style. This concern is certainly to be welcomed. However, the choice made here is intended to respond to an issue – that of the protection of the user and their psychological and physical integrity – which is particularly challenged by AI technologies and XAI systems designed according to the logic of banal deception. From this point of view, ignoring the central concept of human vulnerability, or relegating it to a particular condition, which does not change the application or formulation of the law in general, appears short-sighted and not resolving.

Moreover, embracing a universalistic conception of vulnerability means aligning with the framework outlined by Martha Fineman in her Vulnerability Theory (herenforth VT) [32]. It suggests a theoretical framework of redistribution of responsibility, burdens, support tools, and resilience, starting from the assumption that these measures are functional to the well-being of society, overcoming individual particularisms. This approach seems well suited to the analysis of the interaction between non-specialized users and AI, even when mediated by XAI. In fact, as demonstrated above, the dynamics of banal deception target cognitive and emotional constructs common to humans, not specific categories.

Despite what is sometimes disputed, VT is not a mere argumentative exercise, devoid of practical evidence. It is true that it has never been used holistically as a means of reforming, drafting, or adapting legislation yet. Nevertheless, some of its instances – universalistic interpretation of vulnerability, dependence as transposition of its concept, resilience as its opposite – have been cited or applied without explicit reference in the rulings of the European Court of Justice [39] and the European Court of Human Rights [36], and to address international human rights issues [37]. Moreover, it should not be considered that if vulnerability is no longer diversified in degrees, this implies not having regard to situations of particular fragility. It only means changing the perspective of the investigation – excluding that there may be users completely immune to the possible negative effects of deceptive mechanisms – and to encourage the research and formulation of legal and technological interventions. The latter should aim at creating resources and resilience tools for all, without excluding the possibility that they may be more decisive in some circumstances than in others.

6   Conclusions and Future Works

This study has focused on deception in human-AI interaction, arguing that *banal* deception plays a central role in enhancing the effectiveness of the system's functionalities, raising confidence and appreciation from the users in the given technology. However, the very concept of deception often has a negative connotation, being conceived as a tool of manipulation. Thus, we have pointed out that banal deception is intrinsic in human-AI interaction (HCI, HRI), and it consists of five fundamental elements: ordinary character, functional means, people's obliviousness-centered, low definition, and pre-definition.

Nevertheless, the fact that the banal deception has arisen with the precise intention of encouraging the most efficient interaction possible does not exclude the possibility of harmful side effects—above all, manipulative drifts.

In such a scenario, although XAI can be relevant to counteract such negative effects, the design and content of explanations can also exacerbate the phenomenon described above. This is because individuals are naturally led to attribute human qualities to inanimate objects, even more in the case of AI systems. Such a statement has already been proved by the Media Equation Theory [55], the CASA model [52], and the mechanism of Mindless Behaviour [58].

It has been emphasized here that this tendency is an integral part of an irreducible profile of vulnerability that characterizes human beings as such.

For this reason, this study claims the need to identify a framework to which XAI must refer (or embed) in the design of explanations—to counterbalance the possible harmful effects of banal deception and enhance its benefits. The principles here identified are: (i) the right to the integrity of a person—which consists of the right to both physical and psychological health; (ii) the respect for private and family life—which also includes the protection of personal identity, the inviolability of private space, and the autonomy of one's own choices   ; (iii) right of human dignity – from which the right to inclusion, and the need to enforce a humanisation of the interaction derive. From this perspective, we suggested a new interpretation of the concept of vulnerability as an indirect instrument to evaluate the impact of the phenomenon of banal deception on human dignity.

Such a theoretical framework could represent the essential mean towards a: Vulnerability Impact Assessment and the implementation of related knowledge graphs enabling a semi-automated pre-check and possible handover to humans domain expert if necessary. The first measure could serve as a tool to assess how/how much banal deception impact humans' inherent vulnerability, taking into account the application under analysis and the nature of both the users and the interaction. Thus, it could be possible to address and legally enforce the level of deception to be considered admissible case by case. The second measure would act at a system level, placing a continuous run-time assessment of possible manipulative drift "warnings". Hence, these dynamics could be mitigated and/or limited promptly through the timely intervention of the systems themselves and the human specialist (triggered) intervention if necessary.

Future works will focus on in-depth analysis of the vulnerability concepts and their formalization (from a schematic/systemic perspective) to then enable the design and implementation of semi-automated reasoners bridging data-driven (run-time) generated explanations, legally relevant vulnerability concepts, and the underneath rule-based system vehiculating the overall system dynamics.

Acknowledgments. This work is partially supported by the Joint Doctorate grant agreement No 814177 LAST-JD-Rights of Internet of Everything, and the Chist-Era grant CHIST-ERA19-XAI-005, and by *(i)* the Swiss National Science Foundation (G.A. 20CH21_195530), *(ii)* the Italian Ministry for Universities and Research, *(iii)* the Luxembourg National Research Fund (G.A. INTER/CHIST/19/14589586), *(iv)* the Scientific and Research Council of Turkey (TÜBİTAK, G.A. 120N680).

References

1. Adrienne, K.: Effective enforcement of human rights: the Tysiac v. Poland case. Studia Iuridica Auctoritate Universitatis Pecs Publicata **143**, 186 (2009)
2. AI HLEG: High-level expert group on artificial intelligence (2019)
3. Anjomshoae, S., Najjar, A., Calvaresi, D., Främling, K.: Explainable agents and robots: results from a systematic literature review. In: 18th International Conference on Autonomous Agents and Multiagent Systems, AAMAS 2019, Montreal, Canada, 13–17 May 2019, pp. 1078–1088. International Foundation for Autonomous Agents and Multiagent Systems (2019)
4. UN General Assembly, et al.: Universal declaration of human rights. UN General Assembly **302**(2), 14–25 (1948)
5. Astromskė, K., Peičius, E., Astromskis, P.: Ethical and legal challenges of informed consent applying artificial intelligence in medical diagnostic consultations. AI & Soc. **36**, 509–520 (2021). https://doi.org/10.1007/s00146-020-01008-9
6. Baker, R.S., De Carvalho, A., Raspat, J., Aleven, V., Corbett, A.T., Koedinger, K.R.: Educational software features that encourage and discourage "gaming the system". In: Proceedings of the 14th International Conference on Artificial Intelligence in Education, pp. 475–482 (2009)
7. Banks, J.: Theory of mind in social robots: replication of five established human tests. Int. J. Soc. Robot. **12**(2), 403–414 (2020)
8. Barroso, L.R.: Here, there, and everywhere: human dignity in contemporary law and in the transnational discourse. BC Int'l Comp. L. Rev. **35**, 331 (2012)
9. Beyleveld, D., Brownsword, R.: Human Dignity in Bioethics and Biolaw (2001)
10. Biran, O., Cotton, C.: Explanation and justification in machine learning: a survey. In: IJCAI-17 Workshop on Explainable AI (XAI), vol. 8, pp. 8–13 (2017)
11. Bissoli, L., et al.: A virtual coaching platform to support therapy compliance in obesity. In: 2022 IEEE 46th Annual Computers, Software, and Applications Conference (COMPSAC), pp. 694–699. IEEE (2022)
12. Bradeško, L., Mladenić, D.: A survey of chatbot systems through a Loebner Prize competition. In: Proceedings of Slovenian Language Technologies Society Eighth Conference of Language Technologies, vol. 2, pp. 34–37 (2012)
13. Bublitz, J.C.: The Nascent right to psychological integrity and mental self-determination. In: The Cambridge Handbook of New Human Rights: Recognition, Novelty, Rhetoric, pp. 387–403 (2020)

14. Calvaresi, D., et al.: EREBOTS: privacy-compliant agent-based platform for multi-scenario personalized health-assistant chatbots. Electronics 10(6), 666 (2021)
15. Calvaresi, D., et al.: Ethical and legal considerations for nutrition virtual coaches. AI Ethics, 1–28 (2022). https://doi.org/10.1007/s43681-022-00237-6
16. Calvaresi, D., Cesarini, D., Sernani, P., Marinoni, M., Dragoni, A.F., Sturm, A.: Exploring the ambient assisted living domain: a systematic review. J. Ambient. Intell. Humaniz. Comput. 8(2), 239–257 (2017)
17. Caporael, L.R.: Anthropomorphism and mechanomorphism: two faces of the human machine. Comput. Hum. Behav. 2(3), 215–234 (1986)
18. Carli, R., Najjar, A., Calvaresi, D.: Risk and exposure of XAI in persuasion and argumentation: the case of manipulation. In: Calvaresi, D., Najjar, A., Winikoff, M., Främling, K. (eds.) Explainable and Transparent AI and Multi-Agent Systems, EXTRAAMAS 2022. LNCS, vol. 13283, pp. 204–220. Springer, Cham (2022). https://doi.org/10.1007/978-3-031-15565-9_13
19. Ch'ng, S.I., Yeong, L.S., Ang, X.Y.: Preliminary findings of using chat-bots as a course FAQ tool. In: 2019 IEEE Conference on e-Learning, e-Management & e-Services (IC3e), pp. 1–5. IEEE (2019)
20. Cisek, P.: Beyond the computer metaphor: behaviour as interaction. J. Conscious. Stud. 6(11–12), 125–142 (1999)
21. European Commission: Charter of fundamental rights of the European Union, 2012/c 326/02. Official Journal of the European Union (2012)
22. Coons, C., Weber, M.: Manipulation: Theory and Practice. Oxford University Press (2014)
23. Crevier, D.: AI: The Tumultuous History of the Search for Artificial Intelligence. Basic Books, Inc. (1993)
24. Crowther-Heyck, H.: George A. Miller, language, and the computer metaphor and mind. Hist. Psychol. 2(1), 37 (1999)
25. Dennett, D.C.: The Intentional Stance. MIT Press (1987)
26. Dicke, K.: The founding function of human dignity in the universal declaration of human rights. In: The Concept of Human Dignity in Human Rights Discourse, pp. 111–120. Brill Nijhoff (2001)
27. Druce, J., Niehaus, J., Moody, V., Jensen, D., Littman, M.L.: Brittle AI, causal confusion, and bad mental models: challenges and successes in the XAI program. arXiv preprint arXiv:2106.05506 (2021)
28. Edmonds, B.: The constructibility of artificial intelligence (as defined by the Turing test). In: The Turing test: The Elusive Standard of Artificial Intelligence, pp. 145–150 (2003)
29. Epley, N., Waytz, A., Cacioppo, J.T.: On seeing human: a three-factor theory of anthropomorphism. Psychol. Rev. 114(4), 864 (2007)
30. Fabre-Magnan, M.: La dignité en droit: un axiome. Revue interdisciplinaire d'études juridiques 58(1), 1–30 (2007)
31. Fejes, E., Futó, I.: Artificial intelligence in public administration-supporting administrative decisions. PÉNZÜGYI SZEMLE/Public Finan. Q. 66(SE/1), 23–51 (2021)
32. Fineman, M.A.: Vulnerability: Reflections on a New Ethical Foundation for Law and Politics. Ashgate Publishing, Ltd. (2013)
33. Glocker, M.L., Langleben, D.D., Ruparel, K., Loughead, J.W., Gur, R.C., Sachser, N.: Baby schema in infant faces induces cuteness perception and motivation for caretaking in adults. Ethology 115(3), 257–263 (2009)
34. Graziani, M., et al.: A global taxonomy of interpretable AI: unifying the terminology for the technical and social sciences. Artif. Intell. Rev. 56, 3473–3504 (2022)

35. Guzman, A.L.: Making AI safe for humans: a conversation with Siri. In: Socialbots and Their Friends, pp. 85–101. Routledge (2016)
36. Heri, C.: Responsive Human Rights: Vulnerability, Ill-treatment and the ECtHR. Bloomsbury Academic (2021)
37. Ippolito, F.: La vulnerabilità quale principio emergente nel diritto internazionale dei diritti umani? Ars Interpretandi 24(2), 63–93 (2019)
38. Kim, J., Park, K., Ryu, H.: Social values of care robots. Int. J. Environ. Res. Public Health 19(24), 16657 (2022)
39. Knijn, T., Lepianka, D.: Justice and Vulnerability in Europe: An Interdisciplinary Approach. Edward Elgar Publishing (2020)
40. Kopelman, L.M.: The best interests standard for incompetent or incapacitated persons of all ages. J. Law Med. Ethics 35(1), 187–196 (2007)
41. Korn, J.H.: Illusions of Reality: A History of Deception in Social Psychology. SUNY Press (1997)
42. Lee, S.l., Lau, I.Y.m., Kiesler, S., Chiu, C.Y.: Human mental models of humanoid robots. In: Proceedings of the 2005 IEEE International Conference on Robotics and Automation, pp. 2767–2772. IEEE (2005)
43. Leonard, A.: Bots: The Origin of the New Species. Wired Books, Incorporated (1997)
44. Leonard, T.C.: Richard H. Thaler, Cass R. Sunstein, Nudge: improving decisions about health, wealth, and happiness. Constit. Polit. Econ. 19(4), 356–360 (2008)
45. Magid, B.: The meaning of projection in self psychology. J. Am. Acad. Psychoanal. 14(4), 473–483 (1986)
46. Margalit, A.: Autonomy: errors and manipulation. Jerusalem Rev. Legal Stud. 14(1), 102–112 (2016)
47. Marshall, J.: Personal Freedom Through Human Rights Law? Autonomy, Identity and Integrity under the European Convention on Human Rights. Brill (2008)
48. Massaro, D.W.: The computer as a metaphor for psychological inquiry: considerations and recommendations. Behav. Res. Meth. Instrum. Comput. 18, 73–92 (1986)
49. United States. President's Commission for the Study of Ethical Problems in Medicine and Biomedical and Behavioral Research: Making Health Care Decisions Volume One: Report (1982)
50. Mitnick, K.D., Simon, W.L.: The Art of Deception: Controlling the Human Element of Security. Wiley (2003)
51. Nass, C., Moon, Y.: Machines and mindlessness: social responses to computers. J. Soc. Issues 56(1), 81–103 (2000)
52. Nass, C., Steuer, J., Tauber, E.R.: Computers are social actors. In: Proceedings of the SIGCHI Conference on Human Factors in Computing Systems, pp. 72–78 (1994)
53. Natale, S.: Deceitful Media: Artificial Intelligence and Social Life After the Turing Test. Oxford University Press, USA (2021)
54. Papacharissi, Z.: A Networked Self and Human Augmentics, Artificial Intelligence, Sentience. Routledge, UK (2018)
55. Reeves, B., Nass, C.: Media Equation Theory (1996). Accessed 5 Mar 2009
56. Roberts, T., Zheng, Y.: Datafication, dehumanisation and participatory development. In: Zheng, Y., Abbott, P., Robles-Flores, J.A. (eds.) Freedom and Social Inclusion in a Connected World, ICT4D 2022. IFIP Advances in Information and Communication Technology, vol. 657, pp. 377–396. Springer, Cham (2022). https://doi.org/10.1007/978-3-031-19429-0_23

57. Sabatello, M.: Children with disabilities: a critical appraisal. Int. J. Child. Rights **21**(3), 464–487 (2013)
58. Sætra, H.S.: The parasitic nature of social AI: Sharing minds with the mindless. Integr. Psychol. Behav. Sci. **54**, 308–326 (2020)
59. Sarrafzadeh, A., Alexander, S., Dadgostar, F., Fan, C., Bigdeli, A.: "How do you know that i don't understand?" A look at the future of intelligent tutoring systems. Comput. Hum. Behav. **24**(4), 1342–1363 (2008)
60. Schneider, B.: You are not a gadget: a manifesto. J. Technol. Educ. **23**(2), 70–72 (2012)
61. Schreiber, D.: On social attribution: implications of recent cognitive neuroscience research for race, law, and politics. Sci. Eng. Ethics **18**, 557–566 (2012)
62. Seymour, W., Van Kleek, M.: Exploring interactions between trust, anthropomorphism, and relationship development in voice assistants. Proc. ACM Hum. Comput. Interact. **5**(CSCW2), 1–16 (2021)
63. Switzky, L.: Eliza effects: Pygmalion and the early development of artificial intelligence. Shaw **40**(1), 50–68 (2020)
64. Timmer, A.: A quiet revolution: vulnerability in the European court of human rights. In: Vulnerability, pp. 147–170. Routledge (2016)
65. Trower, T.: Bob and beyond: a Microsoft insider remembers (2010)
66. Turing, A.M.: Computing machinery and intelligence. In: Epstein, R., Roberts, G., Beber, G. (eds.) Parsing the Turing Test, pp. 23–65. Springer, Dordrecht (2009). https://doi.org/10.1007/978-1-4020-6710-5_3
67. White, L.A.: The symbol: the origin and basis of human behavior. Philos. Sci. **7**(4), 451–463 (1940)
68. Yang, Y., Liu, Y., Lv, X., Ai, J., Li, Y.: Anthropomorphism and customers' willingness to use artificial intelligence service agents. J. Hospitality Mark. Manage. **31**(1), 1–23 (2022)
69. Zatti, P.: Note sulla semantica della dignità. Maschere del diritto volti della vita, pp. 24–49 (2009)

Enhancing Wearable Technologies for Dementia Care: A Cognitive Architecture Approach

Matija Franklin[1,2(✉)], David Lagnado[1], Chulhong Min[1,2], Akhil Mathur[1,2], and Fahim Kawsar[2]

[1] Cuasal Cognition Lab, UCL, London, UK
matija.franklin@ucl.ac.uk
[2] Nokia Bell Labs, Cambridge, UK

Abstract. Activities of Daily Living (ADLs) are often disrupted in patients suffering from dementia due to a well-known taxonomy of errors. Wearable technologies have increasingly been used to monitor, diagnose, and assist these patients. The present paper argues that the benefits current and future wearable devices provide to dementia patients could be enhanced with cognitive architectures. It proposes such an architecture, establishing connections between modalities within the architecture and common errors made by dementia patients while engaging in ADLs. The paper contends that such a model could offer continuous diagnostic monitoring for both patients and caregivers, while also facilitating a more transparent patient experience regarding their condition, potentially influencing their activities. Concurrently, such a system could predict patient errors, thus offering corrective guidance before an error occurs. This system could significantly improve the well-being of dementia patients.

Keywords: Cognitive Architecture · Wearable · Dementia

1 Introduction

Activities of Daily Living (ADL), everyday tasks such as making a cup of coffee or calling someone on the phone, are disrupted in patients suffering from dementia. Dementia, a condition affecting over 55 million people worldwide, produces an intention-action gap[1]. The intention-action gap occurs when someone is not able to successfully execute their intentions, often due to a decline in cognitive function. ADLs are good indicators of the level of dementia a person is experiencing. As dementia progresses, patients have increasing difficulty performing ADLs. This decline in function can be gradual or sudden, and its onset can be unpredictable.

---

[1] Intention is the cognitive process by which people decide on and commit to an action. Action is the physical process of executing an intention.

D. Calvaresi et al. (Eds.): EXTRAAMAS 2023, LNAI 14127, pp. 270–280, 2023.
https://doi.org/10.1007/978-3-031-40878-6_15

Numerous assistive devices and technologies are available to aid people with dementia in performing their activities of daily living. These range from simple tools like pillboxes (small containers that remind people to take their medication), to more complex devices such as automated medication dispensers (devices that dispense medication at preset times). The development and utilization of such assistive devices and technologies have gained traction in part because dementia patients report that disruptions to ADLs cause the greatest loss of independence and wellbeing.

Multiple proposals have been put forward for the use of wearable technologies to aid dementia patients [8,36]. In relation to ADLs, wearables equipped with adequate behavioural monitoring would be able to detect patient errors during these activities and provide them with *corrective guidance* - guidance that responds to the error. The present paper proposes an approach that utilizes wearables and cognitive models to create systems that provide patients with *directive guidance* - guidance that predicts errors and aims to prevent them from happening.

2   Dementia Is Diverse and Dynamic

The reason why a one-size-fits-all corrective guidance is not viable for dementia patients lies in the diversity and dynamic nature of dementia. In other words, there are different types of dementia, and each patient's condition changes over time. Furthermore, even among patients with the same type of dementia, there are individual differences in the types of errors they exhibit.

Not all cognitive decline is the result of dementia. From early adulthood onwards, throughout a person's life, thinking speed, reasoning, working memory, and executive function all progressively decline [7]. This age-associated cognitive decline is non-pathological [16] and is an inevitable process of neurological aging. Dementia and mild cognitive impairment (MCI) are relatively rare conditions. Current estimates suggest that less than 20% of adults over the age of 80 have dementia [27].

Dementia is a syndrome characterized by a decline in memory and thinking abilities, as well as deterioration of cognitive abilities [20]. Symptoms of dementia include problems with planning and carrying out tasks, memory loss, mood and personality changes, and confusion. When these symptoms impair ADLs to a point where a person cannot live independently, they are said to have dementia.

Dementia is not a single disease, but an umbrella term for conditions that result in symptoms associated with memory loss, thinking, and communication problems. There are different types of dementia. First, dementia is not synonymous with MCI, although the two display similar symptoms and are thus often confused. MCI is a cognitive impairment that affects 5–20% of seniors [27]. Its symptoms are similar to those of regular brain aging - including decreased processing speed, working memory issues, and difficulties with reasoning and executive function - but more severe. People with MCI often forget names, numbers, and passwords, misplace items, struggle to remember conversations or decisions, and have trouble keeping track of their commitments and plans. As a result, MCI does not always prevent people from living independently, and some cases are even treatable. Approximately one in six people with MCI will develop dementia within a year.

There are many types of dementia, but Alzheimer's disease is the most common, accounting for 60–75% of all cases [27]. Other forms include vascular dementia, frontotemporal dementia, dementia with Lewy bodies, alcohol-related 'dementia', young-onset dementia, and Creutzfeldt-Jakob disease. A comprehensive review of all these conditions is outside the scope of this article. However, it is important to note that each type of dementia has its unique cognitive disruptions, patient errors, and treatment challenges. Additionally, mixed dementia is a condition in which more than one type of dementia occurs simultaneously [5], further complicating the establishment of a homogeneous approach to corrective guidance.

A patient's dementia symptoms also change over time. Dementia is dynamic in that a patient's symptoms worsen over time [24]. The rate at which it progresses and the cognitive functions it impairs will differ even among patients with the same type of dementia [34]. As a result, there are individual differences in the types of errors they display, even within the same type of dementia.

In conclusion, dementia takes different forms, with cases existing where a single patient manifests more than one form. It changes over time, typically worsening by affecting a cognitive modality with greater intensity. However, its progression differs among patients, and different patients exhibit different errors even if they have the same type of dementia. A technology capable of detecting these errors and tracking their changes over time would indeed be beneficial.

3 Wearables: Monitoring, Augmentation, and Guidance

Wearable technology provides a unique advantage for helping dementia patients by monitoring their behavior, identifying errors as they occur, and aiding patients in correcting these errors after they occur - corrective guidance - or even before they occur - directive guidance. Through the use of various sensors and the capacity to communicate with other smart devices, wearable technologies can be context-aware. Multiple examples exist of wearable systems being used for augmented memory, including the wearable remembrance agent [28], which monitors information retrieved from computers; SenseCam [13], a camera that serves as a memory aid; and DejaView [6], a healthcare system comprised of multiple sensors that assist patients in recalling daily activities. Despite potential perceptions of these technologies as intrusive, research suggests that patients support their development [12,31].

There have been proposals for utilizing more minimal setups, using just one wearable device, to monitor user behavior and augment a patient's memory [8,22,36]. Such wearables are made feasible with the aid of tailored machine learning algorithms that can identify dementia type and patient error based on data gathered from wearable technology [17,19]. A systematic implementation incorporating the seven most popular wearables at the time examined whether these devices could serve as suitable dementia patient monitoring devices [26]. The study found that the devices enabled real-time monitoring of dementia patients, but also identified major technological gaps, such as the need for devices with lower power consumption.

Wearables can also serve as diagnostic tools for dementia patients. A recent systematic review assessing wearable technology in dementia found that these devices could effectively monitor patient behavior, highlighting that adults with dementia were less active, had a more fragmented sleep-wake cycle, and exhibited a less varied circadian rhythm [4]. Inertial wearables have been proposed as pragmatic tools for monitoring control and gait, which serve as useful biomarkers in dementia [11]. A more recent paper determined that gait impairment monitoring by wearable technologies combined with machine learning algorithms could differentiate between different types of dementia [21].

Furthermore, wearable technology can provide patients with guidance for ADLs by assisting them through activities, thereby reducing or correcting errors [8,30]. For example, CueMinder reminds patients to perform ADLs using image and vocal cues, aiming to promote patient independence [14]. Other systems are more single-task oriented, such as AWash, which uses a smartwatch to monitor and segment hand-washing actions and prompts users to remind them to wash their hands [2].

We posit that the benefits wearable devices provide to dementia patients could be further enhanced with the use of cognitive architectures in two key ways. First, such models could provide an explanatory layer by identifying the cognitive modality that has been disrupted given the error patterns displayed by a patient. This has implications for the early detection of disruption to cognitive modalities. Apart from the diagnostic benefit, this would also increase transparency with patients. Second, by understanding what cognitive modalities are getting disrupted, wearables could predict the likelihood of future errors, thereby serving as a tool for tailored directive assistance that can enhance a patient's independence and well-being.

4   Cognitive Architectures

A cognitive model is a representation of the mental processes in the mind [1]. It offers an understanding of how the mind takes in, processes, and stores information. Some cognitive models provide a comprehensive description of the mind's operations, while others focus on specific elements such as memory, attention, or decision-making. The goal of cognitive modeling is to emulate human cognition in a way that aligns with observable data, like reaction times or functional magnetic resonance imaging results.

When a cognitive model is constructed to serve as the basis for artificial intelligence, it is referred to as a cognitive architecture [18]. Typically, cognitive architectures consist of various modules, each dedicated to a specific task or set of tasks. For example, one module may be responsible for attention, another for working memory, and another for long-term memory. Each module has its own set of processes and data structures that it uses to fulfill its function. These modules are interconnected, allowing information to flow between them so that the outputs of one module can become the inputs of another. The specific array of modules and their interconnections may vary between different cognitive architectures. Some architectures aim to emulate the operation of the human mind as closely as possible, while others may be more simplified or abstract.

Various approaches have been taken to construct cognitive architectures. However, as John Laird [18] argues, many cognitive architectures share similarities. He proposes a prototypical cognitive architecture consisting of memories and processing units common to other renowned cognitive architectures like SOAR, ACT-R [29], Icarus [3], LIDA [9], Clarion [32], and EPIC [15]. The block diagram of this prototypical cognitive architecture is shown in Fig. 1. This prototypical cognitive architecture will be used in this paper, as it consists of most elements present in other widely-used cognitive architectures.

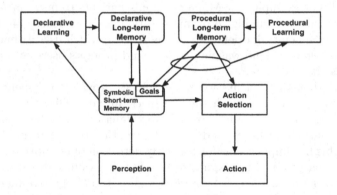

Fig. 1. A Prototypical Cognitive Architecture.

In this architecture, sensory information is initially processed by perception and then transferred to short-term memory. From here, cues from long-term declarative memory can be retrieved to access facts relevant to the current situation. Declarative memory has an associated learning module that uses activity from short-term memory to add new facts to this memory type. Similarly, procedural memory stores knowledge about what actions should be executed and when. Like declarative memory, procedural memory can be cued by activity in short-term memory, and it also includes a learning component. Data from both procedural memory and working memory are used by the action-selection component to determine the most suitable next action. This could involve physical actions in the external environment or deliberate changes in short-term memory. Thus, behavior is generated through a sequence of decisions over potential internal and/or external actions. Complex behavior, including internal planning, arises from actions that create and interpret internally generated structures and respond to the dynamics of the external environment. The architecture of this model supports both reactive behavior in dynamic environments and more deliberate, knowledge-mediated behavior. Lastly, learning mechanisms can be incremental - adding small units of knowledge one at a time - and/or online - learning occurring during performance. As new knowledge is experienced and acquired, it is immediately available. Learning does not require extensive analysis of past behavior or previously acquired knowledge.

5 Implementing a Cognitive Architecture for Dementia

A successful cognitive architecture provides a fixed infrastructure for understanding and developing agents that require only task-specific knowledge to be added, in addition to general knowledge [18]. This allows cognitive modeling to build upon pre-existing theories by using an existing architecture, thereby saving the time and effort spent starting from scratch. For modeling human behavioral data of dementia patients performing ADLs, the proposed cognitive architecture in Fig. 1 can be used to understand how patients observe changes in the environment, interpret them, retrieve other precepts from memory, convert them into actions, and so forth.

[35] have researched and categorized the four most common errors that dementia patients exhibit while performing ADLs. First, *Sequencing errors*, which can be further categorized into: *Intrusion* - the performance of an inappropriate action from a different activity that prevents the completion of the intended activity; *Omission* - the omission of an action required for completing the intended activity; *Repetition* - the repetition of an action that prevents the completion of the intended activity. Second, errors related to finding things; further divided into errors in finding items that are out of view and identifying items that are in view. Third, errors related to the operation of household appliances. Finally, *Incoherence errors*, which can be further divided into *toying* - performing random gestures with no apparent goal - and *inactivity* - not performing any action at all.

With this required task-specific knowledge, the prototypical cognitive architecture can be used to model dementia patients performing ADLs. Sequencing errors thus emerge from disruptions to the action selection or action performance modalities. More specifically, intrusion and repetition are action selection errors while omission could be due to either action selection or action performance. Errors in finding things that are out of view, as well as identifying things that are in view, can result from errors in short-term memory, declarative long-term memory, or declarative learning. Errors in the operation of appliances can emerge from errors in procedural long-term memory and procedural learning. Finally, incoherence errors may be due to disruptions to short-term memory, specifically the ability to hold a goal in mind, or due to errors in action selection and action execution. The block diagram of how patients' errors relate to the cognitive modalities is available in Fig. 2.

6   Evaluating the Cognitive Architecture

Validating a cognitive architecture of patient error would be feasible to conduct at scale with wearables and cognitive assessment batteries. As previously discussed, wearables can be used to detect different types of dementia [21], as well as varying patient behaviors [22], and thus, will be able to discern individual differences in the frequency of patient errors. To identify whether the observed patient errors stem from disruptions to the modalities proposed in Fig. 2, they can be tested against results from cognitive assessments. A cognitive assessment is a set of tests that are administered to evaluate an individual's cognitive abilities. These tests are often used to diagnose cognitive impairments.

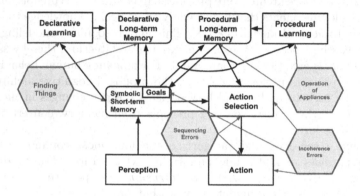

Fig. 2. Cognitive architecture of patient errors.

The Cognitive Assessment Battery (CAB) is one such test [23]. CAB is capable of clearly distinguishing between normal cognitive aging, Mild Cognitive Impairment (MCI), and dementia. It consists of six assessments covering different cognitive domains, namely, language, executive functions, speed, attention, episodic memory, and visuospatial functions. The Mini-Mental State Exam (MMSE), also used to screen for dementia, assesses attention, language, recall, and orientation [10]. These two tests are complimentary and have both been modified and validated using different methodologies. For example, the MMSE has been updated in the Modified MMSE, which includes verbal fluency [33]. Furthermore, there is evidence that such tests can be reliably administered over the phone [25], boding well for the possibility of these tests being administered through wearable technologies. Additionally, when a new type of error appears, patients could be administered new cognitive assessments.

The result of validating the model would be a cognitive architecture that provides an explanatory, causal layer behind patient error; specifically, how different disruptions produce varying distributions of patient error. Aside from being explanatory, such a model would also be predictive of the likelihood of future distributions of error for a patient, as different frequencies of error would result from distinct disruptions to cognition. The functionality of such a model will now be discussed.

7   Functionality

The functionality of a cognitive architecture for wearables assisting dementia patients with ADLs is twofold. First, the architecture would enable such technologies to be diagnostic and transparent with patients. The wearables could continuously track the frequency of patient error, understanding how alterations to these patterns might relate to changes in cognitive modalities. The system would provide a real-time assessment of an individual's performance over time. This information could be used to evaluate an individual's performance, which would be essential for understanding the efficacy of rehabilitation programs.

Provided consent is given, the wearable could also directly send updates to the patient's caregivers or close friends and relatives. This would inform the treatment and care the patient needs to receive. For instance, if a person is going about their daily activities and is not making some of the errors that are typically made, but is making new kinds of errors, this would indicate a deterioration in function. By supplying this information to the caregiver, the caregiver could provide guidance to help prevent further deterioration.

The cognitive architecture would offer a number of advantages for patients with dementia. First, it would provide them with greater transparency over their condition. Second, it would offer a level of support that could assist them in making decisions about their care. Third, it would be highly customized, allowing patients to receive the specific level of support they need. Lastly, it would be non-intrusive, meaning patients could wear the device without feeling as though they are being monitored.

Secondly, the architecture could be used to predict future errors in dementia patients. As the distribution of error will cluster around disruptions to certain cognitive modalities, a wearable with such a cognitive architecture could anticipate patient behavior. This predictive capability could, in turn, be used to provide patients with directive guidance, offering advice on performing ADLs before the predicted error occurs. Arguably, this would be more beneficial than corrective guidance that responds to patient error after it happens.

There are many potential applications for a cognitive architecture, and the specific applications will depend on the unique needs of the dementia patient population. However, the potential benefits of such an architecture are clear. By predicting when errors are likely to occur, a cognitive architecture could provide patients with guidance to help them avoid making errors. In turn, this could help patients stay safe and enhance their quality of life. Future research could explore the communication methods used for talking to patients about changes to their condition, as well as ways of gaining their attention and giving them precise directive guidance.

8   Conclusion

This paper contends that the benefits wearable devices currently offer to dementia patients, and those they could provide in the future, can be amplified by integrating cognitive architectures. It puts forth such an architecture, delineating the connections between modalities within the architecture and patient errors commonly manifested by dementia patients during ADLs. The paper asserts that this model could enable continuous diagnostic monitoring for both patients and caregivers, while also affording patients a more transparent understanding of their condition, which may inform their actions. Furthermore, such a system would have the capacity to predict patient errors, thus offering them corrective guidance before an error occurs. Such a system could greatly enhance the well-being of dementia patients.

References

1. Bermúdez, J.L.: Cognitive Science: An Introduction to the Science of the Mind. Cambridge University Press, Cambridge (2014)
2. Cao, Y., Chen, H., Li, F., Yang, S., Wang, Y.: AWash: handwashing assistance for the elderly with dementia via wearables. In: IEEE Conference on Computer Communications, IEEE INFOCOM 2021, pp. 1–10. IEEE (2021)
3. Choi, D., Langley, P.: Evolution of the ICARUS cognitive architecture. Cogn. Syst. Res. **48**, 25–38 (2018)
4. Cote, A.C., Phelps, R.J., Kabiri, N.S., Bhangu, J.S., Thomas, K.: Evaluation of wearable technology in dementia: a systematic review and meta-analysis. Front. Med. **7**, 501104 (2021)
5. Custodio, N., Montesinos, R., Lira, D., Herrera-Pérez, E., Bardales, Y., Valeriano-Lorenzo, L.: Mixed dementia: a review of the evidence. Dement. Neuropsychologia **11**, 364–370 (2017)
6. De Jager, D., et al.: A low-power, distributed, pervasive healthcare system for supporting memory. In: Proceedings of the First ACM MobiHoc Workshop on Pervasive Wireless Healthcare, pp. 1–7 (2011)
7. Deary, I.J., et al.: Age-associated cognitive decline. Br. Med. Bull. **92**(1), 135–152 (2009)
8. Franklin, M., Lagnado, D., Min, C., Mathur, A., Kawsar, F.: Designing memory aids for dementia patients using earables. In: Adjunct Proceedings of the 2021 ACM International Joint Conference on Pervasive and Ubiquitous Computing and Proceedings of the 2021 ACM International Symposium on Wearable Computers, pp. 152–157 (2021)
9. Franklin, S., Madl, T., D'mello, S., Snaider, J.: LIDA: a systems-level architecture for cognition, emotion, and learning. IEEE Trans. Autonom. Mental Develop. **6**(1), 19–41 (2013)
10. Galea, M., Woodward, M.: Mini-mental state examination (MMSE). Aust. J. Physiotherapy **51**(3), 198 (2005)
11. Godfrey, A., Brodie, M., van Schooten, K., Nouredanesh, M., Stuart, S., Robinson, L.: Inertial wearables as pragmatic tools in dementia. Maturitas **127**, 12–17 (2019)
12. Hassan, L., et al.: Tea, talk and technology: patient and public involvement to improve connected health 'wearables' research in dementia. Res. Involvement Engagem. **3**(1), 1–17 (2017)
13. Hodges, S., et al.: SenseCam: a retrospective memory aid. In: Dourish, Paul, Friday, Adrian (eds.) UbiComp 2006. LNCS, vol. 4206, pp. 177–193. Springer, Heidelberg (2006). https://doi.org/10.1007/11853565_11

14. Kempner, Danielle, Hall, Martha L.: The CueMinder project: patient-driven wearable technology to improve quality of life. In: Gargiulo, Gaetano D., Naik, Ganesh R. (eds.) Wearable/Personal Monitoring Devices Present to Future, pp. 231–238. Springer, Singapore (2022). https://doi.org/10.1007/978-981-16-5324-7_9

15. Kieras, D.E.: A summary of the EPIC cognitive architecture. In: The Oxford Handbook of Cognitive Science, vol. 1, p. 24 (2016)

16. Konar, A., Singh, P., Thakur, M.K.: Age-associated cognitive decline: insights into molecular switches and recovery avenues. Aging Dis. **7**(2), 121 (2016)

17. Kwan, C.L., Mahdid, Y., Ochoa, R.M., Lee, K., Park, M., Blain-Moraes, S.: Wearable technology for detecting significant moments in individuals with dementia. BioMed. Res. Int. **2019**, 6515813 (2019)

18. Laird, J.E.: The Soar Cognitive Architecture. MIT Press (2019)

19. Lim, J.: A smart healthcare-based system for classification of dementia using deep learning. Digit. Health **8**, 20552076221131668 (2022)

20. Livingston, G., et al.: Dementia prevention, intervention, and care: 2020 report of the lancet commission. Lancet **396**(10248), 413–446 (2020)

21. Mc Ardle, R., Del Din, S., Galna, B., Thomas, A., Rochester, L.: Differentiating dementia disease subtypes with gait analysis: feasibility of wearable sensors? Gait Posture **76**, 372–376 (2020)

22. Mohamedali, F., Matoorian, N.: Support dementia: using wearable assistive technology and analysing real-time data. In: 2016 International Conference on Interactive Technologies and Games (ITAG), pp. 50–54 (2016). https://doi.org/10.1109/iTAG.2016.15

23. Nordlund, A., Påhlsson, L., Holmberg, C., Lind, K., Wallin, A.: The cognitive assessment battery (CAB): a rapid test of cognitive domains. Int. Psychogeriatr. **23**(7), 1144–1151 (2011)

24. Raj, A., Kuceyeski, A., Weiner, M.: A network diffusion model of disease progression in dementia. Neuron **73**(6), 1204–1215 (2012)

25. Rapp, S.R., et al.: Validation of a cognitive assessment battery administered over the telephone. J. Am. Geriatr. Soc. **60**(9), 1616–1623 (2012)

26. Ray, P.P., Dash, D., De, D.: A systematic review and implementation of IoT-based pervasive sensor-enabled tracking system for dementia patients. J. Med. Syst. **43**(9), 1–21 (2019)

27. Ray, S., Davidson, S.: Dementia and Cognitive Decline. A Review of the Evidence, vol. 27, pp. 10–12 (2014)

28. Rhodes, B.J.: The wearable remembrance agent: a system for augmented memory. Pers. Technol. **1**(4), 218–224 (1997)

29. Ritter, F.E., Tehranchi, F., Oury, J.D.: ACT-R: a cognitive architecture for modeling cognition. Wiley Interdisc. Rev. Cogn. Sci. **10**(3), e1488 (2019)

30. Siri, S., Divyashree, H., Mala, S.P.: The memorable assistant: an IoT-based smart wearable Alzheimer's assisting device. In: 2021 IEEE International Conference on Computation System and Information Technology for Sustainable Solutions (CSITSS), pp. 1–7. IEEE (2021)

31. Stavropoulos, T.G., et al.: Wearable devices for assessing function in Alzheimer's disease: a European public involvement activity about the features and preferences of patients and caregivers. Front. Aging Neurosci. **13**, 643135 (2021)

32. Sun, R.: Anatomy of the Mind: Exploring Psychological Mechanisms and Processes with the Clarion Cognitive Architecture. Oxford University Press, Oxford (2016)

33. Tombaugh, T., McDowell, I., Kristjansson, B., Hubley, A.: Mini-mental state examination (MMSE) and the modified MMSE (3MS): a psychometric comparison and normative data. Psychol. Assess. **8**(1), 48 (1996)

34. Tschanz, J.T., et al.: Progression of cognitive, functional, and neuropsychiatric symptom domains in a population cohort with Alzheimer dementia: the cache county dementia progression study. Am. J. Geriatr. Psychiatry **19**(6), 532–542 (2011)

35. Wherton, J.P., Monk, A.F.: Problems people with dementia have with kitchen tasks: the challenge for pervasive computing. Interact. Comput. **22**(4), 253–266 (2010)

36. Yang, P., Bi, G., Qi, J., Wang, X., Yang, Y., Xu, L.: Multimodal wearable intelligence for dementia care in healthcare 4.0: a survey. Inf. Syst. Front., 1–18 (2021). https://doi.org/10.1007/s10796-021-10163-3

Author Index

D. Calvaresi et al. (Eds.): EXTRAAMAS 2023, LNAI 14127, p. 281, 2023.
https://doi.org/10.1007/978-3-031-40878-6

Printed in the United States
by Baker & Taylor Publisher Services